McGraw-Hill Ryerson Mathematics 7
Making Connections

Authors

Elizabeth Ainslie
Hon. B.Sc., B.Ed.
Toronto District School Board

Wayne Erdman
B.Math., B.Ed.
Toronto District School Board

Dan Gilfoy
B.Sc. (Agr.), B.Ed.
Halifax Regional School Board

Honi Huyck
B.Sc., B.Ed.
Belle River, Ontario

Stacey Lax
B.A., B.Ed., M.Ed.
York Region District School Board

Brian McCudden
M.A., M.Ed., Ph.D.
Toronto, Ontario

Kelly Ryan
Hon. B.Sc., B.Ed.
Toronto District School Board

Jacob Speijer
B.Eng., M.Sc.Ed., P.Eng.
District School Board of Niagara

Sandy Szeto
B.Sc., B.Ed.
Toronto District School Board

Michael Webb
B.Sc., M.Sc., Ph.D.
Toronto, Ontario

Assessment/Pedagogy Consultants

Elizabeth Ainslie
Toronto District School Board

Brian McCudden
Toronto, Ontario

Combined Grades Consultant

Jonathan Dean
Hamilton-Wentworth District School Board

Special Education Consultants

Pauline Creighton
District School Board of Niagara

Deirdre Gordon
Hastings and Prince Edward District School Board

Technology Consultant

Honi Huyck
Belle River, Ontario

Mental Mathematics Consultant

Joan Manuel
District 10, St. Stephen, New Brunswick

Literacy Consultant

Anne Burnham MacLeod
District 18, Fredericton, New Brunswick

English as a Second Language Consultant

Jane E. Sims
Toronto, Ontario

Advisors

Chris Dearling
OISE - University of Toronto

Catherine Little
Toronto District School Board

Shelley McCurdy
Simcoe Muskoka Catholic District School Board

Tess Miller
Durham District School Board

Troy Parkhouse
District School Board of Niagara

Debbie Price
Greater Essex County District School Board

Mary E. O'Neill
The Halifax Regional School Board, Nova Scotia

McGraw-Hill Ryerson

Toronto Montréal Boston Burr Ridge, IL Dubuque, IA Madison, WI New York
San Francisco St. Louis Bangkok Bogotá Caracas Kuala Lumpur Lisbon London
Madrid Mexico City Milan New Delhi Santiago Seoul Singapore Sydney Taipei

COPIES OF THIS BOOK
MAY BE OBTAINED BY
CONTACTING:

McGraw-Hill Ryerson Ltd.

WEB SITE:
http://www.mcgrawhill.ca

E-MAIL:
orders@mcgrawhill.ca

TOLL-FREE FAX:
1-800-463-5885

TOLL-FREE CALL:
1-800-565-5758

OR BY MAILING YOUR
ORDER TO:
McGraw-Hill Ryerson
Order Department
300 Water Street
Whitby, ON L1N 9B6

Please quote the ISBN and title when placing your order.

Student Text ISBN:
0-07-090950-4

McGraw-Hill
Ryerson Limited
A Subsidiary of The *McGraw-Hill* Companies

McGraw-Hill Ryerson
Mathematics 7: Making Connections

Copyright © 2004, McGraw-Hill Ryerson Limited, a Subsidiary of The McGraw-Hill Companies. All rights reserved. No part of this publication may be reproduced or transmitted in any form or by any means, or stored in a data base or retrieval system, without the prior written permission of McGraw-Hill Ryerson Limited, or, in the case of photocopying or other reprographic copying, a licence from the Canadian Copyright Licensing Agency (Access Copyright). For an Access Copyright licence, visit *www.accesscopyright.ca* or call toll free to 1-800-893-5777.

Any request for photocopying, recording, or taping of this publication shall be directed in writing to Access Copyright.

ISBN 0-07-090950-4

http://www.mcgrawhill.ca

1 2 3 4 5 6 7 8 9 10 TCP 0 9 8 7 6 5 4

Printed and bound in Canada

Care has been taken to trace ownership of copyright material contained in this text. The publishers will gladly accept any information that will enable them to rectify any reference or credit in subsequent printings.

The Geometer's Sketchpad®, Key Curriculum Press, 1150 65th Street, Emeryville, CA 94608, 1-800-995-MATH.

National Library of Canada Cataloging in Publication Data

Mathematics 7 : making connections / authors, Elizabeth Ainslie ... [et al.].

Includes index.
ISBN 0-07-090950-4

1. Mathematics—Textbooks. I. Ainslie, Elizabeth
II. Title: Mathematics seven.

QA107.2.M39 2004   510   C2004-901442-0

PUBLISHER: Diane Wyman
PROJECT MANAGER: Helen Mason
DEVELOPMENTAL EDITORS: Maggie Cheverie, Jean Ford, Tom Gamblin, Jacqueline Lacoursiere
MANAGER, EDITORIAL SERVICES: Linda Allison
SUPERVISING EDITOR: Crystal Shortt
COPY EDITOR: Julia Cochrane
PERMISSIONS EDITOR: Paula Joiner
JUNIOR EDITORS: Scott Rostrop, Darren Scanlan
EDITORIAL ASSISTANT: Erin Hartley
MANAGER, PRODUCTION SERVICES: Yolanda Pigden
PRODUCTION CO-ORDINATOR: Janie Deneau
COVER AND INTERIOR DESIGN: Pronk & Associates
ART DIRECTION: Tom Dart/First Folio Resource Group, Inc.
ELECTRONIC PAGE MAKE-UP: Tom Dart, Greg Duhaney, Claire Milne, Adam Wood of
        First Folio Resource Group, Inc.
COVER IMAGE: David Brooks/CORBIS/MAGMA

Acknowledgements

Reviewers

The authors and editors of McGraw-Hill Ryerson Mathematics 7: Making Connections, wish to thank the following educators for their thoughtful comments and creative suggestions about what would work best in grade 7 classrooms. Their input has been invaluable in making sure that the text and its related Teacher's Resource meet the needs of students and teachers of Ontario.

Rahat Ahmed
Toronto District School Board

Chris Aikman
Hastings and Prince Edward District School Board

Jennifer Anderson
Ottawa-Carleton District School Board

Mary Anderson
York Region District School Board

Dan Antflyck
Toronto District School Board

Vijaya Balchandani
Toronto District School Board

Sarah Barclay
Upper Canada College

Tracey Bates
Ottawa-Carleton Catholic District School Board

Wayne Bechard
St. Clair Catholic District School Board

Matthew Bernstein
York Region District School Board

Michael Blackburn
Limestone District School Board

Andrew Canham
Ottawa-Carleton District School Board

Dennis Caron
Toronto Catholic District School Board

Richard Chaplinsky
Ottawa-Carleton District School Board

Adam Conacher
Ottawa-Carleton District School Board

Gordon Cooke
Upper Canada District School Board

Paul Cornies
Greater Essex County District School Board

Darlene Davison
Kawartha Pine Ridge District School Board

Charmaine Donnelly
Halifax Regional School Board

Angela Esau
District School Board of Niagara

George Fawcett
Hamilton Wentworth District School Board

Jodee Anne Ferdinand
Renfrew County District School Board

Maria Fotias Ginis
Toronto District School Board

Joanne Harris
Halton District School Board

Cheri Heslop
Upper Grand District School Board

Todd Horn
Sudbury Catholic District School Board

Stephen Hua
Hamilton-Wentworth District School

Iwan Jugley
District School Board of Niagara

Marion Kline
Grand Erie District School Board

Sylvia Constancio Kwan
Toronto District School Board

Heather Leonard
York Region District School Board

Dianne Lloyd
St. Clair Catholic District School Board

Stephen MacEachern
Durham Catholic District School Board

Patricia Macey
Peterborough Victoria Northumberland and Clarington Catholic District School Board

Jim Markovski
The Durham District School Board

Sean Marks
Halton District School Board

Christina Maschas-Hammond
Peterborough Victoria Northumberland and Clarington Catholic District School Board

Chester Makischuk
York Catholic District School Board

Mykola Matviyenko
Toronto District School Board

Cindy Terrade Moffat
Ottawa-Carleton District School Board

Suzanne Morrison
Upper Canada District School Board

Stephen Nevills
Durham District School Board

Jeremy Nowiski
Ottawa-Carleton District School Board

Megan Nowiski
Ottawa-Carleton District School Board

Dennis Paré
Ottawa-Carleton District School Board

Christopher Perry
Hamilton Wentworth District School Board

Kenneth Stanley Peterson
Kawartha Pine Ridge District School Board

Marilyn Price
Lower Sackville, Nova Scotia

Anna Przybylo
Durham Catholic District School Board

Lydia Rabenko-Javor
Toronto District School Board

Chad Richard
Toronto District School Board

Sherry St. Denis
Toronto District School Board

William Searle
Ottawa-Carleton District School Board

Jessica Silver
Limestone District School Board

Wuchow Than
Hamilton-Wentworth District School Board

Peter Thompson
Appleby College

Lisa True
Durham Catholic District School Board

Kathy Valiquette
Peterborough Victoria Northumberland and Clarington Catholic District School Board

Theresa Varney
District School Board of Niagara

Cara White
Kawartha Pine Ridge District School Board

Alan Wickens
Toronto District School Board

Student Photos

The authors and editors would like to extend a special thank you to the students of Bowmore Public School, Toronto District School Board, who allowed their photographs to be taken for some of the visuals found in this text.

Field-Test Teachers

The authors and editors of McGraw-Hill Ryerson Mathematics 7: Making Connections, wish to thank these teachers for their thoughtful comments and suggestions about what worked best in their classrooms. Their input has been invaluable in making sure that the text activities, explanations, Tasks, and teacher's resource support meet the needs of students and teachers.

Melanie Allport
Limestone District School Board

Greg Arkwright
Trillium Lakelands District School Board

Andrew Austin
Peterborough Victoria Northumberland and Clarington Catholic District School Board

Wayne Bechard
St. Clair Catholic District School Board

Yolanda Calquhoun
Thames Valley District School Board

Richard Chaplinski
Ottawa-Carleton Catholic District School Board

Pete Cobb
Lambton Kent District School Board

Patrizia DiFabio
Simcoe Muskoka Catholic District School Board

Angela Esau
District School Board of Niagara

Pegita Ghasemi
Toronto District School Board

Maria Fotias Ginis
Toronto District School Board

Gordana Grmuša
Greater Essex County District School Board

Todd Hayward
Lambton Kent District School Board

Cheri Heslop
Upper Grand District School Board

Michael Hill
Simcoe County District School Board

Dana Hopkins
Greater Essex County District School Board

Jan Kidd
Greater Essex County District School Board

Athina Lakoseljac
Toronto District School Board

Gloria Lasovich
Waterloo Catholic District School Board

Cyril Lewin
Toronto District School Board

Chris MacDonald
Toronto District School Board

Mike MacDonald
Dufferin-Peel Catholic District School Board

Stephen MacEachern
Durham Catholic District School Board

Christina Maschas Hammond
Peterborough Victoria Northumberland and Clarington Catholic District School Board

Michael Masse
York Region District School Board

Frank Mercuri
Niagara Catholic District School Board

Cindy Terrade Moffat
Ottawa-Carleton District School Board

Suzanne Morrison
Upper Canada District School Board

Kevin Reid
Durham District School Board

Cathy Renda
Waterloo Catholic District School Board

Elaine Roberts
Algonquin Lakeshore Catholic District School Board

Alex Sikkema
Emmanuel Christian School

Jiiva Somerville
Peel District School Board

Beth Vallance
Avon-Maitland District School Board

Wes Vickers
Greater Essex County District School Board

Rachel Wallage
Brant Haldimand-Norfolk Catholic District School Board

Dawna Wastesicoot
Durham District School Board

Jackie Watson
Niagara Catholic District School Board

Vickie Williams
Hastings and Prince Edward District School Board

Margarita Ziroldo
Ottawa-Carleton Catholic District School Board

Contents

A Tour of Your Textbook — x

Problem Solving — xvi

Get Ready for Grade 7
1. Fractions, Metric Units, Estimation — 2
2. Multiplying and Dividing Decimals, Estimation — 4
3. Patterns With Natural Numbers, Fractions, and Decimals — 6

CHAPTER 1
Measurement and Number Sense — 8
Get Ready — 10
1.1 Perimeters of Two-Dimensional Shapes — 12
1.2 Area of a Parallelogram — 18
1.3 Area of a Triangle — 22
1.4 Apply the Order of Operations — 26
1.5 Area of a Trapezoid — 30
1.6 Draw Trapezoids — 34
Use Technology
    Construct and Manipulate a Trapezoid Using *The Geometer's Sketchpad*® — 37
1.7 Composite Shapes — 40
Review — 46
Practice Test — 48

CHAPTER 2
Two-Dimensional Geometry — 50
Get Ready — 52
2.1 Classify Triangles — 54
2.2 Classify Quadrilaterals — 60
2.3 Congruent Figures — 66
2.4 Congruent and Similar Figures — 70
Use Technology
    Identify Similar Triangles Using *The Geometer's Sketchpad*® — 75
Review — 76
Practice Test — 78
Task: Create a Logo — 81

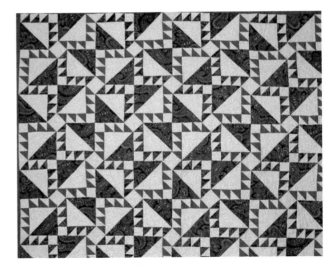

CHAPTER 3
Fraction Operations — 82
Get Ready — 84
3.1 Add Fractions Using Manipulatives — 86
3.2 Subtract Fractions Using Manipulatives — 90
3.3 Find Common Denominators — 94
3.4 Add and Subtract Fractions Using a Common Denominator — 98
3.5 More Fraction Problems — 104
Review — 108
Practice Test — 110

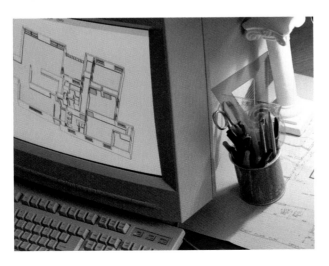

CHAPTER 4

Probability and Number Sense — 112

Get Ready — 114
4.1 Introducing Probability — 116
4.2 Organize Outcomes — 121
4.3 Use Outcomes to Predict Probabilities — 126
4.4 Extension: Simulations — 131
4.5 Apply Probability in Sports and Games — 134
Review — 140
Practice Test — 142
Task: Develop a Fair Game — 145
Chapters 1–4 Review — 146

CHAPTER 5

Fractions, Decimals, and Percents — 148

Get Ready — 150
5.1 Fractions and Decimals — 152
5.2 Calculate Percents — 158
5.3 Fractions, Decimals, and Percents — 162
5.4 Apply Fractions, Decimals, and Percents — 166
Review — 172
Practice Test — 174

CHAPTER 6

Patterning — 176

Get Ready — 178
6.1 Investigate and Describe Patterns — 180
6.2 Organize, Extend, and Make Predictions — 185
6.3 Explore Patterns on a Grid or in a Table of Values — 190
6.4 Express Simple Relationships — 195
Review — 200
Practice Test — 202
Task: Fold Fractals — 205

CHAPTER 7

Exponents — 206

Get Ready — 208
7.1 Understand Exponents — 210
7.2 Represent and Evaluate Square Roots — 214
7.3 Understand the Use of Exponents — 218
7.4 Fermi Problems — 224
Review — 228
Practice Test — 230

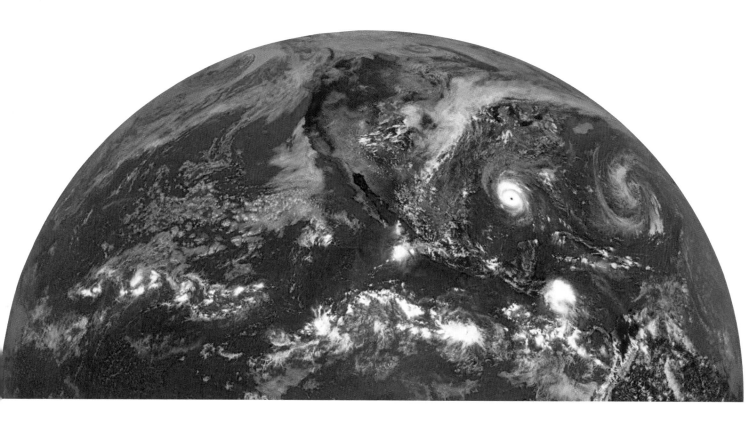

CHAPTER 8

Three-Dimensional Geometry and Measurement — 232

Get Ready — 234
8.1 Explore Three-Dimensional Figures — 236
8.2 Sketch Front, Top, and Side Views — 242
8.3 Draw and Construct Three-Dimensional Figures Using Nets — 247
8.4 Surface Area of a Rectangular Prism — 252
8.5 Volume of a Rectangular Prism — 258
Review — 262
Practice Test — 264
Task: Design a Stage — 267
Chapters 5–8 Review — 268

CHAPTER 9

Data Management: Collection and Display — 270

Get Ready — 272
9.1 Collect and Organize Data — 274
9.2 Stem-and-Leaf Plots — 280
9.3 Circle Graphs — 286
9.4 Use Databases to Find Data — 292
9.5 Use a Spreadsheet to Display Data — 298
Review — 304
Practice Test — 306

CHAPTER 10

Data Management: Analysis and Evaluation — 308

Get Ready — 310
10.1 Analyse Data and Make Inferences — 312
10.2 Measures of Central Tendency — 318
Use Technology
 Find Measures of Central Tendency With a Spreadsheet — 324
10.3 Bias — 326
10.4 Evaluate Arguments Based on Data — 331
Review — 336
Practice Test — 338
Task: Plan a Television Schedule — 341

CHAPTER 11

Integers — 342

Get Ready — 344
11.1 Compare and Order Integers — 346
11.2 Explore Integer Addition — 352
11.3 Adding Integers — 356
11.4 Explore Integer Subtraction — 362
11.5 Extension: Subtracting Integers — 368
11.6 Integers Using a Calculator — 374
Review — 378
Practice Test — 380

CHAPTER 13
Geometry of Transformations 424

Get Ready 426
13.1 Explore Transformations 428
13.2 Investigate Frieze Patterns With *The Geometer's Sketchpad®* 434
13.3 Extension: Translations on a Coordinate Grid 436
13.4 Identify Tiling Patterns and Tessellations 442
Use Technology
Create Tiling Patterns Using *The Geometer's Sketchpad®* 446
13.5 Construct Translational Tessellations 448
Use Technology
Tessellate by Translation Using *The Geometer's Sketchpad®* 450
13.6 Construct Rotational Tessellations 452
Use Technology
Tessellate by Rotation Using *The Geometer's Sketchpad®* 454
Review 456
Practice Test 458

Answers 460
Glossary 481
Index 488

CHAPTER 12
Patterning and Equations 382

Get Ready 384
12.1 Variables and Expressions 386
12.2 Solve Equations by Inspection 392
12.3 Model Patterns With Equations 398
12.4 Solve Equations by Systematic Trial 404
12.5 Model With Equations 410
Review 416
Practice Test 418
Task: Magic Squares 421
Chapters 9–12 Review 422

A Tour of Your Textbook

How is *Mathematics 7: Making Connections* set up?

Each chapter starts off with a **Chapter Problem** that connects math and your world. You will be able to solve the problem using the math skills that you learn in the chapter.

You are asked to answer questions related to the problem throughout the chapter.

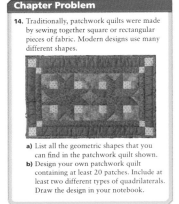

The **Chapter Problem Wrap-Up** is at the end of the chapter, on the second Practice Test page.

The **Get Ready** pages provide a brief review of skills from previous grades that are important for success with this chapter.

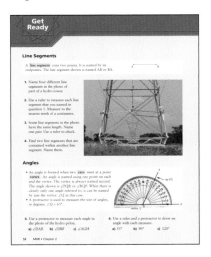

The **numbered sections** often start with a photo to connect the topic to a real setting. The purpose of this introduction is to help you make connections between the math in the section and the real world, or to make connections to previous knowledge.

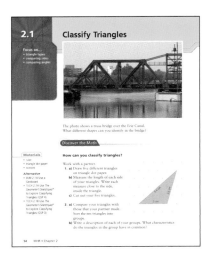

A three-part lesson follows.

Discover the Math

How can you tell if two triangles are congruent?

1. On grid paper, draw any rectangle and one diagonal. Cut out the two triangles formed.

The first part helps you find answers to the key question.
- An activity is designed to help you build your own understanding of the new concept and lead toward answers to the key question.
- **Examples** and **Solutions** demonstrate how to use the concept.

Key Ideas

- The perimeter is the total distance around the outside two-dimensional shape.
- Perimeter is measured in linear units, such as millimetres, centimetres, metres, and kilometres.

- A summary of the main new concepts is given in the **Key Ideas** box.
- Questions in the **Communicate the Ideas** section let you talk or write about the concepts and assess whether you understand the ideas.

Check Your Understanding

Practise

For help with questions 3 to 5, refer to the Example.

3. Zoë drew these parallelograms on centimetre grid paper. Calculate the area of each one.

- **Practise**: these are straightforward questions to check your knowledge and understanding of what you have learned.
- **Apply**: in these questions, you need to apply what you have learned to solve problems.
- **Extend**: these questions may be a little more challenging and may make connections to other lessons.

22. Tania is calculating the perimeter of a shape. She writes:
$P = (2 \times l) + (2 \times w)$
$P = (2 \times 5\ cm) + (2 \times 3\ cm)$
a) Draw and label the shape.
b) Find the perimeter.
c) Explain how you know your answers

The last Apply question in each set of questions is designed to assess your level of success with the section. Everyone should be able to respond to at least some part of each *Try This!* question.

Numbered sections that have a green tab are based on the use of technology such as scientific calculators, spreadsheets, or *The Geometer's Sketchpad®*.

Some numbered sections are followed by a Use Technology feature. This means some or part of the preceding section may be done using the technology shown.

How does *Mathematics 7: Making Connections* help you learn?

Understanding Vocabulary

Key words are listed on the Chapter Opener. Perhaps you already know the meaning of some of them. Great! If not, watch for these terms highlighted the first time they are used in the chapter. The meaning is given close by in the margin.

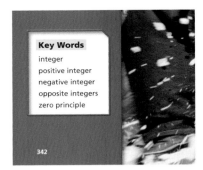

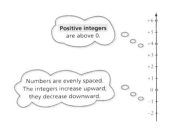

Literacy Connections provide tips to help you read and interpret items in math. These tips will help you in other subjects as well.

Understanding Concepts

The Discover the Math activity is designed to help you construct your own understanding of new concepts. The key question tells you what the activity is about. Short steps, with illustrations, lead you to be able to make some conclusions in the last step, the **Reflect** question.

xii   MHR • A Tour of Your Textbook

The **Examples** and their worked **Solutions** include several tools to help you understand the work.
- Notes in a thought or speech bubble help you think through the steps.
- Sometimes different methods of solving the same problem are shown. One way may make more sense to you than the other.
- **Problem Solving Strategies** are pointed out.
- Calculator key press sequences are shown where appropriate.

Example 2: Identify a Quadrilateral
A certain quadrilateral has two pairs of opposite sides that are equal and parallel. The quadrilateral contains no right angles. Identify and draw the quadrilateral.

Solution
Method 1: Draw a Diagram

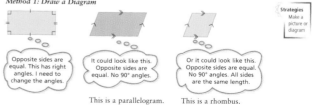

Opposite sides are equal. This has right angles. I need to change the angles.

It could look like this. Opposite sides are equal. No 90° angles.

Or it could look like this. Opposite sides are equal. No 90° angles. All sides are the same length.

This is a parallelogram. This is a rhombus.

Strategies: Make a picture or diagram

The quadrilateral must be either a parallelogram or a rhombus.

Method 2: Work Backward

 The quadrilateral has two pairs of opposite sides parallel. So, it is not a trapezoid or a kite.

Strategies: Work backward

 The quadrilateral contains no right angles. So, it is not a square or a rectangle.

The exercises begin with **Communicate the Ideas**. These two or three short questions focus your thinking on the **Key Ideas** you learned in the section. By discussing these questions in a group, or doing the action called for, you can see whether you understand the main points and are ready to start the exercises.

The first few questions in **Check Your Understanding** can often be done by following one of the worked Examples.

Key Ideas
- The side length of a square represents the square root of a number.
- A perfect square is a number whose square root is a natural number.
- The $\sqrt{}$ symbol indicates the square root of a number.

$A = 25$, $s = 5$

Communicate the Ideas
1. How does the diagram show the square root of 16?
2. How could you use grid paper, tiles, or blocks to show that $\sqrt{36} = 6$?
3. Decide if 49 is a perfect square. Show how you know.
4. Is it possible to find $\sqrt{6.25}$? Explain and justify your answer.

Check Your Understanding

Practise
5. State the side length of each square.
   a)  b)

For help with questions 6 and 7, refer to Example 1.

6. Find the side length of a square with the given area.
   a) 25 m^2 b) 49 cm^2
   c) 100 km^2 d) 9 m^2

7. Use a calculator to find the side length of a square with the given area.

For help with question 8, refer to Example 2.

8. Decide if each number is a perfect square. Show how you know.
   a) 16 b) 24
   c) 58 d) 225

9. Evaluate.
   a) $\sqrt{64}$
   b) $\sqrt{144}$
   c) $\sqrt{400}$

10. Use a calculator to evaluate.
    a) $\sqrt{625}$
    b) $\sqrt{441}$
    c) $\sqrt{10\,000}$

What else will you find in *Mathematics 7: Making Connections*?

Two special sections at the beginning of the book will help you to be successful with the grade 7 course.

Problem Solving

This is an overview of the four steps you can use to approach solving problems. Samples of 12 problem solving strategies are shown. You can refer back to this section if you need help choosing a strategy to solve a problem. You are also encouraged to use your own strategies.

Get Ready for Grade 7

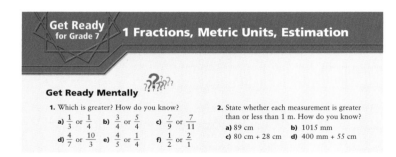

These six pages present a brief review of basic concepts from earlier grades and ways of thinking about the concepts.

Other Special Features

Did You Know?

These are interesting facts related to math topics you are learning.

Making Connections

These activities link the current topic to careers, games, or to another subject.

Internet Connect

You can find extra information related to some questions on the Internet. Log on to **www.mcgrawhill.ca/links/math7** and you will be able to link to recommended Web sites.

Each chapter ends with a **Chapter Review** and a **Practice Test**. The chapter review is organized by section number so you can look back if you need help with a question. The test includes the different types of questions that you will find on provincial tests: multiple choice, short answer, and extended response.

Task
These projects follow each pair of chapters. To provide a solution, you may need to combine skills from multiple chapters and your own creativity.

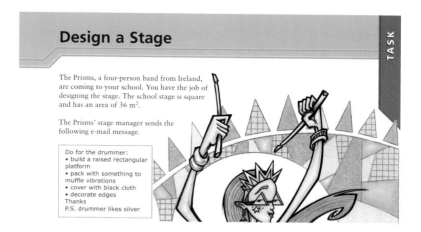

Reviews of the previous four chapters can be found following Chapters 4, 8, and 12.

Answers
Answers are provided to the odd-numbered Practise, Apply, and Extend questions, as well as, Reviews and Practice Tests. Sample answers are given for questions that have a variety of possible answers or that involve communication. If you need help, read the sample and then try to give an alternative response.

Answers are omitted for the Try This and the Chapter Problem questions because teachers may use these questions to assess your progress.

Glossary
Refer to the illustrated Glossary at the back of the text if you need to check the exact meaning of mathematical terms.

Problem Solving

How can you solve problems like the four below? Compare your ideas with the strategies that are shown on the following pages.

Problem 1
Honi has 100 m of fencing. She uses it to fence off a rectangular field for her horse to graze in. The length of the field is 30 m. How wide is the field?

Problem 2
Marja would like to go glow-in-the-dark bowling for her birthday. The bowling alley charges $10 for one lane plus $6 per person. This includes bowling shoe rentals. Marja's mother can afford $40. How many friends can Marja take bowling?

Problem 3
Rani is paid $7 per hour to baby-sit the neighbour's two children. Rani is saving for a new bike. How many hours does he need to baby-sit to earn enough money for the bike?

Problem 4
The corner store has five flavours of ice cream: chocolate, strawberry, bubble gum, rocky road, and orange fizz. How many different two-scoop cones are possible?

People solve mathematical problems at home, at work, and at play. There are many different ways to solve problems. In *Mathematics 7: Making Connections*, you are encouraged to try different methods and to use your own ideas. Your method may be different but it may also work.

A Problem Solving Model

Where do you begin with problem solving? It may help to use the following four-step process.

Understand

Read the problem carefully.
- Think about the problem. Express it in your own words.
- What information do you have?
- What further information do you need?
- What is the problem asking you to do?

Plan

Select a strategy for solving the problem. Sometimes you need more than one strategy.
- Consider other problems you have solved successfully. Is this problem like one of them? Can you use a similar strategy? Strategies that you might use include

  – Make a model
  – Make an assumption
  – Make a picture or diagram
  – Find needed information
  – Choose a formula
  – Solve a simpler problem
  – Act it out
  – Make an organized list
  – Work backward
  – Make a table or chart
  – Use systematic trial
  – Look for a pattern

- Decide whether any of the following might help. Plan how to use them.
  – tools such as a ruler or a calculator
  – materials such as graph paper or a number line

Do It!

Solve the problem by carrying out your plan.
- Use mental math to estimate a possible answer.
- Do the calculations.
- Record each step you are doing.
- Explain and justify your thinking.

Look Back

Examine your answer. Does it make sense?
- Is your answer close to your estimate?
- Does your answer fit the facts given in the problem?
- Is the answer reasonable? If not, make a new plan. Try a different strategy.
- Consider solving the problem a different way. Do you get the same answer?
- Compare your method with that of other students.

Problem Solving Strategies

Here are twelve strategies you can use to help solve problems. The chart shows you different ways to solve the four problems on page xvi. Your ideas on how to solve the problems might be different from any of these.

To see other examples of how to use these strategies, refer to the page references. These show where the strategy is used in other sections of *Mathematics 7: Making Connections*.

Problem 1 Honi has 100 m of fencing. She uses it to fence off a rectangular field for her horse to graze in. The length of the field is 30 m. How wide is the field?

Strategy	Example	Other Examples
Make a model	Use three 30-cm rulers and a piece of string 100 cm long. Assume that each centimetre represents 1 m. $30 + 30 + 20 + 20 = 100$ The width of the field is 20 m.	pages 98, 122, 153, 248, 253
Make a picture or diagram	$30 + 30 = 60$ The two lengths are 60 m. $100 - 60 = 40$ The two widths add to 40 m. $20 + 20 = 40$ One width is 20 m. (distance around is 100 m)	pages 41, 62, 105, 181, 411
Choose a formula	The formula for the perimeter, P, of a rectangle is $P = l + w + l + w$ Substitute $P = 100$ and $l = 30$. $100 = 30 + w + 30 + w$ (Add these two numbers.) $100 = 60 + w + w$ $100 = 60 + 40$ It is a rectangle. The widths are the same size. $20 + 20 = 40$ The width of the field is 20 m.	pages 13, 168, 254

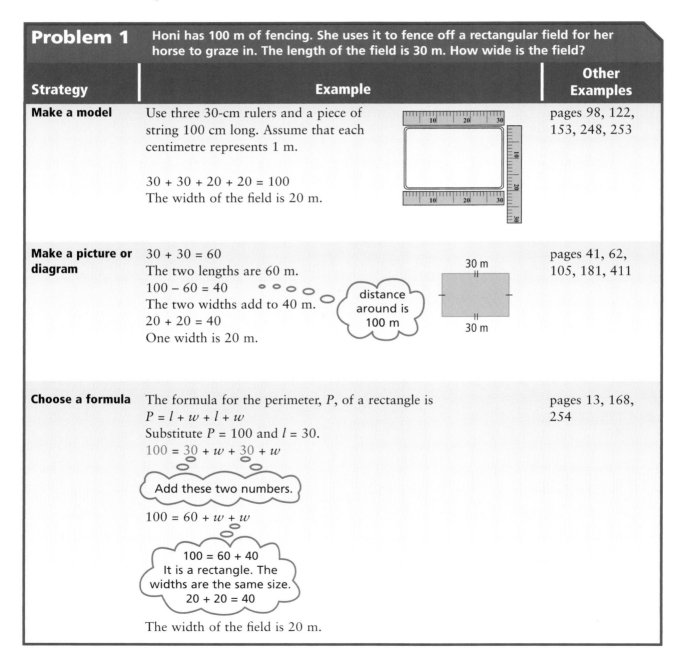

Problem 2

Marja would like to go glow-in-the-dark bowling for her birthday. The bowling alley charges $10 for one lane plus $6 per person. This includes bowling shoe rentals. Marja's mother can afford $40. How many friends can Marja take bowling?

Strategy	Example	Other Examples
Act it out	*Here is $10 for the lane. Add $6 for 1 person. That's $16.* *Four more piles of $6 give $40 altogether.* *I used $10 and then 5 times $6 to make $40.* For $40, five people can go bowling. Marja is one of the people. She can take four friends.	pages 134, 400
Work backward	It costs $10 for the lane. 40 − 10 = 30 This means $30 is left for the people. Each person costs $6. $\frac{30}{6} = 5$ $30 is enough for 5 people. One of these is Marja. She can take four friends.	pages 62, 411
Use systematic trial	The cost is $10 plus $6 per person. Try 3 people: $10 + 3 \times 6$ $= 10 + 18$ $= 28$ *Too low. She can take more friends.* Try 5 people: $10 + 5 \times 6$ $= 10 + 30$ $= 40$ *Right on.* For $40, five people can go bowling. Marja is one of the people. She can take four friends.	pages 215, 406

Problem Solving Strategies

Problem 3 — Rani is paid $7 per hour to baby-sit the neighbour's two children. Rani is saving for a new bike. How many hours does he need to baby-sit to earn enough money for the bike?

Strategy	Example	Other Examples
Make an assumption	*The problem does not say how much Rani's new bike costs. I will assume that he is saving for a racing bike that costs about $350 including taxes.* Find the number of hours to earn $350. $\frac{350}{7} = 50$ Rani needs to baby-sit for 50 h to earn $350.	pages 167, 225
Find needed information	*The problem does not say how much Rani's new bike costs. I found the price of bikes in an advertising flyer. The one I like costs $210.* Find the number of hours to earn $210. $\frac{210}{7} = 30$ Rani needs to baby-sit for 30 h to earn $210.	pages 171, 225

Problem 4

The corner store has five flavours of ice cream: chocolate, strawberry, bubble gum, rocky road, and orange fizz. How many different two-scoop cones are possible?

Strategy	Example	Other Examples
Solve a simpler problem	What if the only two choices were chocolate and strawberry? There are only 3 possible two-scoop cones: chocolate with strawberry, double chocolate, or double strawberry	pages 126, 226

OK, this gets me started. Now I will make an organized list of the possible pairs for five choices. I don't think the order of scoops of different flavours matters.

Make an organized list *I could also show this list in a tree diagram.*	1. chocolate + strawberry 2. chocolate + bubble gum 3. chocolate + rocky road 4. chocolate + orange fizz 5. strawberry + bubble gum 6. strawberry + rocky road 7. strawberry + orange fizz 8. bubble gum + rocky road 9. bubble gum + orange fizz 10. rocky road + orange fizz There are 10 different combinations of two scoops. A person might choose two scoops of the same flavour. That makes 5 more possibilities. Fifteen different two-scoop cones are possible using the five flavours.	pages 127, 135

Make a table or chart		chocolate	strawberry	bubble gum	rocky road	orange fizz
	chocolate	x	x	x	x	x
	strawberry		x	x	x	x
	bubble gum			x	x	x
	rocky road				x	x
	orange fizz					x

Fifteen different two-scoop cones are possible using the five flavours.

Look for a pattern	*loop means double* 1 flavour CC 2 flavours CC — S 3 flavours CC — R, S 4 flavours CC — R, S, B Possibilities 1 double = 1 1 mixed + 2 doubles = 3 3 mixed + 3 doubles = 6 6 mixed + 4 doubles = 10 Look for a pattern: 1, 3, 6, 10, … . Fifteen different two-scoop cones are possible using the five flavours.	pages 95, 105 *From 1 to 3 is add 2, 3 to 6 is add 3, 6 to 10 is add 4. The increase between numbers is one more each time. The next number in the pattern is 10 + 5.*

Problem Solving • MHR

Get Ready for Grade 7

1 Fractions, Metric Units, Estimation

Get Ready Mentally

1. Which is greater? How do you know?
   a) $\dfrac{1}{3}$ or $\dfrac{1}{4}$ b) $\dfrac{3}{4}$ or $\dfrac{5}{4}$ c) $\dfrac{7}{9}$ or $\dfrac{7}{11}$
   d) $\dfrac{4}{7}$ or $\dfrac{10}{3}$ e) $\dfrac{4}{5}$ or $\dfrac{1}{4}$ f) $\dfrac{1}{2}$ or $\dfrac{2}{1}$

2. State whether each measurement is greater than or less than 1 m. How do you know?
   a) 89 cm
   b) 1015 mm
   c) 80 cm + 28 cm
   d) 400 mm + 55 cm

Get Ready by Thinking

1 g

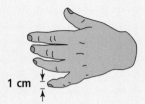
1 cm

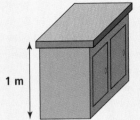
1 m

1 L

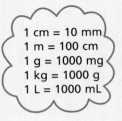
1 cm = 10 mm
1 m = 100 cm
1 g = 1000 mg
1 kg = 1000 g
1 L = 1000 mL

Choose the most reasonable estimate in questions 3 to 9. Share your estimates with a partner. Explain your thinking.

3. The length of a newly sharpened pencil is about
   A 4 cm   **B** 18 cm
   C 74 mm   **D** 18 mm

4. The mass of a calculator is about
   A 25 g   **B** 0.75 kg
   C 40 g   **D** 0.25 kg

5. The length of this textbook is about
   A 8 cm   **B** 120 cm
   C 230 mm   **D** 150 mm

6. The shaded portion is about

   A $\dfrac{2}{3}$ **B** $\dfrac{2}{5}$
   C $\dfrac{9}{10}$ **D** $\dfrac{3}{4}$

7. The volume is about
   A 30 mL   **B** 200 mL
   C 0.5 L   **D** 125 mL

8. The height of your classroom door is about
   A 2.4 m   **B** 360 cm
   C 420 cm   **D** 1.4 m

9. The total of this sale, before tax, is about
   A $17   **B** $20
   C $10   **D** $71

   lined paper $1.99
   binder $3.49
   pens $2.38
   compasses set $2.95
   backpack $9.97

   Estimate to the nearest dollar.

Get Ready by Exploring

Erin's aunt baked a giant cookie for Erin's birthday. Because the cookie was so large, Erin decided to eat it over several days.

Materials
- BLM Get Ready 7A Erin's Cookie

10. If Erin continues in this way, how much of the original cookie will she eat on the eighth day? How much will be left?

11. **a)** What patterns can you see as Erin eats her cookie?
    b) Make a diagram of the different sizes of cookie pieces.

12. A regular cookie has a mass of about 12 g. On the eighth day, Erin's cookie was about the size of a regular cookie. Estimate the mass of Erin's original cookie. Describe how you estimated.

13. About how many regular cookies would make up Erin's cookie?

14. When will Erin finish eating her cookie? Explain using words, manipulatives, or pictures.

Get Ready by Reflecting

15. How would you describe a fraction? Use words or pictures to explain five things you know about fractions.

16. Use words or pictures to describe the relationship between the following units of measure.
    a) millimetres, centimetres, metres, and kilometres
    b) millilitres and litres
    c) milligrams, grams, and kilograms

17. What do you consider to be a good estimate? What advice would you give a classmate who is having difficulty estimating?

Get Ready for Grade 7
2 Multiplying and Dividing Decimals, Estimation

Get Ready Mentally

1. Solve.
   a) 32 × 10
   b) 32 × 100
   c) 32 × 0.1
   d) 32 × 0.01

2. Solve.
   a) 32 ÷ 10
   b) 32 ÷ 100
   c) 32 ÷ 0.1
   d) 32 ÷ 0.01

3. Explain the rules you used in questions 1 and 2.

4. 32 × 6 = 192. Use this fact to find each product.
   a) 32 × 60
   b) 3.2 × 6
   c) 32 × 0.6
   d) 3.2 × 0.6

5. 3205 ÷ 5 = 641. Use this fact to find each quotient.
   a) 320.5 ÷ 5
   b) 3205 ÷ 50
   c) 32.05 ÷ 5
   d) 3205 ÷ 500

Get Ready by Thinking

Choose the most reasonable estimate for questions 6 to 12. Share your estimates with a partner. Explain your thinking.

6. About how many litres of paint are there altogether?
   A 12 L
   B 13 L
   C 12.5 L
   D 14.75 L

7. About how many grams of candy are there altogether?

   (80.5 g) (80.5 g) (80.5 g) (80.5 g) (80.5 g) (80.5 g)

   A 600 g
   B 540 g
   C 480 g
   D 48 g

8. A model train travels seven times around a 6.5-m track. About how many metres does the train travel?
   A 43 m
   B 455 m
   C 46 m
   D 42 m

9. About how many millilitres of juice are there altogether?
   A 1600 mL
   B 1500 mL
   C 1700 mL
   D 1650 mL

10. Heather divides 25 m of rope equally among four people. About how many metres of rope does each person get?
    A 7 m
    B 6 m
    C 21 m
    D 6.3 m

11. Ali's teacher asks him to divide a bucket of centimetre cubes equally into nine plastic bags. There are 758 centimetre cubes. About how many will be in each bag?
    A 80
    B 90
    C 85
    D 7200

12. In a four-person relay, the finish time was 55.3 s. If each person ran for the same length of time, for about how long did each person run?
    A 12 s
    B 13 s
    C 14 s
    D 15 s

Get Ready by Exploring

Coach Doyle held tryouts for the school cross-country team. Two students will make the team. To help decide which ones, Coach Doyle looked at the results from three special events.

Event 1: Students chose their favourite trail and ran as many complete laps as they could without stopping. Only complete laps counted.
- Carriff ran 3 laps of a 2.6-km trail.
- Jeremy ran 2 laps of a 3.2-km trail.
- Len ran 4 laps of a 1.8-km trail.
- Meghan ran 6 laps of a 1.3-km trail.
- Amy ran 3 laps of a 2.1-km trail.

Event 2: Students ran as much of a 1500-m course as they could without stopping and recorded the distance travelled.
- Carriff ran the entire course.
- Jeremy ran 0.75 of the course.
- Len ran 0.65 of the course.
- Meghan ran 0.9 of the course.
- Amy ran 0.8 of the course.

13. How far did each student run in Event 1?

14. How far did each student run in Event 2?

15. Organize the results in a way that would help Coach Doyle decide who should make the team.

16. Which two students would you recommend for the team? Explain your choices.

Get Ready by Reflecting

17. Describe the relationship between the number of decimal places in the numbers you are multiplying or dividing and the number of decimal places in the answer. Use words or diagrams.

18. How can estimation help you in your calculations with decimals?

19. What advice would you give a classmate who is having difficulty multiplying and dividing decimals?

Get Ready for Grade 7
3 Patterns With Natural Numbers, Fractions, and Decimals

Get Ready Mentally

1. Identify the next three numbers in each pattern.
   a) 3, 5, 7, ■, ■, ■
   b) 1, 4, 7, ■, ■, ■
   c) 3, 13, 23, ■, ■, ■
   d) 1, 4, 9, 16, ■, ■, ■
   e) 2, 13, 24, ■, ■, ■
   f) 2, 5, 11, 20, ■, ■, ■

2. Identify the next three numbers in each pattern.
   a) $\frac{1}{4}, \frac{2}{4}, \frac{3}{4}$, ■, ■, ■
   b) 2.7, 3.0, 3.3, ■, ■, ■
   c) $1, \frac{1}{2}, \frac{1}{4}, \frac{1}{8}$, ■, ■, ■
   d) $3.5, 4\frac{1}{2}, 5.5$, ■, ■, ■

Get Ready by Thinking

3. Explain what happens to the input number to get the output number.

   a)
Input	Output
4	7
7	10
12	15

   b)
Input	Output
3	9
5	13
7	17

   c)
Input	Output
2	6
5	15
8	24

   d)
Input	Output
4	9
10	21
3	7

Strategies Look for a pattern

4. Create a pattern of your own and ask a classmate to describe the pattern.

5. Look at the number line.

   a) What number could each letter represent? Explain your reasoning.
   b) Where might 130 be on the number line?
   c) About how far apart are A and D?
   d) Do you think C is greater than or less than 100? Why?

Literacy Connections

Using Problem Solving Strategies
To learn about different problem solving strategies, refer to the Problem Solving section on pages xvi to 1. The orange banner will help you find these pages. Refer to these pages whenever you need help deciding on a strategy to use to solve a problem.

Get Ready by Exploring

Materials
- pencil crayons
- BLM Get Ready 7B Hundred Chart

Sharon was asked to find number patterns on a hundred chart.

6. Describe a general pattern for each column.

7. Describe a diagonal pattern that you see on your hundred chart.

8. Sharon said that the number 48 belonged to only three different number patterns. Is Sharon correct? Explain.

9. How many patterns does the number 65 belong to?

10. Describe a pattern that 84 belongs to.

11. Describe a pattern that 5, 7, 17, and 43 belong to.

12. What do the numbers 8 and 12 have in common?

13. How many patterns could Sharon find containing the number 18? Describe them.

14. Use your own hundred chart to find five patterns. Use pencil crayons to show the numbers that belong to each pattern. Include a legend to identify the pattern for each colour.

15. Work with a partner. Ask each other questions about the patterns each of you found on your hundred charts.

Get Ready by Reflecting

16. Describe the most interesting pattern you found.

17. Describe the most interesting pattern your partner found.

18. What general statements can you make about the patterns you found on your hundred chart?

Measurement
- Estimate and calculate area and perimeter of 2-D shapes, including various trapezoids.
- Develop the formulas for area of a parallelogram, triangle, and trapezoid.
- Define and describe measurement concepts.
- Ask questions about linear measurement and area.
- Research and report on uses of measurement.

Number Sense and Numeration
- Justify the choice of method for calculations.
- Solve problems, using calculators.
- Understand and apply the order of operations, including brackets.

Geometry and Spatial Sense
- Identify and describe geometric figures.

Key Words
parallelogram
base
height
triangle
order of operations
trapezoid
vertex
composite shape

CHAPTER 1

Measurement and Number Sense

Have you ever driven a go-kart or watched go-kart racing?

Look at the go-karts in the picture. What geometric shapes do you recognize?

Participating in and watching go-kart racing can be lots of fun. Many go-karters like to design and build their own go-karts.

By the end of this chapter, you will be able to build your own go-kart model and go off to the races!

Chapter Problem

Think about various parts of the go-karts in the picture. How many pieces would you have to develop to design your own go-kart?

Get Ready

Perimeter

Perimeter is the distance around the outside of a two-dimensional shape or figure. It is measured in linear units.

Common linear units are millimetres (mm), centimetres (cm), metres (m), and kilometres (km).

What is the perimeter of this square?

The markings show which sides have equal lengths.

$P = 7 + 7 + 7 + 7$
$P = 28$

The perimeter of the square is 28 cm.

A **regular polygon** is a closed two-dimensional figure with all sides equal and all angles equal. To find the perimeter of this regular octagon, multiply the side length by 8.

$P = 8 \times 3$
$P = 24$

The perimeter is 24 cm.

Literacy Connections

Reading Diagrams
Short marks across two or more identical sides of a figure show that these sides have equal length.

1. Find the perimeter of each shape.

a) 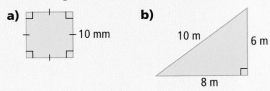 b)

2. For each regular polygon, state the number of sides. Then, find the perimeter.

a) b)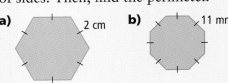

Convert Between Metric Units

To convert between metric units, use the relationships in the table. For example,

$6000 \text{ m} = \dfrac{6000}{1000} \text{ km}$
$\phantom{6000 \text{ m}} = 6 \text{ km}$

I am converting from metres to kilometres. I am converting to a larger unit, so I divide.

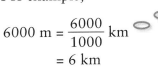

I am converting to a smaller unit, so I multiply.

Metric Units
1 cm = 10 mm
1 m = 100 cm
1 m = 1000 mm
1 km = 1000 m

3. Convert each measure from metres to kilometres.
   a) 9000 m
   b) 18 000 m
   c) 1200 m
   d) 700 m

4. Convert each measure, as described.
   a) 9 m, to centimetres
   b) 12 km, to metres
   c) 150 cm, to metres
   d) 0.5 km, to metres
   e) 2.5 cm, to millimetres

Area

Area measures how much space a two-dimensional shape covers. It is measured in square units. Square units include square centimetres (cm^2), square metres (m^2), and square kilometres (km^2).

One way to measure area is to count the number of square units inside the shape.

This square contains nine square centimetres. The area is 9 cm^2.

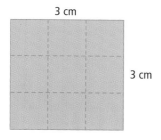

3 cm
3 cm

$A = s \times s$
$A = 3 \times 3$
$A = 9$

This rectangle contains 4 rows of six square centimetres. The area is 24 cm^2.

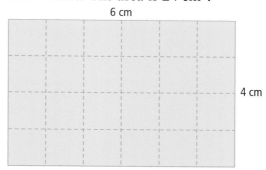

6 cm
4 cm

$A = l \times w$
$A = 6 \times 4$
$A = 24$

5. Find the area of each shape. Use two different methods.

   a) 4 cm, 4 cm
   b) 2 cm, 4 cm
   c) 5 cm, 3 cm

1.1 Perimeters of Two-Dimensional Shapes

Focus on...
- perimeter
- rectangles
- trapezoids
- regular polygons

Ice cream stands like this octagonal one are often found in tourist areas. If the length of each edge is 2 m, what is the distance around the ice cream stand?

Discover the Math

How do you find the perimeter of various shapes?

1. Look at the table. Match the shapes to the photographs. For example, shape 4 goes with photograph A.

Shape and Definition		Mystery Object
1. Quadrilateral	any four-sided figure	A
2. Parallelogram	a quadrilateral with opposite sides parallel	
3. Trapezoid	a quadrilateral with just one pair of opposite sides parallel	B
4. Polygon	a figure with three or more sides	C
5. Regular polygon	a polygon with all sides equal and all angles equal	
6. Regular hexagon	a regular polygon with 6 sides	D
7. Regular octagon	a regular polygon with 8 sides	

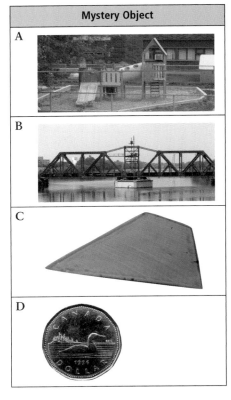

2. Which shapes do not match any photographs? Sketch real-life objects that have each of these shapes.

3. **Reflect** Describe how to calculate the perimeter of each type of shape. What information would you need? How would you use this information?

Example 1: Distance Around a Race Track

Sarah enjoys go-karting. She enters a go-kart race where the track is close to rectangular.
a) Determine the perimeter of the track.
b) How many kilometres must Sarah drive in a 10-lap race?
c) How can you check your answer?

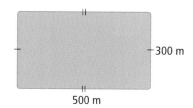

Solution

a) *Method 1: Add Side Lengths*
P = 500 + 300 + 500 + 300
P = 1600
The distance around the track is 1600 m.

I started by labelling the missing side lengths.

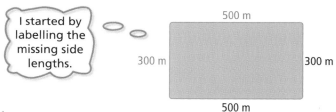

Method 2: Use a Formula
Use the formula for the perimeter of a rectangle.
$P = (2 \times l) + (2 \times w)$
$P = (2 \times 500) + (2 \times 300)$
P = 1000 + 600
P = 1600
The distance around the track is 1600 m. **Add proper units to the final answer.**

Strategies
Choose a formula

b) Find the total distance of a 10-lap race.
1600 m × 10 = 16 000 m
To convert to kilometres, remember that 1000 m = 1 km.
$16\,000 \text{ m} = \frac{16\,000}{1000} \text{ km} = 16 \text{ km}$
Sarah must drive 16 km to complete the race.

I am converting from metres to kilometres. I am converting to a larger unit, so I divide.

c) One kilometre is the distance between Amar's school and the arena. The race distance is 16 times that distance.

What is about 1 km from your school?

1.1 Perimeters of Two-Dimensional Shapes • MHR

Example 2: Trapezoidal Window

Sarah wants to cut out a window in the shape of a trapezoid for the back of her go-kart. What total length of cut must she make?

One length is not given. The markings show that the missing side is 60 cm.

Solution
$P = 1.2$ m $+ 60$ cm $+ 50$ cm $+ 60$ cm
$P = 120$ cm $+ 60$ cm $+ 50$ cm $+ 60$ cm
$P = 290$ cm

Sarah must cut 290 cm to make the window.

1.2 m = 1.2 × 100 cm
     = 120 cm

Example 3: Cost of a Hot Tub

This hot tub is in the shape of a regular octagon. It needs new padding around its edge.
a) What length of padding is needed?
b) Padding costs $4.50 per metre. What is the total cost of padding the sides of the hot tub?

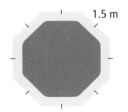

Solution
a) side length = 1.5 m
all sides are equal
number of sides = 8
$P = 8 \times 1.5$
$P = 12$

The length of padding is 12 m.

Remember to include units.

Strategies
What other strategy might you use?

The length of my driveway is 12 m. What does 12 m mean to you?

b) cost of padding: $4.50 per metre
length of padding: 12 m
$4.50 \times 12 = 54$ **Multiply cost per metre by length.**

The padding around the hot tub costs $54.

Key Ideas

- The perimeter is the total distance around the outside of a two-dimensional shape.
- Perimeter is measured in linear units, such as millimetres, centimetres, metres, and kilometres.

Communicate the Ideas

1. Write definitions in your own words for these terms: perimeter, distance, two-dimensional figure, linear units.

2. Matthew and Sonja calculated the perimeter of this figure. Matthew said the perimeter is 20 cm. Sonja said the perimeter is 20 cm². Who is right? How do you know?

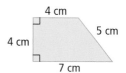

Literacy Connections

Reading Diagrams
The ⌐ symbol means that the lines are at right angles, or 90°, to each other.

3. $P = 6 \times 1.5$ m
   $P = 9$ m
   Draw and label this shape.

Check Your Understanding

Practise

For help with questions 4 and 5, refer to Example 1.

4. Find the perimeter of each rectangle.

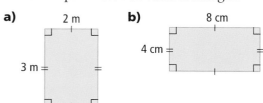

5. Find the perimeter of each rectangle.

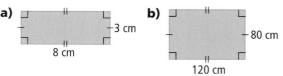

For help with questions 6 to 8, refer to Example 2.

6. You want to cut out this window. What length of cut should you make?

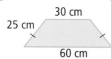

7. What length of cut is needed for this shape?

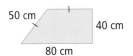

8. Find the perimeter of each trapezoid.

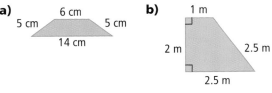

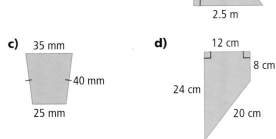

For help with questions 9 to 11, refer to Example 3.

9. Find the perimeter of each regular polygon.

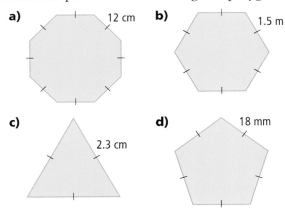

10. Find each perimeter.

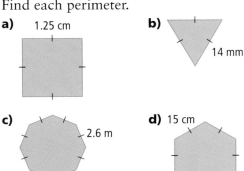

11. a) Look around the classroom, or your home. Sketch five different polygons. Identify where you found each one.
  b) Measure the perimeter of each polygon. Choose the best measurement unit for each polygon.

12. What projects at home, at a workplace, or in the community might involve finding a perimeter?

Apply

13. You need to find the perimeter of this rectangle. What steps must you take before adding lengths?

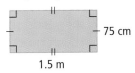

14. Why is perimeter measured in linear units? Use a diagram to help you explain.

For help with question 15, refer to Example 1.

15. A rectangular go-kart track has length 450 m and width 150 m.
  a) How long is one lap, in metres?
  b) How long is a 20-lap race, in kilometres?

16. Describe any relationships you see between the right triangle and the rectangle.

Making Connections

You will learn more about right triangles in Chapter 2.

17. Anders found the perimeter of this rectangle. His friend Sasha says that his solution is wrong.

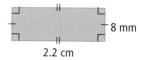

P = (2 × l) + (2 × w)
P = (2 × 2.2) + (2 × 8)
P = 4.4 + 16
P = 20.4

Who is right, Anders or Sasha?

Chapter Problem

18. Sarah wants to put glow-in-the-dark tape around the nose of her go-kart.

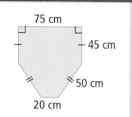

- Regular glow-in-the-dark tape costs $6.00 per metre.
- Sparkly purple tape costs $7.50 per metre.

Design a colour scheme for Sarah's go-kart tape. How much will the tape for your design cost? Explain your calculations

19. What kind of geometric shape is a loonie? Determine the perimeter of a loonie.

20. Leila skates around the perimeter of an ice rink.

a) How far does Leila skate on each lap?
b) How far does Leila skate in 8 laps?
c) How many laps should Leila skate to travel 3 km?

21. A parallelogram is a quadrilateral that has opposite sides parallel and opposite sides equal in length.

a) Measure the length of each side of this parallelogram.

b) Find the perimeter.
c) Describe another way to find the perimeter of a parallelogram.

Literacy Connections

Reading Diagrams
Arrows such as > or >> on two sides of a figure show that these sides are parallel.

22. Tania is calculating the perimeter of a shape. She writes:

P = (2 × l) + (2 × w)
P = (2 × 5 cm) + (2 × 3 cm)

a) Draw and label the shape.
b) Find the perimeter.
c) Explain how you know your answers to parts a) and b) are correct.

Extend

23. The gazebo shown in the plan has a perimeter of 18 m. Can you fit a 2.5-m bench along one side? Justify your answer mathematically.

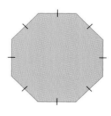

24. Heidi wants to use stones to surround a rectangular garden she is creating. How many different ways can she design her garden so that it has a perimeter of 26 m? Show each different design.

1.2 Area of a Parallelogram

Focus on...
- area
- parallelograms
- base and height

parallelogram
- four-sided figure with both pairs of opposite sides equal and parallel

Materials
- centimetre grid paper
- ruler

Look at this building. Which windows are **parallelograms**? The formula for the area of a rectangle is $A = l \times w$. Will this work for parallelograms?

Discover the Math

What is the area formula for a parallelogram?

1. Jadzia wanted to find the area of a parallelogram. Check the method she used.

18 MHR • Chapter 1

2. How is the area of Jadzia's rectangle related to the area of her parallelogram? Explain your answer. Hint: Compare the length and width of the rectangle to the **base** and **height** of the parallelogram.

3. Try Jadzia's method with your own parallelogram. Do you get the same result?

4. Reflect Brainstorm the steps you need to find the accurate area of a parallelogram. Develop a formula to explain what you are doing.

base
- a side of a polygon
- short form is *b*

height
- distance from the base to the opposite side or vertex, measured at right angles to the base
- short form is *h*

Example: Calculate Parallelogram Areas

Robert drew these parallelograms on centimetre grid paper. Use the formula $A = b \times h$ to calculate the area of each parallelogram.

a) 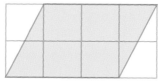   **b)**    **c)**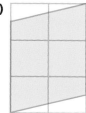

Solution

a) $A = b \times h$
$A = 3 \times 2$
$A = 6$
The area of the parallelogram is 6 cm².

Use square centimetres for area.

The centimetre grid tells me base = 3 cm height = 2 cm

b) $A = b \times h$
$A = 1 \times 3$
$A = 3$
The area of the parallelogram is 3 cm².

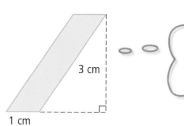

This time, I have to extend the line of the base to measure the height.

c) $A = b \times h$
$A = 2.5 \times 2$
$A = 5$
The area of the parallelogram is 5 cm².

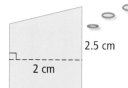

There is no horizontal side. But I can use one vertical side as the base, and measure the height from that.

Key Ideas

- You can find the area of a parallelogram if you know its base and height. Use the formula
  area = base × height   or $A = b \times h$

  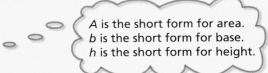
  *A is the short form for area.
  b is the short form for base.
  h is the short form for height.*

- The height of a parallelogram is always at right angles to its base.

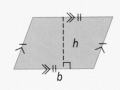

Communicate the Ideas

1. In your journal, compare the formulas for the area of a parallelogram and the area of a rectangle.
   How are they similar? How are they different?

2. What's wrong? Natasha calculated the area of a parallelogram with base 5 cm and height 4 cm.
   Area = 5 × 4
       = 20
   The area is 20 cm.

   Explain the error in Natasha's solution. Why is correcting this error important?

Check Your Understanding

Practise

For help with questions 3 to 5, refer to the Example.

3. Zoë drew these parallelograms on centimetre grid paper. Calculate the area of each one.

   a) 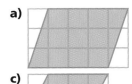   b)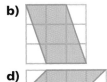

   c)   d)

4. Sameh drew these parallelograms on centimetre grid paper. Calculate the area of each one.

   a)   b)

   c)   d)

5. Measure the base and height of each parallelogram. Then, find the area.

a) 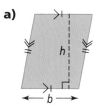   b)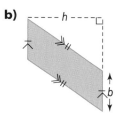

Apply

6. a) Measure the base and height of this parallelogram. Then, find the area.
b) What was unusual about measuring the parallelogram? Explain.

7. a) Can you find the area of parallelogram A with the given information? Explain why or why not.
b) Explain how you can find the height of parallelogram B.

Parallelogram A

Parallelogram B

8. Madra is choosing a garden layout.

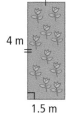 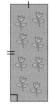    
Rectangular layout   Parallelogram layout

a) Predict which layout gives a greater area for Madra's flowers.
b) Calculate the area of each layout. Compare to your prediction.

9. Joel measured the parallelogram shown and then calculated the area.

A = b x h
A = 8 x 4
A = 32

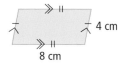

The area of the parallelogram is 32 cm².

There is an error in Joel's work. Find the error and describe what Joel should have done.

10. Victor planned a decal with his first initial. He used centimetre grid paper. What is the area of Victor's decal?

 Try This!

Extend

11. a) Who do you agree with, Monica or Michel? Explain why.
b) Make a statement that shows why the *other* person could also be right.

You know, you can calculate the area of a rectangle using the parallelogram formula.

You can't mix the two formulas up because rectangles are not like parallelograms.

12. Katie is constructing a shelf for her videos and DVDs. She is designing parallelogram-shaped wooden dividers to sort the movies.

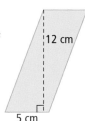

a) How much wood will Katie need for 8 dividers?
b) Create a plan for Katie to make her 8 dividers. Research the shapes and sizes of wood available. How can Katie conserve wood?

1.3 Area of a Triangle

Focus on...
- area
- triangles
- base and height

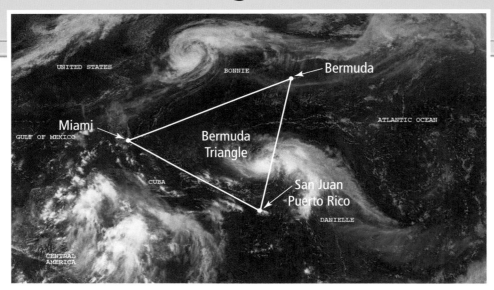

The Bermuda Triangle is one of the most mysterious regions on Earth. A number of strange things have occurred there over the years, including the unexplained disappearances of planes and ships.

How large do you think the Bermuda Triangle is? Can you use a formula you already discovered to answer this question?

Discover the Math

Materials

Optional:
- centimetre grid paper
- ruler

triangle
- three-sided closed figure

Making Connections

Shapes that are identical, but in different positions, are congruent. Learn more about congruence in Chapter 2.

How can you find the area of a triangle?

1. Stefan was trying to calculate the area of a **triangle**. He used what he knew about the area of a parallelogram to help him. Find the total area of Stefan's parallelogram.

2. Next, Stefan cut out the two triangles.
   a) Do the triangles have equal areas? Explain.
   b) What else about the two triangles could be equal?

3. a) How is the area of the parallelogram from step 1 related to the area of each triangle? Explain.
   b) Use this information to calculate the area of each triangle.

22 MHR • Chapter 1

4. a) How is the area of a triangle related to the area of a parallelogram with the same base and height?
   b) Write down the formula for finding the area of a parallelogram.

5. Reflect Modify your answers from step 4. Make them into a formula for finding the area of a triangle.

Example: Apply the Triangle Area Formula

Find the area of the Bermuda Triangle.
Use the formula for the area of a triangle:
area = base × height ÷ 2
or
$A = b \times h \div 2$

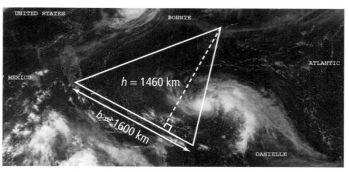

Solution
base = 1600 km
height = 1460 km
$A = b \times h \div 2$
$A = 1600 \times 1460 \div 2$
$A = 2\ 336\ 000 \div 2$
$A = 1\ 168\ 000$
The area of the Bermuda Triangle is 1 168 000 km².

Did You Know?

1 168 000 km² is large enough to swallow all of the Great Lakes!

Key Ideas

- The area of a triangle is related to the area of a parallelogram.
- The area of a triangle can be found by using the formula
  $A = b \times h \div 2$ or $A = \dfrac{b \times h}{2}$
- The base and height always form a right angle.

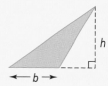

Communicate the Ideas

1. Your classmate missed this lesson. How could you explain to your classmate how to measure the height of this triangle?

2. Sketch or trace each triangle. Label the base and height of each triangle. In each case, explain how you know which measures to identify.

   a)   b)   c)

3. a) Draw a triangle to match this solution:

   Area = base × height ÷ 2
   = 5 × 4 ÷ 2
   = 10

   The area is 10 cm².

   b) Exchange with a partner. Did you draw the same triangles? Are different triangles possible? Explain.

Check Your Understanding

Practise

4. Identify the base and height of each triangle.

   a)   b)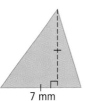

   c) 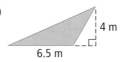   d)

For help with questions 5 to 7, refer to the Example.

5. Find the area of each triangle in question 4.

6. Find the area of each triangle.

   a)    b)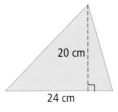

7. Find the area of each triangle.

   a) 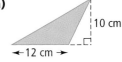   b)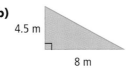

   c)   d)

24   MHR • Chapter 1

Apply

8. Kristin is painting a mural to cover one end wall of her attic room. What area will Kristin have to paint?

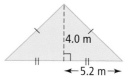

9. a) Where does the $b \times h$ part come from in the formula $A = b \times h \div 2$?
   b) Where does the "$\div 2$" come from in the formula $A = b \times h \div 2$?
   c) How is the area of a triangle related to the area of a parallelogram with the same base and height? Use pictures, numbers, and symbols to explain.

10. Mac is a defender for his soccer team. The coach has assigned Mac to cover the region shown. What area is Mac defending?

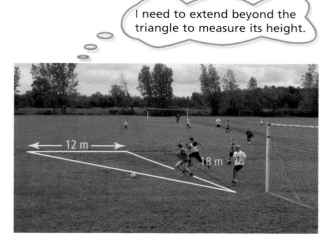

I need to extend beyond the triangle to measure its height.

11. If two triangles have the same area, do they also have an equal perimeter? Find out. Hint: Draw several triangles with the same area. Measure their perimeters.

Chapter Problem

12. Sarah is adding two triangular reflectors to the back panel of her go-kart.

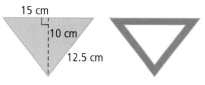

   a) What total perimeter must Sarah cut out for the reflectors?
   b) Sarah is painting one side of the reflectors with a reflective paint. What is the total area that must be painted?

13. a) Without measuring or calculating, rank the triangles in order from least area to greatest.

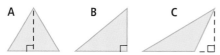

   b) Measure the base and height of each triangle.
   c) Calculate the area of each triangle. Compare these results to your prediction. What do you observe?

 14. Create a short triangle quiz. Your quiz should reflect the types of questions asked in this section. Include triangles of different kinds and in different positions.

Extend

15. Lindsay is building a tree house. The floor of the tree house needs to be a triangle that joins three large branches. The base of the triangle needs to be 4.0 m and the total area is to be 5.0 m².

   a) Draw several floor plans that meet Lindsay's needs.
   b) Lindsay also wants to paint the edge of the floor a bright orange. Will each design require the same amount of paint for the edge? Explain.

1.3 Area of a Triangle • MHR   25

1.4 Apply the Order of Operations

Focus on...
- order of operations
- perimeter and area

In math, you often have to do several operations in one calculation.

How does the order of the operations affect the answer? What can you do to change the order?

Discover the Math

How do brackets work in the order of operations?

1. How many different answers can you get for this expression if you do the operations in different orders?

   $4 + 5 - 3 \times 2 + 7 \div 7$

2. What is the *correct* answer? How do you know?

3. If brackets appear in an expression, the operations in the brackets are done first. Insert brackets in the expression from step 1 to get several different answers. How many different answers can you get by inserting brackets?

4. **Reflect** Think about steps 1 to 3.
   a) What is the advantage of inserting brackets in expressions?
   b) What is the advantage of having a specific order of operations?

Example: The Order of Operations

Cora has found a winning ticket for a music CD in her box of cereal. To claim her prize, she must answer this skill-testing question:
$3 \times 3 - (2 - 0.2) \div 0.3 + 4 \times 2$
Use the **order of operations** to find the answer.

Solution

Method 1: Pencil and Paper

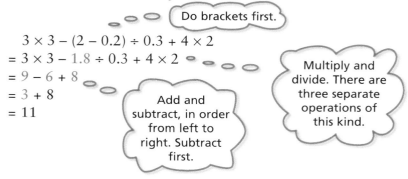

Do brackets first.

$3 \times 3 - (2 - 0.2) \div 0.3 + 4 \times 2$
$= 3 \times 3 - 1.8 \div 0.3 + 4 \times 2$
$= 9 - 6 + 8$
$= 3 + 8$
$= 11$

Multiply and divide. There are three separate operations of this kind.

Add and subtract, in order from left to right. Subtract first.

order of operations
- correct sequence of steps for a calculation

B Brackets, then
O Order
D ⎫ Division and
M ⎭ Multiplication, in order from *left to right*
A ⎫ Addition and
S ⎭ Subtraction, in order from *left to right*

Method 2: Calculator

$3 \times 3 - (2 - 0.2) \div 0.3 + 4 \times 2$
$= 3 \times 3 - 1.8 \div 0.3 + 4 \times 2$

$= 9 - 6 + 8$

$= 3 + 8$
$= 11$

For the bracket: ⓒ 2 ⊖ 0.2 ⊜ 1.8
For the multiplications and division:
ⓒ 3 ⓧ 3 ⊜ 9., 1.8 ⊘ 0.3 ⊜ 6., 4 ⓧ 2 ⊜ 8.
Now, add and subtract, *in order from left to right.* **Subtract first.**

Technology Tip
- You can use brackets with some calculators. For this example, key in
ⓒ 3 ⓧ 3 ⊖ ⓘ 2
⊖ 0.2 ⓘ ⊘ 0.3 ⊕ 4
ⓧ 2 ⊜ 11.

Key Ideas

- Some mathematical expressions include brackets and +, −, ×, and ÷ operations.

- When evaluating expressions, the correct order of operations must be followed. The term "BODMAS" can be used to help remember the order.

In the expression $7 + 4 \times (5 - 3)$, do the brackets first.

Communicate the Ideas

1. How is it useful to have a single, standard order of operations?

2. In the skill-testing question $3 \times 3 - (2 - 0.2) \div 0.3 + 4 \times 2$, what if you add before subtracting in the third step?
$9 - 6 + 8 = 9 - 14 = \blacksquare$
Can you claim the prize? Why or why not?

3. Use a flow diagram to show the correct order for the operations in this expression.
$12 \div (3 + 1) - 18 \div (2 \times 3)$

Check Your Understanding

Practise

For help with questions 4 to 9, refer to the Example.

4. Answer this skill-testing "warm-up" question:
$3 + 4 \times 2$
Show all your steps.

5. Answer this skill-testing "challenge" question:
$3 \times 7 - (6 - 1) \div 4 + 12$
Show all your steps.

6. Evaluate each expression.
   a) $6 + 2 \times 5$ b) $12 \div 3 - 2$
   c) $4 \times 10 \div 5$ d) $54 \div 9 \div 3$
   e) $(7 - 2) \times 8$ f) $3 \times (8 - 1)$

7. Evaluate.
   a) $25 \div 5 + 3 \times 2$ b) $4 + (23 - 7) \div 8$
   c) $(6 + 2) \times 3 - 9$ d) $18 - 2 \times (4 - 2)$

8. Evaluate.
   a) $27 \div (9 \div 3) - 8$
   b) $(14 - 2) \div 3 + (12 - 2) \times 3$

9. Which operation do you perform first in each expression? Explain why.
   a) $7 + 2 \times 5$ b) $4 \times (5 - 2)$ c) $12 \div 4 \times 2$

10. Is each statement true or false? Explain.
    a) "According to BODMAS, division is always done before addition."
    b) "According to BODMAS, addition is always done before subtraction."

Apply

11. What's wrong? To claim a prize, Vanya answers a skill-testing question:
$64 \div 16 \div 4 + 3 - 2 \times 2$
$= 64 \div 4 + 3 - 2 \times 2$
$= 16 + 3 - 2 \times 2$
$= 16 + 1 \times 2$
$= 16 + 2$
$= 18$

   a) Find two errors in Vanya's solution.
   b) Give a correct solution.

12. You can use this expression to calculate the perimeter of Leon's patio:
$(7 \times 3) + (2 \times 7.5 + 3)$

   a) Explain what the expression in each set of brackets means.
   b) Simplify the expression.
   c) What is the advantage of using this expression to find the perimeter?

13. Evaluate.
   a) $(2.1 + 3.6) \times 2$
   b) $(3.4 + 7.1) \div (3.6 + 1.4)$
   c) $8 \div (0.4 \div 0.2) \div 2 - 3 + 5 - 4$
   d) $(8 + 12 - 3 - 1) \div [7 - (4 + 1)]$
   e) $2 + 9 \div [1.5 \times (5 - 3)]$

14. a) Put brackets into each expression to make the equation true.
   $4 + 6 - 5 \times 2 = 6$
   $4 + 6 - 5 \times 2 = 0$
   $4 + 6 - 5 \times 2 = 10$
   b) Look at the three expressions in part a). Why is it important that everyone follow the same order of operations when evaluating expressions? Explain.
   c) For one of these expressions, you can leave the brackets out and still get the given value, if you follow BODMAS correctly. Which expression is it? Explain.

15. Don and Phil are organizing a karate tournament. They are painting several large wooden arrows to show directions. Phil suggests finding the area from the single expression
$80 \times 20 + 60 \times 40 \div 2$

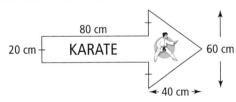

   a) Where do you think the first part of this expression, 80×20, comes from?
   b) Where does the second part of the expression come from?
   c) Twelve arrow signs are needed for the directions. Use brackets to create a new expression for the total area that needs painting. Evaluate this expression.

Extend

16. A five-sided shape can be split into a square and an isosceles triangle. You can find the total area by simplifying this expression:
$6 \times 6 + 6 \times 4 \div 2$
   a) Simplify the expression to find the total area, in square centimetres.
   b) Draw a diagram of this composite shape. Can it be drawn a different way? Explain.
   c) In what different way could you split your shape to find its area?
   d) Use your splitting from part c) to write down a new expression for the area. Check that this gives the same value as the original expression.

17. Roland's Aeronautical School has an airplane crest as a logo.

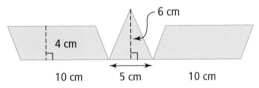

   a) Write an expression that you can use to find the total area of the crest. Evaluate your expression.
   b) What facts did you have to assume about the shape to find its area?

Making Connections

What's My Order?

Use these simple rules to play the game *What's My Order?*
- Create a secret order of operations (different from BODMAS).
- Use it to do a calculation.
- Challenge a classmate to discover what your order is, based on your calculation and result.
- Switch roles with your classmate and play again.

1.5 Area of a Trapezoid

Focus on...
- area
- trapezoids
- base and height

The wave pool at Wild Waterworks in Stoney Creek, Ontario, is the largest of its type in Canada. It creates waves over a metre high.

The pool is roughly in the shape of a **trapezoid**. How can you measure the swimming area? How would it help to split the area up into triangles?

trapezoid
- four-sided figure with one pair of opposite sides parallel

Materials
- ruler
- protractor
- centimetre grid paper and pencil

vertex
- point on a figure where two sides meet

Which measures do I need for each triangle? Which measure is the same for both?

Discover the Math

How can you find the area of a trapezoid?

1. On a blank sheet of centimetre grid paper, draw a large trapezoid. Mark the two parallel sides a and b with parallel markings. Mark the height h.

   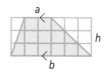

2. **a)** Estimate the area by counting squares.
   b) Measure the lengths of the parallel sides of the trapezoid and label them on your shape. Also measure and label the height of the trapezoid.

3. **a)** Draw a diagonal line from one **vertex** of the trapezoid to the opposite vertex to form two triangles.

   b) Explain how you can find the area of these triangles.
   c) Find the area of the two triangles.
   d) How can you use your answers from part c) to calculate the area of the trapezoid?
   e) Use this method to find the area of the trapezoid.
   f) Compare this result to your estimate from step 2. How close are they?

4. **a)** Try to write a formula that gives the area of a trapezoid, using the symbols A, a, b, and h.
   b) Test your formula. Substitute the measures and calculate the area.

5. **Reflect** Describe the steps you need to find the area of a trapezoid.

Example: Backyard Area

Elvira is replacing the sod in one part of her backyard. Determine the area that must be covered. Round your answer to the nearest square metre.

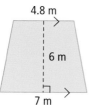

Solution

Method 1: Split Into Triangles
Split the trapezoid into two triangles:
Area of triangle 1 = 4.8 × 6 ÷ 2
            = 14.4
Area of triangle 2 = 7 × 6 ÷ 2
            = 21
Area of trapezoid = Area of triangle 1 + Area of triangle 2
            = 14.4   + 21
            = 35.4
Elvira needs about 36 m² of sod.

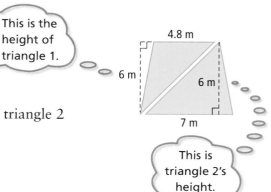

This is the height of triangle 1.

This is triangle 2's height.

Method 2: Use a Formula
Use this formula for the area of a trapezoid:
$A = (a + b) \times h \div 2$
$A = (4.8 + 7) \times 6 \div 2$ **Remember BODMAS. Do brackets first.**
$A = 11.8 \times 6 \div 2$ **Do multiplication and division, from left to right.**
$A = 35.4$

© 4.8 + 7 = × 6 ÷ 2 = 35.4
or © (4.8 + 7) × 6 ÷ 2 = 35.4

Elvira needs about 36 m² of sod.

The grassy area in front of my school is 3 m by 12 m. This is 36 m², about the same as Elvira's backyard.

Key Ideas

- The area of a trapezoid can be found by
  - splitting the trapezoid into two triangles
  - finding the area of each triangle
- The two triangles that make up the trapezoid have the same height.
- The formula for the area of a trapezoid is
  $A = (a + b) \times h \div 2$ or $A = \dfrac{(a + b) \times h}{2}$

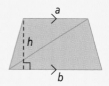

Communicate the Ideas

1. Use pictures and diagrams to explain where the trapezoid formula comes from.

2. Use an organizer to describe the ways in which parallelograms and trapezoids are
   a) alike
   b) different

3. Mya tells Joe that to use the trapezoid area formula, you have to know BODMAS. Joe does not believe that there is any connection. Who is right? Explain.

4. Draw a trapezoid to match this calculation.
   $A = (a + b) \times h \div 2$
   $A = (4 + 6) \times 3 \div 2$
   $A = 10 \times 3 \div 2$
   $A = 15$
   The area of the trapezoid is 15 cm².

Literacy Connections

Making Tables
A table is a form of organizer. You can use a table like this to show likenesses and differences.

	Parallelograms	Trapezoids
Alike		
Different		

Check Your Understanding

Practise

5. Identify the values of a, b, and h in each trapezoid.
   a)   b)
   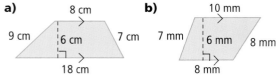
   c)   d)
   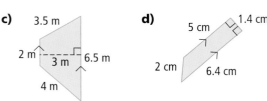

6. For each trapezoid, measure a, b, and h with a ruler.
   a)   b)
   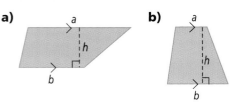

7. Measure a, b, and h.
   a) 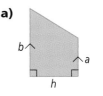   b)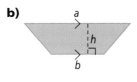

For help with questions 8 to 12, refer to the Example.

8. Find the area of each trapezoid in question 5.

9. Find the area of each trapezoid in questions 6 and 7.

10. Calculate the area of the label on the bottle of sunscreen.
    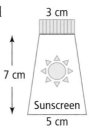

11. Have a friend use a ruler and centimetre grid paper to draw a trapezoid. Find its area.

12. Karsten is replacing the sod in one part of his backyard. Determine the area that must be covered. Round your answer to the nearest square metre.

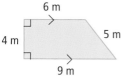

Apply

13. Regional Road signs in Ontario are in the shape of a trapezoid.

How much metal is needed to make 50 signs?

14. Research to find different uses of measurement in technology, the arts, and everyday life. Do any of these involve trapezoids? What about other shapes? Go to your local reference library, or search the Internet. Go to www.mcgrawhill.ca/links/math7 for a place to start.

15. Study these trapezoids. Without calculating, which ones have the same area? How do you know?

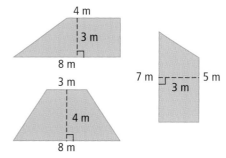

16. a) Use this plan to calculate an approximate area of the wave pool in Wild Waterworks.

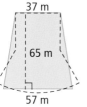

b) How accurate do you think this answer is? Justify your answer.
c) How could you improve the accuracy?
d) Why is the wave pool such an unusual shape?

 17. A quilt pattern uses green and yellow trapezoids. Each green piece is the same shape and size.

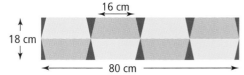

a) Find the area of each trapezoid.
b) Find the total green area.
c) Find the total blue area.
d) Explain how you solved each part.

Extend

18. Alex is designing a sign in the shape of a trapezoid.

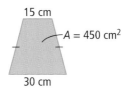

a) What is the height of Alex's sign?
b) Explain how you found your answer.

19. There is another way to discover the formula for the area of a trapezoid. You can use a parallelogram instead of two triangles.

a) Draw and cut out two identical trapezoids. Measure the lengths of the parallel sides, a and b, and the height, h.
b) Use the two trapezoids to create a parallelogram.
c) How can you now find the area of each trapezoid?

1.6 Draw Trapezoids

Focus on...
- perimeter and area of trapezoids
- constructing trapezoids

Trapezoids are common shapes in large structures like bridges and buildings. Computer software, such as *The Geometer's Sketchpad®*, makes it easy to create and adapt drawings and design models.

Even with pencil and paper, you can explore properties of trapezoids, such as area and perimeter.

Discover the Math

Materials
- ruler
- centimetre grid paper
- scissors
or
- centimetre linking cubes

Optional:
- BLM 1-6A Use a Geoboard to Construct Trapezoids

How can you construct a trapezoid, given its perimeter or area?

Part 1: Draw a trapezoid, given its perimeter

1. On a sheet of centimetre grid paper, try to draw a trapezoid that has a perimeter of 20 cm. Draw lightly, since you may need to make changes to your figure.

2. a) Measure the length of each side.
   b) Add the lengths to get the perimeter. How close to 20 cm were you able to get?
   c) Describe the process you used to draw this trapezoid.

3. a) Use 24 centimetre cubes to create a single, long strip.
   b) Split the strip into four pieces. Choose your own lengths.
   c) Using the four pieces, try to form a trapezoid with perimeter 24 cm. You can change your lengths if you need to.
   d) When you have a good trapezoid model, copy it on a piece of centimetre grid paper. Hint: Measure the sloping sides.

34   MHR • Chapter 1

4. **Reflect** Review your methods and difficulties in completing steps 1 to 3. How can you improve at drawing trapezoids when the perimeter is given?

Part 2: Draw a trapezoid, given its area

1. **a)** Design a trapezoid that has an area of 40 cm². Use centimetre cubes or strips of centimetre grid paper to create your design.
   b) Draw your trapezoid on centimetre grid paper.
   • Draw and label the height, *h*.
   • Verify that *a* and *b* are parallel. Adjust your diagram if necessary.

2. **a)** Measure *a*, *b*, and *h*.
   b) Calculate the area of your trapezoid.
   c) How close to 40 cm² were you able to get?
   d) Describe the process you used to draw this trapezoid.

3. **a)** Try to create a trapezoid that has a perimeter of 40 cm and an area of 60 cm². Describe the method you used to construct the trapezoid.
   b) Calculate the perimeter and area. How close were you?

4. Create a problem similar to the one in step 3. Trade with a partner. Try to solve each other's problem.

5. **Reflect** Review your methods and difficulties in both parts of the activity. How can you improve at drawing trapezoids, when you are given

   a) the perimeter   **b)** the area   **c)** the perimeter *and* the area?

Key Ideas

- You can construct a trapezoid by creating a quadrilateral with one pair of opposite sides parallel.

Where should I put the sides?

- You can change the area or perimeter of a trapezoid by changing its side lengths and/or its height.

The sides need to be longer. I'll put the parallel sides farther apart.

Communicate the Ideas

1. a) Describe a method you prefer to draw a trapezoid with a given perimeter.
   b) What difficulties might you have?

2. a) Describe a method you prefer to draw a trapezoid with a given area.
   b) What difficulties might you have?

3. a) Describe how you can construct a trapezoid with a given perimeter and area.
   b) What strategies or shortcuts did you need to learn to solve this type of problem? Compare your strategies with a classmate's.

Check Your Understanding

Practise

For questions 4 to 6, use centimetre grid paper or centimetre linking cubes.

4. Draw a trapezoid with each perimeter. Use whole numbers for the side lengths.
   a) 18 cm   **b)** 26 cm
   c) 10 cm   **d)** 40 cm

5. Draw a trapezoid with each area.
   a) 20 cm^2 **b)** 80 cm^2
   c) 36 cm^2 **d)** 72 cm^2

6. Draw a trapezoid with perimeter 30 cm and area 30 cm^2.

Apply

7. a) Draw a trapezoid with perimeter 40 cm.
   b) Calculate the area of the trapezoid you drew in part a).
   c) Explore making other trapezoids with perimeter 40 cm, but different areas. Hint: Use grid paper for sketches.

8. Describe your methods for questions 4 to 7.

 9. a) Draw a trapezoid with perimeter 50 cm. Can you do this in more than one way? Explain.
   b) Draw another trapezoid with perimeter 50 cm and area 80 cm^2. Describe your techniques.

Extend

For question 10, use centimetre grid paper.

10. a) Draw a triangle with vertical side 4 cm and horizontal side 3 cm. Measure the sloping side. How long is it? Does your measure appear exact?
   b) Use your triangle from part a) to create trapezoids that fit exactly on centimetre grid paper. Label dimensions on your trapezoids.
   c) Look for, and describe, any patterns in the perimeters of your trapezoids.
   d) Look for, and describe, any area patterns.

Use Technology

Focus on...
- constructing trapezoids

This is another way to do the investigation on pages 34 and 35.

Materials
- computers
- *The Geometer's Sketchpad®* software, Version 4.0

Alternatives:
- TECH1.6A Construct and Manipulate a Trapezoid (GSP 4)
- TECH1.6B Construct and Manipulate a Trapezoid (GSP 3)

Technology Tip
- Holding the **Shift** key while dragging makes it easier to draw a horizontal segment.

Construct and Manipulate a Trapezoid Using *The Geometer's Sketchpad®*

Part 1: Construct a Trapezoid

1. Open *The Geometer's Sketchpad®* and begin a new sketch.

2. From the **Edit** menu, choose **Preferences**. Set preferences as shown.

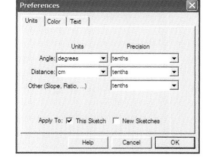

3. **a)** Choose the **Straightedge Tool** by clicking on [/.] on the Toolbar. The Toolbar is at the left of the screen.

   b) Using the **Straightedge Tool**, draw a horizontal line segment near the top of the sketch.

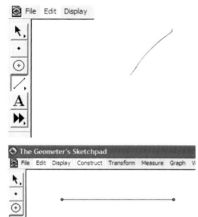

4. Label one endpoint A.
   a) Choose the **Text Tool** from the Toolbar.
   b) Move the cursor to one endpoint until the cursor's hand turns dark.
   c) Click on the point. A letter label should appear.
   d) To change the letter to "A," check that the cursor is still dark, and double-click. This panel should appear.
   Type "A" in the panel, and click **OK**.

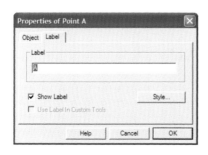

5. Repeat step 4 to label the other point B.

6. a) Choose the **Selection Arrow Tool** from the Toolbar.
b) Click in a blank part of the screen. This is called **deselecting**.
c) Use the **Selection Arrow Tool** to try selecting and deselecting points A and B and the line segment AB, singly and in combinations.
d) Try moving a label around an object by clicking and dragging it. Then, put it back neatly.
e) Deselect again.

> **Technology Tip**
> • Before you select a new object, make sure that you deselect first, by clicking somewhere in the white space. Then, select the objects you want.

7. a) Choose the **Point Tool** from the Toolbar. Place a point below and to the right of the line segment.
b) Label the new point C.

8. a) Choose the **Selection Arrow Tool**. Keeping point C selected, select the line segment AB.
b) From the **Construct** menu, choose **Parallel Line**. A line will appear across the screen. The line should go through point C. It should be parallel to line segment AB.
c) Choose the **Point Tool** from the Toolbar. Place a second point on the line through C. Label this point D.

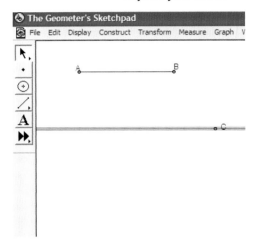

9. a) Deselect, and select the line through C and D. From the **Display** menu, choose **Hide Parallel Line**. Your sketch should look like this.
If you accidentally hide some points as well, choose **Undo Hide** from the Edit menu.
b) Deselect, and select the points A, B, C, and D, *in clockwise order*.

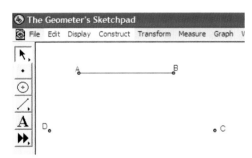

c) From the **Construct** menu, choose **Segments**. A trapezoid should appear.
If your diagram looks like a bow tie, you probably selected the points in the wrong order. Choose **Undo Construct Segments** from the **Edit** menu. Then, select the points in the correct order and construct the trapezoid.

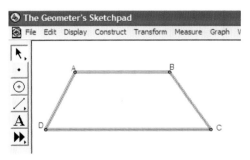

10. Click on and drag any corner of the trapezoid. Does the shape stay a trapezoid? Why or why not? Hint: What is true about sides AB and CD?

Part 2: Measure and Manipulate the Trapezoid

1. Begin with your trapezoid from part 1.
   a) To make sure nothing is selected, deselect.
   b) Select points A, B, C, and D again, in clockwise order.
   c) From the **Construct** menu, choose **Quadrilateral Interior**. The inside of the trapezoid should become coloured. You are now ready to measure your trapezoid.

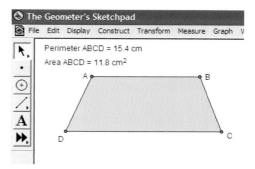

2. a) From the **Measure** menu, choose **Perimeter**. Then, deselect the measure.
   b) Select the trapezoid interior. From the **Measure** menu, choose **Area**.

3. a) Click and drag one corner and describe what happens to the trapezoid.
   b) Repeat part a) for the other corners.

4. Move the corners to create a trapezoid with a perimeter of 30 cm. Can you do this by moving just one point? Explain.

5. Move the corners to create a trapezoid with an area of 40 cm². (The perimeter need not be 30 cm.) Can you do this by moving just one point? Explain.

6. a) Try to create a trapezoid that has a perimeter of 40 cm and an area of 60 cm². Can you do this by moving just one point? Which point is best to move, and why?
   b) Describe what you did.

7. **Reflect** Review your methods and problems in both parts of the activity. Explain how you created trapezoids with
   a) given perimeters
   b) given areas
   c) given perimeters and areas

1.7 Composite Shapes

Focus on…
- splitting and combining shapes
- perimeter
- area

Earlier, you found the area of a trapezoid by splitting the shape up into two triangles.

Suppose that you want to lay sod in this backyard. You also want to fence the perimeter. Can you split the backyard into simpler shapes to help you find its perimeter and area?

Discover the Math

composite shape
- two-dimensional shape that can be split into two or more simpler shapes

What strategies can I use to measure composite shapes?

Example 1: Area of a Composite Shape

Nina's backyard needs to be covered with fresh sod. Sod costs $8.99 per square metre. How much sod will Nina need? How much will it cost?

Solution

- What is the area of the backyard?
- How much will it cost to sod?

Plan

Do It!

1. Identify the missing dimensions. Then, add them to the diagram.
2. To find the area, split the composite shape into two rectangles.
3. Find the area of each rectangle. Add to get the total area.
4. Multiply the area by the cost of sod per square metre to find the total cost.

1. The missing dimensions are 4 m and 12 m.

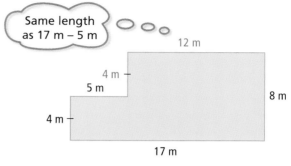

2. Split the backyard like this:

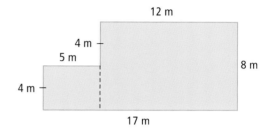

3. Area of small rectangle = length × width
   = 4 × 5
   = 20
   Area of large rectangle = length × width
   = 12 × 8
   = 96
   Add the areas of the two rectangles.
   Total area = Area of small rectangle + Area of large rectangle
   = 20 + 96
   = 116
   The area of Nina's backyard is 116 m².

4. Cost of sod = 116 × 8.99
   = 1042.84
   The total cost for sod is $1042.84.

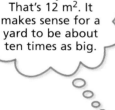

Look Back

- The large rectangle looks about five times the area of the small one. 20 × 5 = 100. That's close to 96 m².
- For the cost, estimate: 100 m² × $10 per square metre is $1000. This is close to the calculated cost.

Example 2: Cost of a Fence

Nina's family wants to build a fence around their backyard. Fencing costs $19 per metre. Nina offers to use her math skills to calculate the cost. Model her solution.

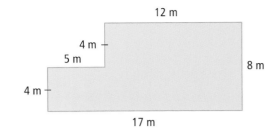

Solution

$P = 4 + 5 + 4 + 12 + 8 + 17$ **Add lengths, going around the outside of the shape.**
$P = 50$
The perimeter of Nina's backyard is 50 m.
Cost of fencing = 50×19
= 950
The total cost for the fence is $950.

> I can't just add the rectangle perimeters. If I did, I'd include lengths that go through the *inside* of the yard. Nina doesn't want a fence through the middle of her yard!

Example 3: Go-Kart Side Panels

Rupau is constructing two side panels for his go-kart. He also wants to paint the outside of each panel.
a) Find the perimeter that must be cut.
b) Find the area that must be painted.

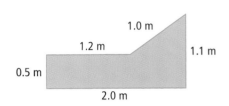

Solution

a) $P = 1.1 + 2.0 + 0.5 + 1.2 + 1.0$
$P = 5.8$
Rupau must cut a perimeter of 5.8 m.

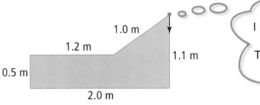

> I need to pick a starting point on the shape. Then, I can add lengths as I go all the way around.

b) To find the total area, split the composite shape into a rectangle and a triangle.

Shape	Diagram	Calculation	Area
Rectangle	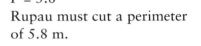	Area of rectangle = $l \times w$ $= 2.0 \times 0.5$ $= 1$	1 m²
Triangle	0.6 m, 0.8 m	Area of triangle = $b \times h \div 2$ $= 0.8 \times 0.6 \div 2$ $= 0.24$	0.24 m²
		Area of each side panel	1.24 m²

Strategies
What strategy is used here?

Area of both side panels = 2×1.24
= 2.48
Rupau needs to paint an area of 2.48 m².

Key Ideas

- You can find the area of a composite shape by dividing it into simpler shapes. Add the areas of the simpler shapes to get the total area.

- The perimeter of a composite shape is the total distance *around the outside*.

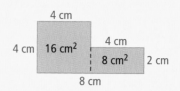

Communicate the Ideas

1. Draw an example of a composite shape that contains two simpler shapes.

2. Identify three or more examples of composite shapes that appear at home, in school, or elsewhere.

3. Identify the simple shapes that combine to make the composite shapes in questions 1 and 2. Find the area of each composite shape:
   a) Split the shape.
   b) Find the area of each part. Then, find the total area.

Check Your Understanding

Practise

For questions 4 to 6, refer to Example 1.

4. Find the area of this composite shape.
   a) Split the shape.
   b) Find the area of each part. Then, find the total area.

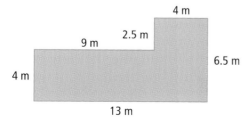

5. Find the missing dimensions of each composite shape.

   a)

   b)

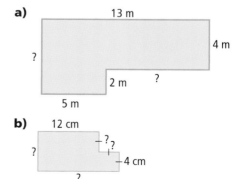

6. Find the area of each shape in question 5.

For questions 7 and 8, refer to Example 2.

7. Calculate the perimeter of this shape.

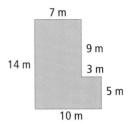

8. Find the perimeter of each figure in questions 4 and 5.

9. Describe how to split each composite shape into simpler shapes.

a)

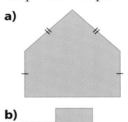

b)

For questions 10 and 11, refer to Example 3.

10. Find the perimeter and the area of each shape in question 9. Measure any dimensions you need.

11. Naveed is building a frame for the front of his shed. Determine the total length of wood needed.

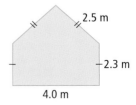

12. To find the area of a composite shape, you can add areas of simpler shapes. Why does this not work for the perimeter?

13. Sabra is the manager of a rock band called M-pathy. She is planning to order concert T-shirts with the band's logo.

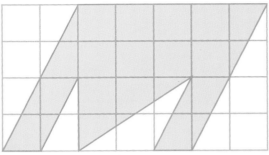

Embroidery costs $0.25 per square centimetre. Sabra needs to find the total cost to embroider 300 shirts.

a) Which strategy will you use to solve this problem? Why?
b) Use your strategy to solve the problem.
c) Show how you could have used a different strategy.

Apply

Chapter Problem

14. Sarah is constructing two side panels for her go-kart. She wants to paint the outside of each panel.

a) Find the perimeter that must be cut.
b) Find the area that must be painted.

15. Choose a room in your home or school. Suppose you decided to wallpaper the walls of this room.

a) What total area would you need to wallpaper?
b) You want to add a wallpaper border. Where will you place it? How much border will you need?
c) Research the cost of wallpaper and borders. How much will it cost to redecorate the room?

16. Example 1 showed one way to split Nina's backyard into simpler shapes.

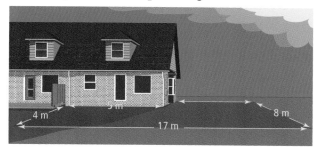

a) Find another way to split Nina's backyard. Use this method to find the total area.
b) Compare this answer to the one found in Example 1. Does this make sense? Explain.

17. a) Josh solved Nina's problem a different way: "I thought of the shape as a large rectangle with a small rectangle removed." How do you think Josh calculated the area?

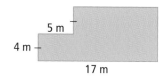

b) Find the area using Josh's method.
c) Compare this answer to the one found in Example 1. Does this make sense? Explain.

18. The square shown was created on a geoboard. The horizontal and vertical distance between pegs is 1 cm.

a) How can you find the area of this square?
b) Describe another way that you could solve this problem.

19. You can estimate the swimming area in this wave pool by splitting up the pool into a rectangle and a trapezoid.

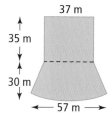

a) Find the area of the rectangle.
b) Find the approximate area of the whole pool. Compare this result to the one you found earlier (page 33, question 16).
c) Which result do you think is more accurate? Explain why.
d) Can you find a more accurate answer? If so, describe how. If not, explain why not.

Extend

20. Repeat question 18 for this geoboard square.

21. Blue County is planning to paint 1320 new stop signs. Each stop sign is a regular octagon. The diagram shows the dimensions.

a) Each can of red paint covers 2 m² and costs $5. How much will all the red paint for this project cost? Justify any estimates you made to find your answer.
b) Extend your budget estimate to include other costs. Consider the white paint for the edging and the word STOP, labour costs, gas for the trucks, and so on.

CHAPTER 1 Review

Key Words

Match the key words to the correct descriptions.

1. the perpendicular distance from the base of a shape to its opposite side or vertex
2. a shape that can be split into two or more simpler shapes
3. these contain operations to be done first
4. calculated by splitting into two different-shaped triangles
5. a measure of how much space a two-dimensional shape covers

A area
B area of a parallelogram
C area of a trapezoid
D height
E brackets
F composite shape

1.1 Perimeters of Two-Dimensional Shapes, pages 12–17

6. Find the perimeter of each shape.

a)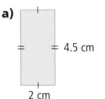
4.5 cm
2 cm

b)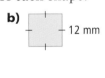
12 mm

c)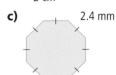
2.4 mm

d)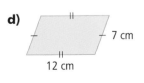
7 cm
12 cm

7. What length of fence is needed to surround the yard shown?

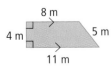
8 m
4 m   5 m
11 m

1.2 Area of a Parallelogram, pages 18–21

8. What is the area of the top of the machine part?

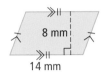
8 mm
14 mm

9. Draw a parallelogram to match this calculation.
$A = b \times h$
$A = 6 \times 3$
$A = 18$
The area of the parallelogram is 18 cm².

1.3 Area of a Triangle, pages 22–25

10. A park is bounded by a river and two roads.

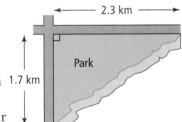
2.3 km
1.7 km
Park

a) Find the area of the park.
b) Describe your method.
c) How accurate do you think your answer is? Explain.

11. Karsten is designing a flag to fly at the back of his go-kart. Determine the area of Karsten's flag.

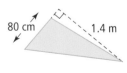
80 cm   1.4 m

1.4 Apply the Order of Operations, pages 26–29

12. Evaluate each expression.
a) $13 - 9 \div 3$
b) $3 \times (16 \div 2) - 5$
c) $8 \times 3 \div 6 \div 3$
d) $20 + (12 - 2) \div 5 \times 3$

13. What's wrong? Find each error and explain how to correct it.

   a) $4 \times 4 + 6 \div 2$
   $= 16 + 6 \div 2$
   $= 22 \div 2$
   $= 11$

   b) $81 \div 9 \div 3$
   $= 81 \div 3$
   $= 27$

14. Frieda is designing a logo for her hockey team, the Bulls.

    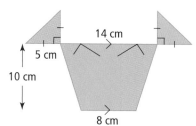

    a) Write an expression for the area of each ear.
    b) Write an expression for the area of the face.
    c) Find the total area.

1.5 Area of a Trapezoid, pages 30–33

For questions 15 and 16, refer to this plan of the side panel of a CD storage case.

15. a) What shape is the side panel? Explain how you know.
    b) Determine the area of the side panel.

    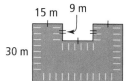

16. Split the side panel shape into simpler shapes. Find the area of each part and add them together to get the total area. Compare this to your answer to question 15b). Does this make sense? Explain.

1.6 Draw Trapezoids, pages 34–36

17. a) Can all trapezoids be split into a rectangle and a triangle? Support your answer with diagrams.
    b) Can you draw a trapezoid that cannot be split into two triangles? Explain.

18. a) Draw a trapezoid that has a perimeter of 36 cm.
    b) Explain your method.

19. a) Draw a trapezoid that has an area of 48 cm².
    b) Explain your method.
    c) Calculate the area of the trapezoid you have drawn. How close is it to 48 cm²?

1.7 Composite Shapes, pages 40–45

20. a) Find the missing dimensions of the parking lot.
    b) Find the area.
    c) Find the perimeter.

    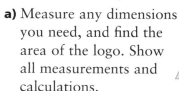

21. The Bulls' archrivals are the Electric, whose logo is shown.
    a) Measure any dimensions you need, and find the area of the logo. Show all measurements and calculations.
    b) Measure any dimensions you need, and find the perimeter of the gold trim. Show all measurements and calculations.

Chapter 1 Practice Test

Strand	NSN	MEA	GSS	PA	DMP
Questions	1–8, 11–14	1–7, 11–14	9, 10		

Multiple Choice

For questions 1 to 5, choose the best answer.

1. What is the perimeter of this shape?

   A 3.5 cm **B** 7 cm
   C 14 cm **D** 21 cm

2. Matt is adding a piece of wood to the side of a ladder. What is the area of wood that Matt must cut?

   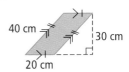

   A 300 cm² **B** 400 cm²
   C 600 cm² **D** 800 cm²

3. What is the area of the blue region of the flag?

   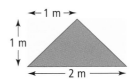

   A 0.5 m² **B** 1 m²
   C 2 m² **D** 4 m²

4. The perimeter of the trapezoid is

   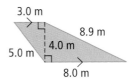

   A 12 m **B** 20.9 m
   C 24.9 m **D** 28.9 m

5. Look at the trapezoid in question 4. The area is
   A 6 m² **B** 22 m²
   C 23 m² **D** 32 m²

Short Answer

6. Identify each figure. Then, find its area.

   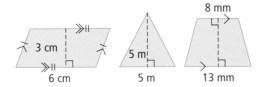

7. Simplify each expression, following the correct order of operations. Show all steps.
   a) $5 + 9 \div 3$
   b) $12 - (6 - 3)$
   c) $3 \times (4 - 2 + 5)$

8. Simplify each expression. Show all steps.
   a) $6 + 12 \div 3 - 4 \div 2$
   b) $2.4 + 3 \times 1.1 + 4.8 \div (4 \div 0.2)$

9. a) Draw a trapezoid that has a perimeter of 26 cm.
   b) Explain how you drew the trapezoid.
   c) Draw a different trapezoid with the same perimeter. Compare the areas of the two trapezoids.

10. a) Draw a trapezoid that has an area of 38 cm².
    b) Calculate the area of the trapezoid you have drawn. How close is it to 38 cm²?
    c) Draw a different trapezoid with the same area. Compare the perimeters of the two trapezoids.

11. a) Draw a two-dimensional shape to match this area calculation.
$A = (a + b) \times h \div 2$
$A = (15 + 9) \times 4 \div 2$
$A = 48$
The area is 48 cm².
b) Find the perimeter of your shape.

Extended Response

12. The layout of an outdoor fairground is shown.

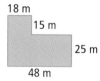

a) Copy the composite shape. Show how you can split it into simpler shapes.
b) Calculate the area of the fairground.
c) Find the length of fencing needed to surround the whole perimeter.
d) Fencing costs $15 per metre. What will it cost for the entire fence?

13. a) Find the perimeter of the building shown in the floor plan.

b) Copy the composite shape. Show two different ways you can split it.
c) Find the area using each way to split the shape. Are your answers the same? Explain.

14. The formula for the perimeter of a rectangle is $P = (2 \times l) + (2 \times w)$. Can this also be written as $P = 2 \times (l + w)$? Use the example shown here and at least one other to explain.

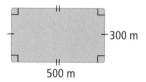

Chapter Problem Wrap-Up

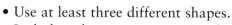

1. Use these shapes to design a model go-kart.
   - Use at least three different shapes.
   - Include at least one composite shape.
   - Decide on the sizes of the shapes you will use.
   - If you use round wheels, do not include them in any calculations.
   Include sketches of your design.

2. Calculate
   a) the area of each shape
   b) the total area of material you will need
   c) the total length of all cuts

Geometry and Spatial Sense
- Identify, describe, compare, and classify geometric figures.
- Identify congruent and similar figures.
- Use mathematical language effectively to describe geometric concepts, reasoning, and investigations.
- Identify two-dimensional shapes that meet certain criteria.
- Identify and explain why two shapes are congruent.
- Create and solve problems involving congruence.

Measurement
- Demonstrate an understanding of and apply accurate measurement strategies.

Key Words

equilateral triangle
isosceles triangle
scalene triangle
acute triangle
right triangle
obtuse triangle
quadrilateral
congruent
similar

CHAPTER 2

Two-Dimensional Geometry

Geometric shapes are used in the design of buildings, bridges, vehicles, clothing, toys, and more. What two-dimensional shapes do you recognize in the pattern used on the material in the photograph?

Artists and designers learn a lot about shapes. They explore ways of putting shapes together to make attractive designs.

Chapter Problem

Patterns on items such as fabric, wallpaper, and floor tiles often repeat. Each part of the pattern is called a pattern block. Look at the designs on items around you. Choose one that includes geometric shapes. Sketch a pattern block that you like from the design.

- Describe what you like about the pattern.
- Explain how the design is created.
- Describe what happens to the pattern block on each repetition. Is the pattern block turned through 90° or flipped upside-down? Are the colours changed?

In this chapter you will explore how shapes are used to create attractive patterns. You will make your own patterns and designs.

Get Ready

Line Segments

A **line segment** joins two points. It is named by its endpoints. The line segment shown is named AB or BA.

1. Name four different line segments in the photo of part of a hydro tower.

2. Use a ruler to measure each line segment that you named in question 1. Measure to the nearest tenth of a centimetre.

3. Some line segments in the photo have the same length. Name one pair. Use a ruler to check.

4. Find two line segments that are contained within another line segment. Name them.

Angles

- An angle is formed when two **rays** meet at a point **vertex**. An angle is named using one point on each and the vertex. The vertex is always named second. The angle shown is ∠PQR or ∠RQP. When there is clearly only one angle referred to, it can be named by just the vertex: ∠Q in this case.
- A protractor is used to measure the size of angles, in degrees. ∠Q = 65°.

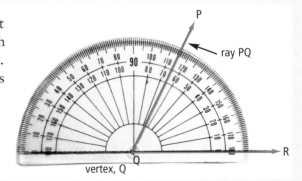

5. Use a protractor to measure each angle in the photo of the hydro pylon.
   a) ∠DAB   b) ∠DBF   c) ∠AGH

6. Use a ruler and a protractor to draw an angle with each measure.
   a) 55°   b) 90°   c) 120°

Classify Angles

Angles are classified by their size.

An **acute angle** measures less than 90°.

A **right angle** measures 90°.

An **obtuse angle** measures more than 90° but less than 180°.

7. Classify each angle that you drew for question 6.

8. Find an example of each type of angle in the hydro tower photo.

Transformations

The diagram shows a translation of 2 units to the right and 3 units up. △KLM is the translation image of △ABC.

The diagram shows a reflection in the reflection line, *m*. △PQR is the reflection image of △DEF.

The diagram shows a rotation of 90° clockwise about the turn centre I. △JKI is the rotation image of △GHI.

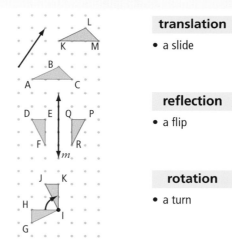

translation
• a slide

reflection
• a flip

rotation
• a turn

9. Copy △PQR onto dot or grid paper. Use different colours to show the image for each translation.

a) 5 units to the right and 3 units up
b) 3 units to the left and 4 units down

10. Make a new copy of △PQR. Show the image for each reflection.

a) in a horizontal reflection line that passes through P
b) in a vertical reflection line that passes through R

11. Make another copy of △PQR. Show the image for each rotation.

a) 90° clockwise about P
b) 180° counterclockwise about Q

2.1 Classify Triangles

Focus on...
- triangle types
- comparing sides
- comparing angles

The photo shows a truss bridge over the Erie Canal. What different shapes can you identify in the bridge?

Discover the Math

Materials
- ruler
- triangle dot paper
- scissors

Alternative
- BLM 2.1A Use a Geoboard
- TECH 2.1A Use *The Geometer's Sketchpad* ® to Explore Classifying Triangles (GSP 4)
- TECH 2.1B Use *The Geometer's Sketchpad* ® to Explore Classifying Triangles (GSP 3)

How can you classify triangles?

Work with a partner.
1. a) Draw five different triangles on triangle dot paper.
   b) Measure the length of each side of your triangles. Write each measure close to the side, inside the triangle.
   c) Cut out your five triangles.

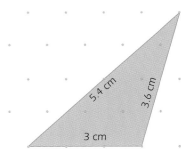

2. a) Compare your triangles with those that your partner made. Sort the ten triangles into groups.
   b) Write a description of each of your groups. What characteristics do the triangles in the group have in common?

3. Use the **Glossary** at the back of this text to find a definition of each of the following:
   - **equilateral triangle**
   - **isosceles triangle**
   - **scalene triangle**

   Classify each of your groups of triangles.

4. **a)** Take each of your five triangles and flip them over.
   b) Recall the different types of angles: acute, right, and obtuse. What type are the angles in each of your triangles? Write the angle type inside each vertex.

5. **a)** Compare your triangles with those that your partner has.
   Sort the ten triangles into groups by their angle types.
   b) Write a description of each group. What characteristics do the triangles have in common?

6. Use the **Glossary** to find a definition of each of the following:
   - **acute triangle**
   - **right triangle**
   - **obtuse triangle**

   Classify each of your groups of triangles.

7. **Reflect** Triangles are classified in two different ways. Write a brief summary of the two ways.

Literacy Connections

Using a Glossary
The Glossary starts on page 481. It lists mathematical terms in alphabetical order. Each word is defined.

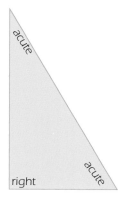

Example 1: Classify a Triangle

a) Measure the side lengths and the angles of △ABC.
b) Classify the triangle in two ways. Give reasons for your answers.

Solution

a) AB = 3 cm   ∠A = 90°
AC = 4 cm   ∠B = 53°
BC = 5 cm   ∠C = 37°

∠A looks like the corner of a sheet of paper which is 90°. Measure with a protractor to check.

b) △ABC is scalene, because the side lengths are all different.
△ABC is a right triangle, because ∠A measures 90°.

2.1 Classify Triangles • MHR 55

Example 2: Name and Classify Triangles

a) Identify all the triangles in the diagram.
b) Classify each triangle by its angle measures.

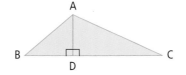

Literacy Connections

Reading Diagrams
To interpret some diagrams, it may help to cover parts with your finger or with a piece of paper. Do this so that you can look at one shape at a time.

Solution

a)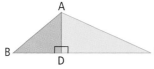

There are three different triangles: △ABC, △ABD, and △ACD.

b)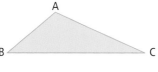

△ABC is an obtuse triangle, because ∠BAC is an obtuse angle.

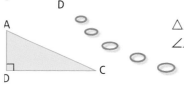

△ABD is a right triangle, because ∠ADB is 90°.

△ACD is a right triangle, because ∠ADC is 90°.

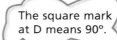
The square mark at D means 90°.

Key Ideas

- Triangles can be classified by their side lengths.

equilateral triangle	**isosceles triangle**	**scalene triangle**
• three equal sides	• two equal sides	• no equal sides

- Triangles can be classified by the size of their angles.

acute triangle	**right triangle**	**obtuse triangle**
• three acute angles	• one right angle	• one obtuse angle

   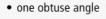

Communicate the Ideas

1. What parts of a triangle do you compare to decide whether a triangle is equilateral, isosceles, or scalene?

2. How do you decide whether a triangle is acute, right, or obtuse?

3. Which one of the four triangles shown does not belong in the same group as the other three? Give reasons.

4. Sketch an acute scalene triangle. Justify your sketch.

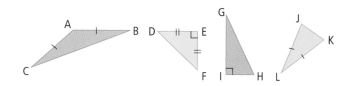

Check Your Understanding

Practise

For help with Questions 5 to 9, refer to Example 1.

5. Classify each triangle as equilateral, isosceles, or scalene. Explain your choice.

   a) 3 cm, 3 cm, 4 cm
   b) 2 cm, 7 cm, 8 cm

6. Classify each triangle by its side lengths. Explain your choice.

   a) 5 cm, 12 cm, 13 cm
   b) 5 cm, 5 cm, 5 cm

7. Classify each triangle as acute, right, or obtuse. Explain your choice.

   a) 60°, 30°, right angle
   b) 80°, 60°, 40°

8. Classify each triangle by its angle measures. Explain your choice.

   a) 30°, 120°, 30°
   b) 70°, 70°, 40°

9. Classify each triangle in two ways. Give reasons for your answers.

   a) 5 m, 4 m, 3 m (right angle)
   b) 4 cm, 100°, 5 cm, 45°, 35°, 7 cm

For help with questions 10 and 11, refer to Example 2.

10. a) Name all the triangles in the figure.
    b) Classify each triangle by its angle measures.

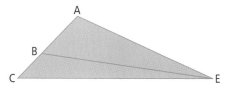

11. a) Name all the triangles in the figure.
    b) Classify each triangle in two ways.

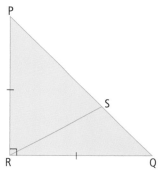

Apply

12. Look at the objects in the pictures.

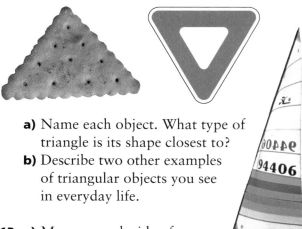

   a) Name each object. What type of triangle is its shape closest to?
   b) Describe two other examples of triangular objects you see in everyday life.

13. a) Measure each side of △PQR, to the nearest tenth of a centimetre.
   b) Measure each angle.
   c) Classify the triangle in two ways.

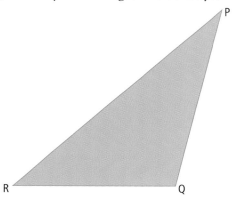

14. Draw each triangle using a ruler and a protractor. Measure and record each side and angle that is not given. Classify each triangle in two ways. Hint: First sketch a triangle and mark the given information. Plan your steps. Draw and label the first side. Then measure and mark the first angle.

   a) △ABC with AB = 5 cm, ∠A = 60°, AC = 5 cm
   b) △DEF with DE = 4 cm, ∠D = 60°, ∠E = 60°

15. Draw each triangle using a ruler and a protractor. Measure and record each side and angle that is not given. Classify each triangle in two ways.

   a) △KLM with KL = 8 cm, KM = 8 cm, ∠K = 40°
   b) △PQR with PQ = 6 cm, ∠P = 80°, ∠Q = 50°

16. Use a ruler and a protractor to draw each triangle. Then, classify the triangle in two ways.

   a) one angle of 65° between sides measuring 5 cm and 5 cm
   b) one side measuring 6 cm between angles of 45° and 45°

17. a) How do you know that △XYZ is an isosceles triangle?

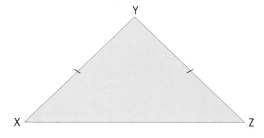

   b) Are any of the angles in △XYZ equal? If so, which ones?

18. Use a ruler and a protractor to draw an isosceles triangle that has
   a) only one 30° angle
   b) two 30° angles
   Classify each triangle by its angle measures.

19. Use a ruler and a protractor to draw △RST. In the triangle RT = 5 cm, ∠R = 60°, and ∠T = 60°.
   a) Measure and record each side and angle that is not given.
   b) Classify △RST by its side measures and by its angle measures.

20. Use diagrams to support your answers.
  a) When one angle in a triangle is a right angle, what type of angle are the other two angles?
  b) When one angle in a triangle is obtuse, what type of angle are the other two angles?
  c) In an equilateral triangle, what is the measure of each angle?

21. There are many paper airplane models. Here is one example.

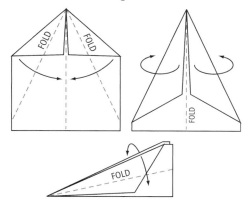

  a) What types of triangles do you see in the model shown?
  b) Draw two other designs for paper airplanes. What types of triangles do they involve? Go to **www.mcgrawhill.ca/links/math7** and follow the links if you need some suggestions.

22. a) Draw a triangle with one acute angle.
  b) What type(s) of angle are the other two angles in your triangle? What type of triangle have you drawn?
  c) Can you draw one or more triangles, with one acute angle, that give a different answer to part b)? Demonstrate and explain.

Extend

23. Why do you often see triangles in bridge designs? What type of triangle occurs most? Conduct Internet research to find out more about the different types of truss bridges and the role of triangle in their design.

Go to **www.mcgrawhill.ca/links/math7** and follow the links to some helpful sites with information on bridges.

24. Compare the sum of any two side lengths of a triangle with the length of the third side. What relationship is true?

Making Connections

Symmetry

Draw a line from any vertex of an equilateral triangle through the middle of the opposite side. This is a line of symmetry. You can fold the triangle along this line and the sides match.

1. How many lines of symmetry does an equilateral triangle have?
2. How many lines of symmetry does an isosceles triangle have? Draw a diagram to illustrate your answer.
3. How many lines of symmetry does a scalene triangle have?

2.2 Classify Quadrilaterals

Focus on…
- quadrilaterals
- side lengths
- angle measures
- parallel sides

A tangram is a geometric puzzle that was invented in China. In the puzzle, a square is divided into seven geometric shapes. Many other figures can be made by rearranging the seven pieces.

Try to use all seven pieces of a tangram to make each of the **quadrilaterals** shown.

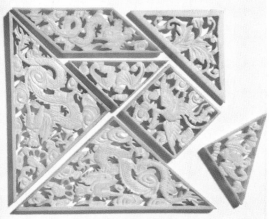

quadrilateral
- a closed shape with four straight sides

Discover the Math

Materials
- ruler
- protractor

Optional
- BLM 2.2A Quadrilaterals

How can you distinguish quadrilaterals?

1. There are six special types of quadrilaterals. How would you sort them into two groups? Which shapes would you group together? Why?

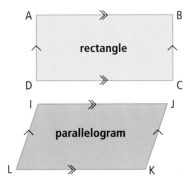
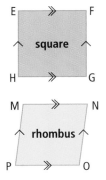

60 MHR • Chapter 2

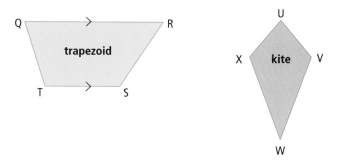

2. Copy and complete the table to compare the types of quadrilaterals. The first one has been done for you.

Quadrilateral Type	Side Lengths	Angle Measures	Parallel Sides?
Rectangle	opposite sides are equal	all angles are 90°	two pairs of opposite sides are parallel
Square			
Parallelogram			
Rhombus			
Trapezoid			
Kite			

3. **Reflect** How do you classify quadrilaterals? Write a short description of the features you need to look at to name the type of quadrilateral.

Example 1: Classify Quadrilaterals

Classify each quadrilateral. Give reasons.

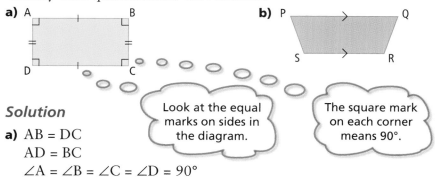

Look at the equal marks on sides in the diagram.

The square mark on each corner means 90°.

Solution

a) AB = DC
   AD = BC
   ∠A = ∠B = ∠C = ∠D = 90°
   Two pairs of opposite sides have equal lengths. All four angles are 90°.
   Quadrilateral ABCD is a rectangle.

b) No sides are marked as equal.
   One pair of opposite sides is parallel.
   Quadrilateral PQRS is a trapezoid.

Example 2: Identify a Quadrilateral

A certain quadrilateral has two pairs of opposite sides that are equal and parallel. The quadrilateral contains no right angles. Identify and draw the quadrilateral.

Solution

Method 1: Draw a Diagram

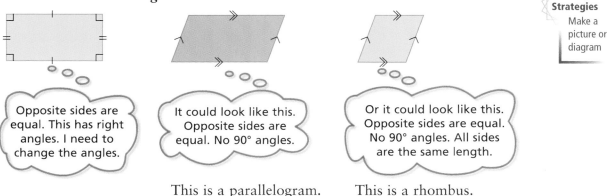

Opposite sides are equal. This has right angles. I need to change the angles.

It could look like this. Opposite sides are equal. No 90° angles.

Or it could look like this. Opposite sides are equal. No 90° angles. All sides are the same length.

This is a parallelogram.   This is a rhombus.

> **Strategies**
> Make a picture or diagram

The quadrilateral must be either a parallelogram or a rhombus.

Method 2: Work Backward

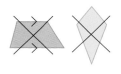

The quadrilateral has two pairs of opposite sides parallel. So, it is not a trapezoid or a kite.

The quadrilateral contains no right angles. So, it is not a square or a rectangle.

> **Strategies**
> Work backward

The quadrilateral must be either a parallelogram or a rhombus.

Check: A parallelogram has two pairs of opposite sides that are equal and parallel. It has no right angles. There is no information on whether all four sides are the same length. If they are, the quadrilateral is the special type of parallelogram called a rhombus.

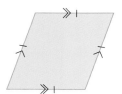

> **Literacy Connections**
>
> To learn about different problem solving strategies, refer to the Problem Solving section on pages xvi to xxi. The orange banner will help you find these pages. Refer to these pages whenever you need help deciding on a strategy to use to solve a problem.

Key Ideas

- Quadrilaterals are closed shapes with four sides. They are formed by joining four line segments and contain an angle at each vertex.

- Quadrilaterals are classified according to their side and angle properties.

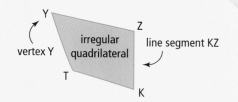

Communicate the Ideas

1. Compare a rectangle and a parallelogram. Use a chart to show how they are the same and how they differ.

   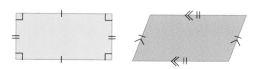

2. How is a rhombus like a square? How is it like a parallelogram? Take turns explaining to your partner.

3. Draw a quadrilateral. Draw a shape that is not a quadrilateral. Compare your drawings with a partner's. List the criteria a shape must have to be a quadrilateral.

Check Your Understanding

Practise

For help with Questions 4 to 8, refer to Example 1.

4. Classify each quadrilateral. Give reasons.

   a) 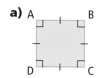   b)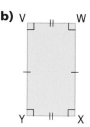

5. Classify each quadrilateral. Give reasons.

   a) 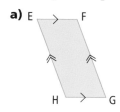   b)

6. Describe each figure, and then classify it.

   a)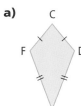

   b)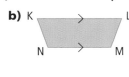

7. Name and classify the two quadrilaterals found in the figure.

   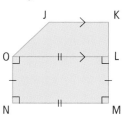

8. Name and classify the three quadrilaterals found in the figure.

   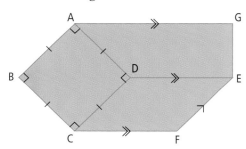

For help with Question 9, refer to Example 2.

9. What shape am I? Match my description with my name.

   Description
   a) I have four equal sides, but no right angles.
   b) I have no equal sides, but I do have one pair of parallel sides.
   c) I have two pairs of equal sides, but no parallel sides.
   d) I have four equal sides and four right angles.

   Name
   A square **B** kite
   C rhombus **D** trapezoid

Apply

10. The quadrilateral shown is sometimes called an isosceles trapezoid. Explain why this name is appropriate.

11. a) Two of the seven tangram pieces are quadrilaterals. What type are they?
    b) One other type of quadrilateral is formed by three pairs of neighbouring pieces in the completed tangram. What pieces? What type of quadrilateral do they form?

Did You Know?

The Chinese legend of the tangram's origin tells of a man who accidentally broke a pane of glass while carrying it up a mountain. While trying to put the pieces back together, he realized that the pieces could be arranged to form many other shapes. His bad luck led to the invention of a new game. ☯

Literacy Connections

Reading Special Shapes
☯ This is a yin yang symbol. It is a Chinese symbol with many meanings. One meaning is that things have both positive and negative sides. What negative thing happened in the legend? What was the positive result?

12. a) Name the types of quadrilaterals found in the tile pattern shown.
   b) Describe the pattern in which the tiles are laid.

13. Draw and label a quadrilateral that matches each description. Then, classify it.
Hint: First sketch a quadrilateral and mark the given information.
   a) AB is parallel to CD, AB is twice as long as CD, ∠A = 90°
   b) all sides measure 3 cm, DE is parallel to GF, DG is parallel to EF, DE is not at right angles to DG

Chapter Problem

14. Traditionally, patchwork quilts were made by sewing together square or rectangular pieces of fabric. Modern designs use many different shapes.

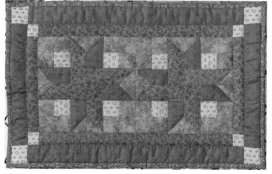

   a) List all the geometric shapes that you can find in the patchwork quilt shown.
   b) Design your own patchwork quilt containing at least 20 patches. Include at least two different types of quadrilaterals. Draw the design in your notebook.

Making Connections

You will explore more tiling patterns in Chapter 13.

15. One diagonal of a square divides the shape into two right isosceles triangles.

What types of triangles can be formed when one diagonal of each of the following quadrilaterals is drawn? Use diagrams to support your answers.
   a) a rectangle
   b) a kite

Extend

16. A Venn diagram uses nested and/or overlapping shapes to show relationships. The Venn diagram below can be used to show the relationships among types of triangles.

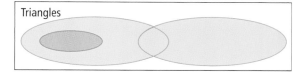

   a) Make a larger copy of the diagram. Add labels to your diagram to show equilateral, isosceles, and right triangles.
   b) What type of triangle is represented by the overlap of the two ovals?

17. Draw and label a Venn diagram to show the relationships among the different types of quadrilaterals. Hint: Do question 16 first.

2.3 Congruent Figures

Focus on...
- matching figures
- comparing side lengths and angle measures
- congruence

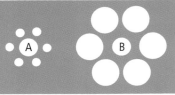

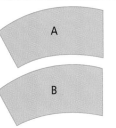

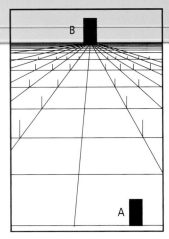

Look carefully at the three diagrams. In which one(s) is figure A identical to figure B? What properties of the figures do you compare to decide?

Figures that have the same shape and size are **congruent**.

congruent
- same shape and size

Materials
- grid paper
- scissors
- ruler
- protractor

Discover the Math

How can you tell if two triangles are congruent?

1. On grid paper, draw any rectangle and one diagonal. Cut out the two triangles formed. Compare the triangles. Are they congruent? How can you tell?

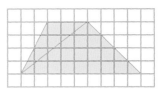

2. On grid paper, draw any trapezoid and one diagonal. Cut out the two triangles formed. Compare the triangles. Are they congruent? How can you tell?

3. **a)** On grid paper, draw any kite and one diagonal. Compare the two triangles formed. Are they congruent? How can you tell without cutting them out?
   b) Draw a copy of your kite. This time draw the other diagonal. Compare the two triangles formed. Are they congruent? How can you tell without cutting them out?

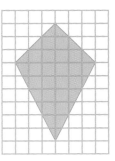

4. **Reflect** What must match if two triangles are congruent? What do you need to do to be certain?

Example 1: Identify Congruent Figures

Are the figures in each group congruent? Explain your answers.

a)  b)  c)

Solution

a) The figures are not congruent because their shapes are different from each other.

b) The figures are all equilateral triangles. Their sides are marked as being the same size. The second triangle is a rotation image. The three triangles are congruent.

c) The figures are all the same shape, circles. The circles are not congruent because they are different sizes.

Example 2: Match Parts of Congruent Triangles

Compare △ABC and △DEF.
a) List the corresponding equal angles and sides.
b) Are the two triangles congruent? Give reasons.

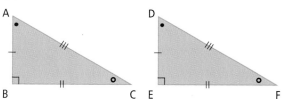

Solution

a) Corresponding angles: Corresponding sides:
∠A = ∠D   AB = DE
∠B = ∠E   AC = DF
∠C = ∠F   BC = EF

b) The corresponding angles and side lengths are equal. So, △ABC and △DEF are congruent.

Key Ideas

- Two figures are congruent if they have the same shape and the same size. Rotations and reflections are allowed.

  Both rectangular, both measure 2 by 3, so they are congruent rectangles.

- Corresponding angles and sides of congruent figures have the same measures.

  AB and PQ are corresponding sides. They are the same length. ∠B and ∠Q are corresponding angles. They contain the same mark. This means that the angles are of equal size.

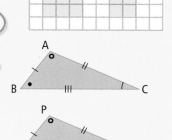

Communicate the Ideas

1. Which two of the rectangles shown are congruent? Why is the third rectangle not congruent?

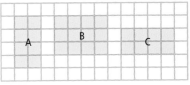

2. Explain why the two triangles shown are congruent.

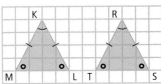

Check Your Understanding

Practise

For help with questions 3 to 6, refer to Example 1.

3. Are the figures in each group congruent? Justify your answer.

   a)

   b)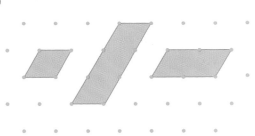

   c)

4. Which two rectangles are congruent? Explain.

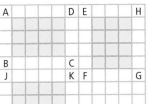

5. Which triangles are congruent? Explain.

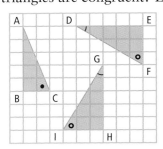

6. Are any of these figures congruent? Justify your answer.

For help with questions 7 and 8, refer to Example 2.

7. △ABC and △DEF are congruent. List the corresponding sides and the corresponding angles.

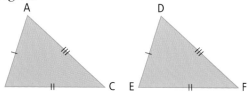

8. △LMN and △PQR are congruent. List the corresponding sides and the corresponding angles.

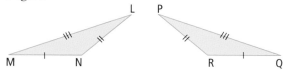

Apply

9. Examine the triangular end of the house roof frame.

a) Name two congruent right triangles.
b) Name two congruent acute triangles.
c) Name two congruent obtuse triangles.

10. Use a ruler and a protractor. Draw a triangle that is congruent to △KLM.

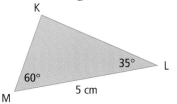

11. If two figures are the same shape, must they be congruent? Draw diagrams to illustrate your answer.

12. Are two rectangles with the same perimeter always congruent to each other? Explain, using diagrams to illustrate your answer.

13. If two figures have the same area, are the figures congruent? Use diagrams to illustrate your answer.

14. If you place two congruent equilateral triangles together with two sides aligned, you always get a rhombus.

Draw and name the geometric figure(s) that you can create by placing the following triangles with two equal sides aligned.

a) two congruent isosceles triangles
b) two congruent right scalene triangles

Extend

15. Draw three equilateral triangles. Add line segments to divide

a) the first triangle into two congruent triangles
b) the next triangle into three congruent triangles
c) the next triangle into four congruent triangles

2.4 Congruent and Similar Figures

Focus on...
- comparing angles
- patterns in side lengths
- congruent figures
- similar figures

Compare the dolls in the photograph. How are they alike? How do they differ?

Are the figures congruent? Are they **similar**?

similar
- same shape but different size

Discover the Math

Materials
- grid paper
- ruler
- protractor

Optional
- BLM 2.4A Recording Sheet
- BLM 2.4B Large Triangles

How are similar figures related?

1. Three similar rectangles are shown.

   a) Compare the length of rectangle EFGH with the length of rectangle ABCD. Compare the widths of these two rectangles. What do you notice?

   b) Compare the length of rectangle MJKL with the length of rectangle ABCD. Compare the widths of these two rectangles. What do you notice?

   c) Copy and complete the table.

Rectangle	ABCD	EFGH	MJKL
$\dfrac{\text{length}}{\text{width}}$			

   d) What do you notice about the value of $\dfrac{\text{length}}{\text{width}}$ for the three rectangles? What does this tell you about the shape of the three rectangles?

2. Two similar triangles are shown.
   a) Make larger copies of the triangles on centimetre grid paper.
   b) Use a protractor to measure the angles in each triangle. What do you notice?
   c) Copy and complete to compare the lengths of corresponding sides of the triangles.
   $$\frac{PQ}{XY} = \frac{\blacksquare}{\blacksquare} \qquad \frac{PR}{XZ} = \frac{\blacksquare}{\blacksquare} \qquad \frac{RQ}{ZY} = \frac{\blacksquare}{\blacksquare}$$
   d) Compare the three ratios in part c). What do you notice?

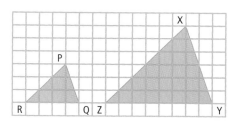

3. **Reflect** How are the angles in similar figures related? How are the sides of similar figures related?

Example 1: Identify Congruent or Similar Figures

Examine the figures shown.
a) Explain why there are no congruent pairs among them.
b) Are there any similar figures? Give reasons.

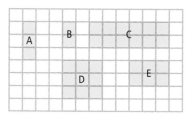

Solution

a) To be congruent, figures must have the same shape and size. The dimensions of the rectangles are
   A: 3 by 1   B: 2 by 1   C: 6 by 2
   D: 3 by 3   E: 3 by 2
   None of the rectangles are identical, so none of them are congruent.

   I wrote the length first, then the width.

b) Compare rectangle A and rectangle C.
   $$\frac{\text{length of C}}{\text{length of A}} = \frac{6}{2} \qquad \frac{\text{width of C}}{\text{width of A}} = \frac{3}{1}$$

   For both dimensions, rectangle C is three times rectangle A.
   So, rectangles A and C are similar.

   A and C look to be the same shape. I checked the corresponding sides.

   Check whether any other figures are similar.
   D is the only square shown.
   Compare rectangle B and rectangle E.
   $$\frac{\text{length of E}}{\text{length of B}} = \frac{3}{2} \qquad \frac{\text{width of E}}{\text{width of B}} = \frac{2}{1}$$

   These ratios are not the same.
   Rectangles A and C are the only similar figures shown.

Example 2: Match Parts of Similar Triangles

Compare the corresponding angles and sides of △ABC and △DEF. Are the two triangles similar? Give reasons.

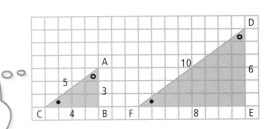

Solution
Compare corresponding angles:
∠A = ∠D   ∠B = ∠E   ∠C = ∠F

∠B and ∠E are both 90° because of the grid lines.

Compare corresponding sides:
$\dfrac{AB}{DE} = \dfrac{3}{6}$ $\dfrac{BC}{EF} = \dfrac{4}{8}$ $\dfrac{AC}{DF} = \dfrac{5}{10}$

Each side of △DEF is double the length of the corresponding side of △ABC.

△ABC and △DEF are similar because they are the same shape. Each side of △DEF is twice the corresponding side of △ABC.

Key Ideas

- Similar figures have the same shape but may be different in size.

- Corresponding angles in similar figures are equal.

∠R = ∠U, ∠S = ∠V, ∠T = ∠W

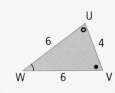

- The lengths of corresponding sides in similar figures are in proportion.

Each side of △UVW is double the length of the corresponding side of △RST.

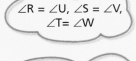

Communicate the Ideas

1. Compare the rectangles shown. Justify your answers to the following.
   a) Which are congruent?
   b) Which are similar?

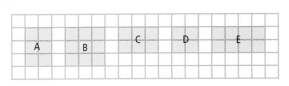

2. Compare the three triangles shown. Which two are similar? Why is the third not similar?

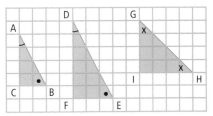

Check Your Understanding

Practise

For help with Questions 3 to 6, refer to Example 1.

3. Are the figures in each pair similar? Give reasons for your answers.

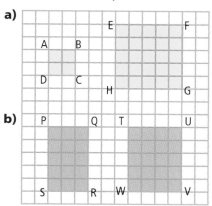

4. Are the triangles in each pair congruent? If not, are they similar? Explain your answers.

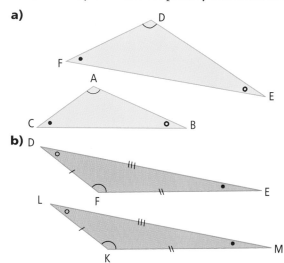

5. Which of the rectangles shown are similar? Explain why.

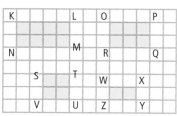

6. List the similar quadrilaterals. List the congruent ones. Give reasons for your choices.

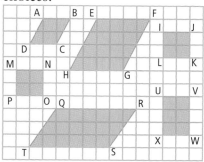

For help with questions 7 and 8, refer to Example 2.

7. Compare the corresponding angles and sides of △KLM and △XYZ. Are the two triangles similar? Give reasons.

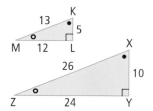

8. Compare the corresponding angles and sides of △PQR and △STV. Are the triangles similar? Give reasons.

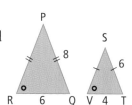

Apply

9. Compare the figures.

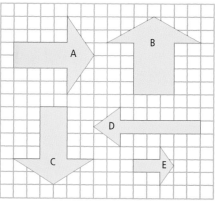

a) Which arrow is congruent to A?
b) Which arrow is similar to A? Explain.

2.4 Congruent and Similar Figures • MHR   73

Chapter Problem

10. What congruent and similar shapes are used in the design of this quilt?

11. Consider the seven pieces of the tangram.

    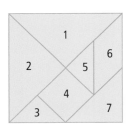

    a) Are any pieces congruent? Explain.
    b) Are any pieces similar? Explain.

12. Is each statement true? Explain your decision.
    a) All squares are similar to each other.
    b) All rectangles are similar to each other.
    c) All right isosceles triangles are similar to each other.
    d) All rhombi are similar to each other.

13. You can tell if two rectangles are similar by using the "diagonal test." Place the smaller rectangle on top of the larger one as shown. If the diagonals align, as in figure A, then the rectangles are similar. If the diagonals do not align, as in B, then the rectangles are not similar. Draw several rectangles on grid paper. Use the diagonal test to check which are similar.

14. a) Use grid paper to draw a triangle that is similar to the one shown.

    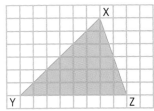

    b) Compare your similar triangle with those drawn by other students. Are the triangles that you have drawn similar to each other? Are they congruent? Explain.

15. Make a list of the congruent figures found in the diagram. Then, make a list of all the similar figures that you can find.

    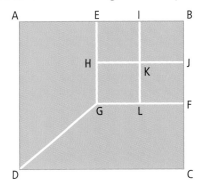

Extend

16. In △ABC, a line segment DE is drawn parallel to BC. How are △ABC and △ADE related? Justify your answer.

    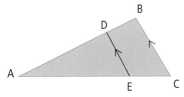

17. The national flag of Canada is twice as long as it is wide. The two red rectangles are similar to the complete flag. What can you deduce about the shape of the white rectangle? Explain.

Use Technology

Identify Similar Triangles Using *The Geometer's Sketchpad®*

Focus on...
- Exploring the properties of similar triangles

Materials
- The Geometer's Sketchpad® software
- computers

Optional
- TECH 2.4A Identify Similar Triangles (GSP4)
- TECH 2.4B Identify Similar Triangles (GSB3)
- BLM 2.4C Identify Similar Triangles Without Technology

1. Use *The Geometer's Sketchpad®* to construct and label a triangle ABC.

2. **a) Select** line segment AB. **Construct** its **Midpoint**.
   b) Repeat the previous step to construct the midpoint for line segment BC.
   c) Select points D and E. **Construct** line segment DE.

3. If △ABC and △DBE are similar triangles, then what is true about the measures of their corresponding angles? Check your answer by measuring corresponding pairs of angles.

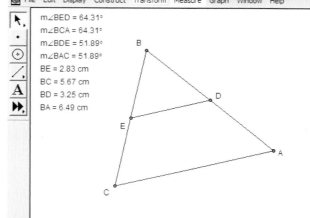

4. **a)** Make a conjecture about how the lengths of corresponding sides of △ABC and △DBE are related.
   b) Check by measuring the lengths of corresponding sides.

5. Extend your investigation by changing the shape of △ABC. Do the patterns still hold?

6. **Reflect** How can you use technology to check whether two triangles are similar?

CHAPTER 2 Review

Key Words

1. Draw and label diagrams to show the meaning of each word.
   a) isosceles triangle
   b) obtuse triangle
   c) rhombus
   d) trapezoid

2. Compare the two shapes shown. How are the triangles related? Explain.

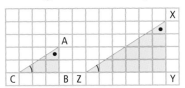

2.1 Classify Triangles, pages 54–59

3. Classify each triangle in two ways. Give reasons for your answers.

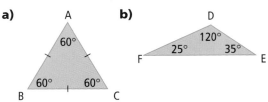

4. A support wire from the top of a 12-m flag pole reaches the ground 3 m away from the foot of the flag pole. What type of triangle is formed?

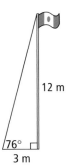

5. Use a ruler and a protractor to draw each triangle. Then, classify the triangle in two ways.
   a) one angle of 50° between sides measuring 4 cm and 4 cm
   b) one side measuring 7 cm between angles of 25° and 40°

2.2 Classify Quadrilaterals, pages 60–65

6. Classify each quadrilateral. Explain your choice.

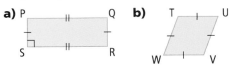

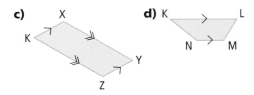

7. Look at the spider's web. What types of quadrilaterals can you find?

2.3 Congruent Figures, pages 66–69

8. Draw two congruent shapes. Explain why they are congruent.

9.

  a) Examine the picture of the provincial flag of Ontario. Find two different pairs of congruent figures.
  b) Find three other flags that use different shapes in their design. Sketch the flags and list the congruent figures.

10. In each part, are the shapes congruent? Explain your answer. If the shapes are congruent, list their equal sides and angles.

  a)

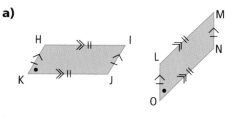

  b)

  c)

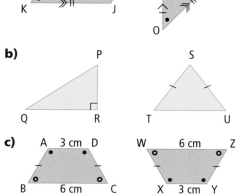

11. If two triangles have the same perimeter, are they congruent? Draw diagrams to illustrate your answer.

2.4 Congruent and Similar Figures, pages 70–74

12. Draw two similar figures. Explain how you can tell that they are similar.

13. Make a list of the similar figures found in the diagram. Are there any congruent figures? Explain.

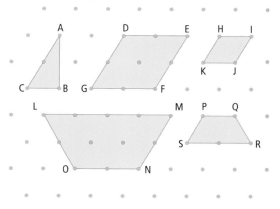

14. Decide whether each image is similar to the original. Can the image also be congruent? Explain.
  a) a photocopy of a figure
  b) a photograph of a figure

15. Both triangles contain a 60° angle.

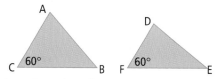

  a) Are the triangles congruent? Give reasons for your answer.
  b) Are the triangles similar? Give reasons for your answer.

CHAPTER 2 Practice Test

| Strand Questions | NSN | MEA 5, 8, 9 | GSS 1–11 | PA | DMP |

Multiple Choice

For questions 1 to 6, select the correct answer.

1.

   △ABC can be classified as
   A an obtuse scalene triangle
   B an acute isosceles triangle
   C a right scalene triangle
   D an acute equilateral triangle

2.

   △RST can be classified as
   A an obtuse isosceles triangle
   B an acute equilateral triangle
   C an obtuse scalene triangle
   D a right isosceles triangle

3.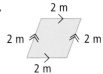

   The quadrilateral can be classified as
   A a rectangle
   B a square
   C a trapezoid
   D a rhombus

4.

   The quadrilateral can be classified as
   A a parallelogram
   B a square
   C a kite
   D a trapezoid

5.

   The two shapes are
   A isosceles triangles
   B congruent triangles
   C similar angles
   D congruent angles

6.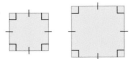

   The two shapes are
   A similar triangles
   B congruent squares
   C similar squares
   D congruent rectangles

Short Answer

7. Use a ruler and a protractor to draw each triangle. Then, classify the triangle in two ways.
   a) In △XYZ, ∠Y is a right angle. Sides XY and YZ are each 5 cm.
   b) △ABC with AB = 5 cm, BC = 7 cm, and ∠B = 60°

8. Compare the three photo frames shown. Which are similar rectangles? Explain your reasoning.

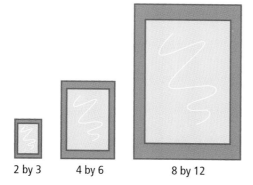

2 by 3   4 by 6   8 by 12

10. Which triangles are congruent? Which are similar? Explain why.

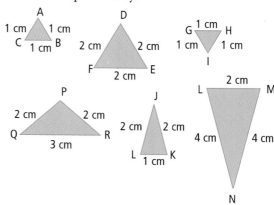

Extended Response

9. Explain why a right triangle can never be similar to an obtuse triangle. Draw a sketch to help in your explanation.

11. The height of square ABCD is half the height of square EFGH. ABCD has a perimeter of 16 cm. Use a ruler to draw the two quadrilaterals. Label the dimensions of both. Are the figures congruent? similar? Explain.

Chapter Problem Wrap-Up

Patterns that use a variety of shapes are more interesting. Design a pattern for the front of your binder, or for another similar purpose. You may draw it on paper, and then create it using pieces of coloured tissue paper, fabric, wood, or other materials you choose.

Your pattern block should include
- two different quadrilaterals
- two different triangles
- some congruent figures
- some similar figures

Write an e-mail to a friend giving a brief description of your design. List its geometric properties.

Making Connections

Transformations and Congruence

The diagram shows a variety of shapes on grid paper.

Materials
- grid paper
- scissors

Optional
- tracing paper
- pattern blocks
- Mira

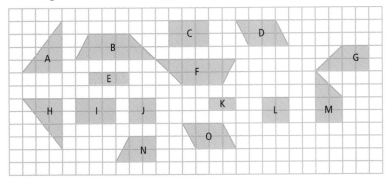

1. Identify pairs of shapes that are related by a translation, rotation, or reflection. To help you decide, you can cut the shapes out of a piece of grid paper. Tracing paper, or a Mira, may also be helpful.
2. Describe the transformation that relates each pair. Draw diagrams as necessary. For a translation, show the translation arrow. For a reflection, show the mirror line. For a rotation, show the turn centre and the angle of rotation.
3. List pairs of congruent shapes in the diagram. Compare this list with your answers in step 1. What do you notice? How can you explain this?

Making Connections

What's math got to do with sports?

Geometry is found on all sports playing surfaces. Squares, rectangles, and circles are the most common shapes used in marking out the playing areas.

The plan of a soccer pitch is shown. What shapes can be seen when a soccer game is being played?

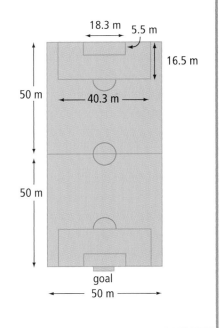

80   MHR • Chapter 2

Create a Logo

New School Logo!

Create a new logo that might be used on a crest for team shirts, on school banners, and on the opening page of the school's web site.

The student council wants a logo with
- at least two congruent shapes
- at least two similar shapes
- at least two different shapes

1. Create a logo. Explain how your logo meets the requirements set out by the student council.

2. In order to make a school crest, your logo is going to be sewn onto material. Every line in the logo has to be sewn. Sewing costs $0.20 per centimetre. How much will it cost to sew one crest of your logo? Explain.

3. The school is creating a large banner. Adding colour costs $4 per square metre. Design and price the banner. Consider the following:
    - How large a banner will your school need?
    - Wll the banner have a large logo?
    - What else will be on the banner?

Number Sense and Numeration

- Generate, compare, and order multiples.
- Understand and explain operations with fractions using manipulatives.
- Add and subtract fractions with simple denominators using concrete materials, drawings, and symbols.
- Relate the repeated addition of fractions with simple denominators to the multiplication of a fraction by a whole number.
- Ask "what if" questions, pose problems involving simple fractions, and investigate solutions.
- Solve problems involving fractions using appropriate strategies and calculation methods.

Key Words

equivalent fractions

common denominator

multiple

CHAPTER 3

Fraction Operations

Fractions are all around us. When Rebecca and Vishal were helping to clean up the kindergarten room, Rebecca dropped an armload of puzzles. How do these puzzles relate to fractions? Consider how you can use the puzzles to talk about numbers, colours, and seasons. For example:
- There are four seasons in a year. What fraction of a year does each season cover?
- When you were 5 years old, what fraction of your life did 1 year make? 2 years?
- Sets of markers or paints often come in eight colours. Which colours are your favourites? What fraction of your wardrobe uses these colours?

Understanding fractions is important outside of the classroom. Describe the last time you used fractions in your life.

In this chapter, you will add and subtract fractions. You will find a common denominator and learn how to multiply a fraction by a whole number.

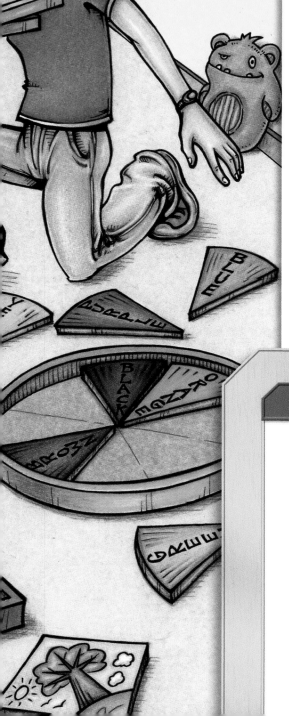

Chapter Problem

How can you use fraction puzzles to help others learn more about fractions?

Get Ready

Writing Fractions

A fraction is a number that represents a part of a whole or a part of a group.

$\frac{3}{8}$ means that 3 parts out of a group of 8 equal parts are shaded.

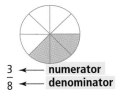

$\frac{3}{8}$ is a **proper fraction**. Its denominator is greater than its numerator.

$\frac{5}{3}$ is an **improper fraction**. Its numerator is greater than its denominator.

A **mixed number** is made up of a whole number and a fraction.
$\frac{5}{3}$ can be written as the mixed number $1\frac{2}{3}$.

1. Write the fraction shaded in each diagram. Show each as an improper fraction and as a mixed number.

   a) b) c)

2. Draw a diagram to represent each fraction.

   a) $\frac{1}{4}$ b) $\frac{1}{3}$ c) $\frac{2}{5}$

   d) $\frac{8}{5}$ e) $1\frac{3}{4}$ f) $2\frac{1}{2}$

Comparing and Ordering Fractions

Which is greater, $\frac{1}{4}$ or $\frac{1}{2}$?

Use diagrams that are the same size to compare the two fractions.

$\frac{1}{2}$ is greater than $\frac{1}{4}$.

A number line can help you order fractions.
$1\frac{1}{2}, \frac{1}{4}, 1,$ and $\frac{7}{4}$ are ordered from least to greatest on the number line.

3. Use diagrams that are the same size to compare the fractions. Which is greater?
   a) $\frac{1}{4}$ and $\frac{1}{8}$
   b) $\frac{1}{4}$ and $\frac{1}{3}$
   c) $\frac{2}{3}$ and $\frac{5}{8}$

4. Use a number line to order $\frac{3}{4}, \frac{3}{2}, \frac{1}{2},$ and $1\frac{1}{4}$ from least to greatest.

Multiples

The first five **multiples** of 3 are 3, 6, 9, 12, and 15.

Each multiple is the product of 3 and a natural number.

$3 \times 1 = 3 \quad 3 \times 2 = 6 \quad 3 \times 3 = 9 \quad 3 \times 4 = 12 \quad 3 \times 5 = 15$

5. List the first five multiples of each number.
   a) 2
   b) 4
   c) 5

6. Which of the following numbers is not a multiple of 8?
   8   24   40   15   32

Equivalent Fractions

Equivalent fractions represent the same part of the whole or group.

$\frac{4}{8}$ and $\frac{1}{2}$ are equivalent fractions.

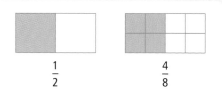

$\frac{1}{2}$ $\frac{4}{8}$

7. Identify the fraction shaded in each diagram. Show the fraction in another way using equivalent fractions.
   a)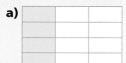
   b)

8. Draw a diagram to show an equivalent fraction for each fraction.
   a) $\frac{3}{4}$
   b) $\frac{4}{6}$
   c) $\frac{10}{15}$

3.1 Add Fractions Using Manipulatives

Focus on...
- representing fractions using concrete materials
- using patterns to add fractions

 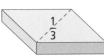 

The blue rhombus is one third of the yellow hexagon. What other pattern block relationships are there?

Discover the Math

Materials
- pattern blocks
- pencil crayons
- BLM 3.1A Pattern Block Worksheet

How can you add fractions using manipulatives?

1. Allison used pattern blocks to make 1. Here is how she started.

2. There are at least 8 different ways to make 1 whole hexagon using yellow hexagon, red trapezoid, blue rhombus, and green triangle pattern blocks. How many ways can you find?

3. Reflect How can concrete materials and diagrams help you represent and add fractions?

Example: Add Fractions Using Manipulatives

Add $\frac{1}{2} + \frac{1}{3}$.

Solution

Method 1: Use Concrete Materials
$\frac{1}{2} + \frac{1}{3} = \frac{5}{6}$

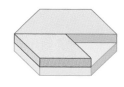

Method 2: Use a Diagram
$\frac{1}{2} + \frac{1}{3} = \frac{5}{6}$

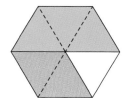

Key Ideas

- Concrete materials or diagrams can be used to represent fractions.
- Concrete materials or diagrams can be used to show equivalent fractions.

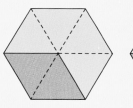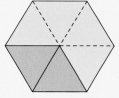

 =

- To add fractions using concrete materials or diagrams, each fraction can be shown using parts of equal size.

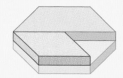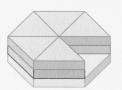

Communicate the Ideas

1. Show why $\frac{1}{6} + \frac{1}{6} + \frac{1}{6} = \frac{3}{6}$.

2. Use words or diagrams to show that $\frac{3}{6} = \frac{1}{2}$.

3. Describe how you would explain your answers to questions 1 and 2 to a friend.

Check Your Understanding

Practise

For help with questions 4 to 9, refer to the Example.

4. What fraction of each hexagon is covered?
   a)  b)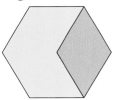

5. What fraction of each hexagon is covered?
   a)  b)

6. Write an addition sentence to represent the total fraction of each hexagon that is covered. State the total fraction covered.
   a)  b)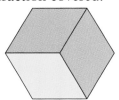

7. Write an addition sentence to represent the total fraction of each hexagon that is covered. State the total fraction covered.
   a) 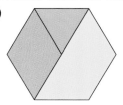 b)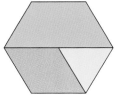
   c)  d)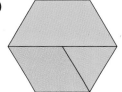

8. Add. Use concrete materials or diagrams.
   a) $\frac{1}{3} + \frac{1}{3}$ b) $\frac{1}{6} + \frac{1}{6}$ c) $\frac{2}{6} + \frac{3}{6}$

9. Add. Use concrete materials or diagrams.
   a) $\frac{4}{6} + \frac{1}{3}$ b) $\frac{1}{2} + \frac{3}{6}$ c) $\frac{2}{3} + \frac{1}{6}$

Chapter Problem

10. Write an addition sentence to describe this puzzle.

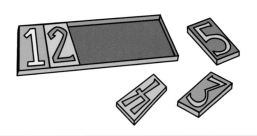

Apply

11. Write an addition sentence to describe how many hexagons are covered in each of the following.

   a)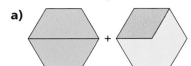

   b)

   c)

12. Suppose 1 red trapezoid = 1 whole.

   a) What fraction of the trapezoid is represented by this diagram?

   b) What fraction of the trapezoid is represented by this diagram?

   c) Write an addition statement to show the sum of the fractions in parts a) and b).

   d) Find the sum.

13. In Chapter 2, you used a tangram to classify geometric shapes.

   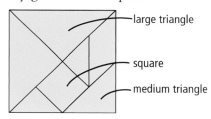

   a) Suppose 1 medium triangle = 1 whole. Write an addition statement to represent this diagram.

   b) Suppose 1 square = 1 whole. Write an addition statement to represent this diagram.

   c) Suppose 1 large triangle = 1 whole. Write an addition statement to represent this diagram.

   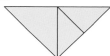

14. Some of the tangram pieces are placed in the tangram square. Write an addition statement to show the fraction of the completed tangram that is placed in the square.

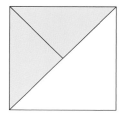

15. Go to www.mcgrawhill.ca/links/math7 and follow the links to a Web site where you can build pattern block diagrams on-screen.

   a) Create three addition diagrams using yellow, red, green, and blue pattern blocks.

   b) Ask a classmate to write an addition sentence to represent your diagrams.

16. Use concrete materials of your own choice to make up two addition questions. Draw diagrams to show your questions. Challenge a classmate to solve your questions using the materials you chose.

Extend

17. Use pattern blocks to simplify $\frac{1}{3} + \frac{1}{3} + \frac{1}{6}$. Show your answer visually and using a number sentence.

18. Suppose 2 hexagons = 1 whole. You have 1 triangle, 2 rhombuses, and 2 trapezoids.

   a) Draw a diagram to represent this situation.

   b) What fraction of 1 whole is covered?

19. Suppose 1 hexagon = 1 whole. You have 2 trapezoids, 1 rhombus, and 3 triangles. How many hexagons can you make? Show how you found your answer.

3.2 Subtract Fractions Using Manipulatives

Focus on...
- representing fractions using concrete materials
- using patterns to subtract fractions

The hexagon represents 1. What fraction of the hexagon is still covered?

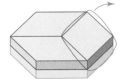

Discover the Math

Materials
- pattern blocks
- pencil crayons
- BLM 3.1A Pattern Block Worksheet

Strategies
What strategies are you using?

How can you use pattern blocks to represent subtracting fractions?

1. Use pattern blocks to show subtracting $\frac{1}{2}$. How many different ways can you take away $\frac{1}{2}$? Record your answers.

3. **Reflect** How can pattern blocks help you represent subtracting fractions?

Example: Subtract Fractions

Subtract $\frac{1}{2} - \frac{1}{6}$.

Solution

Method 1: Use Concrete Materials

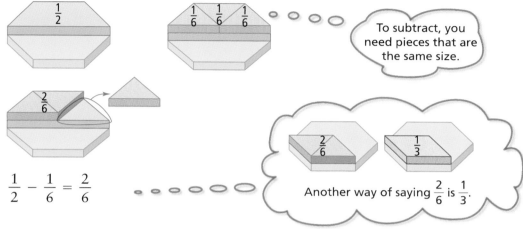

$$\frac{1}{2} - \frac{1}{6} = \frac{2}{6}$$

Method 2: Use a Diagram

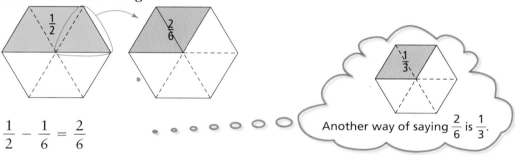

$$\frac{1}{2} - \frac{1}{6} = \frac{2}{6}$$

Key Ideas

- When subtracting fractions using concrete materials,
  - represent each fraction using parts of equal size
  - remove the blocks represented by the fraction that is being subtracted
  - identify the fraction that remains

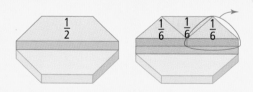

Communicate the Ideas

1. Show $\frac{2}{3}$ in as many ways as you can.

2. Show visually how to subtract $\frac{2}{3} - \frac{1}{6}$.

Check Your Understanding

Practise

For help with questions 3 to 10, refer to the Example.

3. Write a subtraction sentence to represent each diagram.

   a)

   b)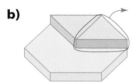

   c)

4. Draw a diagram to show the answer to each part in question 3.

5. Write a subtraction sentence to represent each diagram.

   a)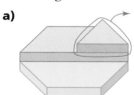

   b)

   c)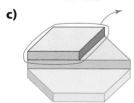

6. Draw a diagram to show the answer to each part in question 5.

7. Use concrete materials or diagrams to show each subtraction.

   a) $1 - \dfrac{1}{2}$ b) $\dfrac{5}{6} - \dfrac{1}{6}$

   c) $\dfrac{2}{3} - \dfrac{1}{3}$ d) $1 - \dfrac{2}{3}$

8. Write a subtraction sentence to represent each diagram.

   a)

   b)

9. Draw a diagram to show the answer to each part in question 8.

10. Use concrete materials or diagrams to show each subtraction.

    a) $\dfrac{1}{2} - \dfrac{2}{6}$ b) $\dfrac{1}{2} - \dfrac{1}{3}$

    c) $\dfrac{1}{3} - \dfrac{1}{6}$ d) $\dfrac{5}{6} - \dfrac{1}{3}$

Apply

11. Use concrete materials or diagrams to show each subtraction.

    a) $\dfrac{5}{6} - \dfrac{1}{2}$ b) $\dfrac{2}{3} - \dfrac{1}{2}$

12. One of the tangram pieces is removed from a finished tangram. Write a subtraction sentence to show the fraction of the completed tangram that remains.

13. Use the tangram diagram to help you answer the following questions.

a) Suppose 1 parallelogram = 1 whole. Write a subtraction sentence to represent this diagram.

b) Suppose 1 large triangle = 1 whole. Write a subtraction sentence to represent this diagram.

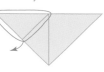

Chapter Problem

14. Write a subtraction sentence to represent each puzzle. Explain what each sentence represents. Solve it.

15. Use concrete materials of your own choice to make up two subtraction questions. Draw diagrams to show your questions. Challenge a classmate to solve your questions using the materials you chose.

Extend

16. Suppose 2 hexagons = 1 whole.

a) What fraction of the whole does one red trapezoid represent?

b) Write a subtraction sentence to represent this diagram.

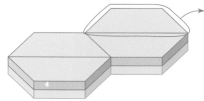

c) Show the answer to the subtraction sentence.

17. Suppose 2 hexagons = 1 whole.

a) Show $\frac{1}{2} - \frac{1}{3}$ visually.

b) Explain how this representation differs from the representation when 1 hexagon = 1 whole.

c) How is the answer to part a) related to the answer when 1 hexagon = 1 whole?

3.3 Find Common Denominators

Focus on...
- common denominators

Estimate how much of the rectangle is covered. Is it more or less than $\frac{1}{2}$? Is it more or less than 1?

Discover the Math

How can you find a common denominator?

Materials
- paper
- pencil crayons

1. Fold a square piece of paper one way into quarters. Open it up and colour $\frac{1}{4}$.

2. Now, fold the paper into thirds the other way.

3. Open up the paper.
   a) How many parts is the paper divided into now?
   b) How many sections are shaded? Name an **equivalent fraction** for $\frac{1}{4}$.

equivalent fractions
- two or more fractions that represent the same part of a whole or a group

common denominator
- a number that is a common multiple of the denominators of a set of fractions
- a common denominator for $\frac{1}{2}$ and $\frac{1}{5}$ is 10

4. Use a fresh piece of paper. Fold the piece of paper into thirds one way. Open it up and colour $\frac{1}{3}$.

5. Now, fold the paper into quarters the other way.

6. Open up the paper. Name an equivalent fraction for $\frac{1}{3}$.

7. **Reflect** How can paper folding help you find a **common denominator**?

94 MHR • Chapter 3

Example: Find a Common Denominator

Find a common denominator for $\frac{1}{2}$ and $\frac{1}{3}$.

Method 1: Use Paper Folding
Fold a square piece of paper in half one way and in thirds the other way.

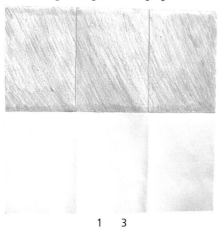
$\frac{1}{2} = \frac{3}{6}$

$\frac{1}{3} = \frac{2}{6}$

A common denominator for $\frac{1}{2}$ and $\frac{1}{3}$ is 6.

Method 2: Use a Diagram
Divide a square in half one way and in thirds the other way.

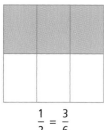

$\frac{1}{2} = \frac{3}{6}$ $\frac{1}{3} = \frac{2}{6}$

A common denominator for $\frac{1}{2}$ and $\frac{1}{3}$ is 6.

Method 3: Use Multiples
The denominator of $\frac{1}{2}$ is 2. **Multiples** of 2 are 2, 4, ⑥, 8, … .

The denominator of $\frac{1}{3}$ is 3. Multiples of 3 are 3, ⑥, 9, 12, … .

The multiple 6 is in both lists.

A common denominator for $\frac{1}{2}$ and $\frac{1}{3}$ is 6.

Strategies
Look for a pattern

Literacy Connections

Reading the Symbol …
When you see …, this means "and so on." The list continues.

multiple
- the product of a given number and a natural number
- multiples of 2 are 2, 4, 6, 8, … .

3.3 Find Common Denominators • MHR 95

Key Ideas

- Paper folding, diagrams, and multiples can be used to find a common denominator.

Paper Folding

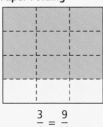 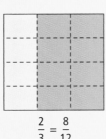

$\frac{3}{4} = \frac{9}{12}$ $\frac{2}{3} = \frac{8}{12}$

Common denominator: 12

Diagrams

$\frac{3}{4} = \frac{9}{12}$ $\frac{2}{3} = \frac{8}{12}$

Common denominator: 12

Multiples

The denominator of $\frac{3}{4}$ is 4. Multiples of 4 are 4, 8, ⑫, 16, 20, …

The denominator of $\frac{2}{3}$ is 3. Multiples of 3 are 3, 6, 9, ⑫, 15, 18, …

Common denominator: 12

Communicate the Ideas

1. Mei is trying to find a common denominator for $\frac{1}{5}$ and $\frac{2}{8}$. She says that a common denominator is 13. Do you agree with her? What might you say to Mei?

2. Brian found both 24 and 42 as common denominators for $\frac{1}{3}$ and $\frac{5}{6}$.

   Explain how this is possible. What other common denominators are there? Brian's work:

   Multiples of 3 are 3, 6, 9, 12, 15, 18, 21, ㉔, 27, 30, 33, 36, 39, ㊷
   Multiples of 6 are 6, 12, 18, ㉔, 30, 36, ㊷
   Two common denominators are 24 and 42.

3. A common denominator for $\frac{1}{2}$ and $\frac{1}{3}$ is 6.

   What other common denominators could be used? Which denominator would be best to use? Explain why.

Check Your Understanding

Practise

For help with questions 4 and 5, refer to the Example.

4. Use paper folding or a diagram to find a common denominator for each pair of fractions.
   a) $\frac{1}{3}$ and $\frac{1}{4}$
   b) $\frac{1}{2}$ and $\frac{1}{8}$
   c) $\frac{1}{6}$ and $\frac{1}{4}$
   d) $\frac{1}{5}$ and $\frac{1}{2}$

5. Use multiples to find a common denominator for each pair of fractions.
   a) $\frac{1}{5}$ and $\frac{1}{3}$
   b) $\frac{1}{7}$ and $\frac{1}{3}$
   c) $\frac{1}{4}$ and $\frac{1}{10}$
   d) $\frac{1}{6}$ and $\frac{1}{8}$

6. Write two equivalent fractions for the shaded part of each diagram in each pair.

   a)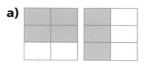
   b)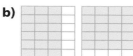
   c)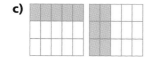
   d)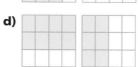

7. State a common denominator for each pair of diagrams in question 6.

8. State a common denominator for each pair of diagrams.

   a)
   b)

Apply

9. Find two common denominators for each pair of fractions.
   a) $\frac{1}{3}$ and $\frac{1}{5}$
   b) $\frac{1}{4}$ and $\frac{1}{6}$

10. Find three common denominators for each pair of fractions.
    a) $\frac{2}{3}$ and $\frac{1}{2}$
    b) $\frac{1}{2}$ and $\frac{3}{8}$

11. Name all the common denominators between 1 and 40 for the fractions $\frac{1}{4}$ and $\frac{2}{3}$.

12. a) Name two common denominators between 10 and 20 for the fractions $\frac{1}{2}$ and $\frac{1}{6}$.
    b) Describe how you found the common denominators.

13. Find a common denominator for each pair of fractions. Use equivalent fractions to rewrite each pair of fractions with the common denominator.
    a) $\frac{3}{5}$ and $\frac{1}{2}$
    b) $\frac{5}{8}$ and $\frac{1}{4}$

14. Which fraction in each pair is greater? Show at least one method.
    a) $\frac{5}{6}$ and $\frac{2}{3}$
    b) $\frac{2}{5}$ and $\frac{1}{4}$

Extend

15. Find a common denominator for each group of fractions. Describe your method.
    a) $\frac{1}{2}, \frac{1}{3}$, and $\frac{1}{4}$
    b) $\frac{1}{3}, \frac{1}{4}$, and $\frac{1}{5}$

3.4 Add and Subtract Fractions Using a Common Denominator

Focus on...
- common denominator
- adding and subtracting fractions

Suppose you invite a friend over for pizza. You order a six-slice pizza. One slice is left after you both eat. What fraction of the pizza has been eaten? How can you show this using numbers?

Discover the Math

How can you add and subtract fractions using a common denominator?

Example 1: Add Fractions
How much pizza is left?

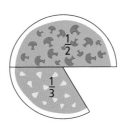

Solution

Method 1: Use Manipulatives
To add fractions, the pieces have to be the same size.

Each green triangle represents $\frac{1}{6}$.

Count the number of pieces.

$$\frac{1}{2} + \frac{1}{3} = \frac{5}{6}$$

$\frac{5}{6}$ of the pizza is left.

Strategies
Make a model

98 MHR • Chapter 3

Method 2: Use a Common Denominator

Fold a piece of paper in half one way and then in thirds the other way. Colour $\frac{1}{2}$ of the page. Three sections are coloured.

Fold another piece of paper in thirds one way and then in half the other way. Colour $\frac{1}{3}$ of the page. Two sections are coloured.

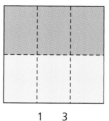
$\frac{1}{2} = \frac{3}{6}$

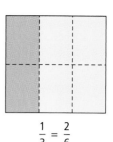
$\frac{1}{3} = \frac{2}{6}$

A common denominator is 6.

$\frac{1}{2} + \frac{1}{3} = \frac{3}{6} + \frac{2}{6}$ **Add the numerators**

$= \frac{5}{6}$

$\frac{5}{6}$ of the pizza is left.

Literacy Connections

Adding and Subtracting Fractions

$\frac{3}{6}$ ← number of shaded pieces
← number of pieces in whole or group

When you add or subtract fractions with the same denominator, add or subtract only the number of shaded pieces (numerator) in each fraction. The number of pieces in the whole or group (denominator) stays the same.

Example 2: Subtract Fractions

Subtract $\frac{5}{6} - \frac{1}{3}$.

Solution

Method 1: Use Manipulatives
To subtract fractions, the pieces have to be the same size.

$\frac{5}{6} - \frac{1}{3} = \frac{3}{6}$

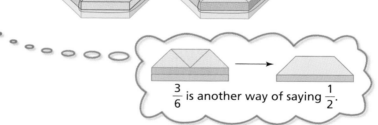

$\frac{3}{6}$ is another way of saying $\frac{1}{2}$.

Method 2: Use Multiples
Multiples of 6 are ⑥, 12, 18, ….
Multiples of 3 are 3, ⑥, 9, 12, ….
The first common denominator in the two lists is 6.
Write equivalent fractions with 6 as a denominator.

$\frac{5}{6} = \frac{5}{6}$ $\frac{1}{3} = \frac{1 \times 2}{3 \times 2}$

$= \frac{2}{6}$

To find equivalent fractions, multiply the numerator and denominator by the same number.

$\frac{5}{6} - \frac{1}{3} = \frac{5}{6} - \frac{2}{6}$ **Subtract the numerators.**

$= \frac{3}{6}$

Literacy Connections

Remembering Common Denominators
People who have friends in common share the *same* friends. Fractions that have a common denominator share the *same* denominator.

$\frac{1}{2} = \frac{3}{⑥}$
$\frac{1}{3} = \frac{2}{⑥}$ common denominator

Example 3: Add the Same Fraction

How much pizza is left over?

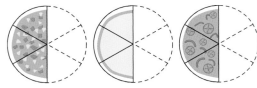

Solution

Method 1: Use Manipulatives

Strategies
Make a model

Each red trapezoid represents $\frac{1}{2}$.

There are 3 trapezoids.

$\frac{1}{2} + \frac{1}{2} + \frac{1}{2} = \frac{3}{2}$

$\frac{3}{2}$ can be written as the mixed number $1\frac{1}{2}$.

There are $\frac{3}{2}$ or $1\frac{1}{2}$ pizzas left over.

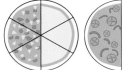

This means that there are 3 halves of pizza left over.

Method 2: Multiply

Each strip shows $\frac{1}{2}$.

There are 3 half strips.

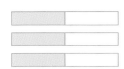

3 halves = $3 \times \frac{1}{2}$

Multiply the whole number 3 by the numerator.

$3 \times \frac{1}{2} = \frac{3}{2}$

$\frac{3}{2}$ can be written as the mixed number $1\frac{1}{2}$.

There are $\frac{3}{2}$ or $1\frac{1}{2}$ pizzas left over.

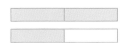

Key Ideas

- To add or subtract fractions with different denominators, use a common denominator.
- Repeated addition of the same fraction can be written as multiplication.

Communicate the Ideas

1. Use diagrams to subtract $\frac{1}{2} - \frac{1}{3}$.

2. Use a common denominator to subtract $\frac{1}{2} - \frac{1}{3}$. Explain why you chose your common denominator.

3. Add $\frac{5}{6} + \frac{1}{2}$ in two different ways. Explain which method you prefer and why.

4. Describe two ways to represent the diagram using numbers.

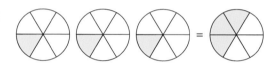

Check Your Understanding

Practise

For help with question 5, refer to Example 1.

5. Write an addition sentence to represent the fraction of each figure that is shaded?

   a) 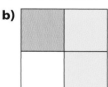   b)

   c) 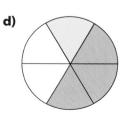   d)

For help with question 6, refer to Example 2.

6. What fraction of each figure remains?

   a)    b)

   c) 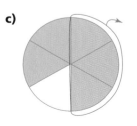   d)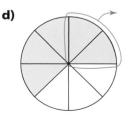

For help with question 7, refer to Example 3.

7. In each diagram, what fraction of a whole is shaded?

a)

b)

c)

8. Rewrite each expression with a common denominator. Subtract.

a) $\dfrac{2}{3} - \dfrac{1}{6}$ b) $\dfrac{3}{4} - \dfrac{1}{2}$

c) $\dfrac{7}{8} - \dfrac{1}{4}$ d) $\dfrac{3}{5} - \dfrac{1}{2}$

9. Rewrite each expression with a common denominator. Add.

a) $\dfrac{1}{5} + \dfrac{1}{2}$ b) $\dfrac{1}{4} + \dfrac{5}{6}$

c) $\dfrac{2}{5} + \dfrac{2}{3}$ d) $\dfrac{1}{6} + \dfrac{1}{2}$

10. Write each repeated addition as a multiplication and evaluate. Show your answer as an improper fraction and as a mixed number.

a) $\dfrac{1}{2} + \dfrac{1}{2} + \dfrac{1}{2}$

b) $\dfrac{1}{3} + \dfrac{1}{3} + \dfrac{1}{3} + \dfrac{1}{3} + \dfrac{1}{3}$

Apply

11. a) Evaluate each of the following.

$2 \times \dfrac{1}{2} = \blacksquare$ $3 \times \dfrac{1}{3} = \blacksquare$

$4 \times \dfrac{1}{4} = \blacksquare$ $5 \times \dfrac{1}{5} = \blacksquare$

b) Describe the pattern you see.

c) Predict the value of $20 \times \dfrac{1}{20}$.

Chapter Problem

12. Use multiplication to describe the number of pieces in this puzzle.

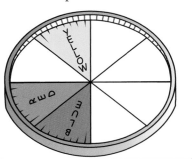

13. Write each repeated addition as a multiplication and evaluate. Show your answer as an improper fraction and as a mixed number.

a) $\dfrac{3}{2} + \dfrac{3}{2} + \dfrac{3}{2} + \dfrac{3}{2} + \dfrac{3}{2}$

b) $\dfrac{4}{3} + \dfrac{4}{3} + \dfrac{4}{3}$

14. Cheryl and Monica volunteered to shovel the snow from Monica's driveway.

If the two girls shovelled the whole driveway, were they both correct? Explain.

15. Kayla used the following diagram and explanation to add $\frac{1}{4} + \frac{3}{8}$.

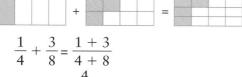

$$\frac{1}{4} + \frac{3}{8} = \frac{1+3}{4+8}$$
$$= \frac{4}{12}$$

a) Explain the error in Kayla's solution.
b) Redraw the diagram to show the correct answer. Show the correct equation.

16. Jamal ate $\frac{1}{3}$ of a pizza for lunch. His friend Kevin ate $\frac{1}{4}$ of the same pizza.

a) What fraction of the pizza did the two friends eat?
b) What fraction of the pizza was left over? Draw a diagram to show your answer.

17. Which is greater, $\frac{2}{5} + \frac{1}{2}$ or $\frac{2}{3} + \frac{1}{6}$? Show how you know.

18. Which is greater, $1 - \frac{2}{3}$ or $1 - \frac{5}{8}$? Show how you know.

19. Add. Hint: Use multiples to find a common denominator.

a) $\frac{1}{2} + \frac{1}{3} + \frac{1}{4}$
b) $\frac{1}{4} + \frac{2}{5} + \frac{3}{10}$

 20. a) Draw a diagram to show $\frac{1}{4} + \frac{1}{2}$.
b) Draw a diagram to show $\frac{2}{3} + \frac{1}{6}$.
c) Which sum is larger? How do you know?

Extend

21. Use diagrams to show how you would add $1 + \frac{1}{2} + \frac{1}{4}$.

22. Evaluate.

a) $1 + \frac{1}{3} + \frac{3}{4}$ b) $2 - \frac{2}{5} + \frac{1}{2}$

23. Three students were hired to clean the school windows before the start of the new school year.

Should the group be paid the full amount for cleaning the windows? Explain.

Making Connections

Order of Operations and Fractions

You learned about the order of operations in Chapter 1. The order of operations also applies to fractions. Add or subtract the fractions in brackets first.

Evaluate.

a) $\left(\frac{1}{4} + \frac{1}{2}\right) - \frac{1}{3}$ b) $\frac{1}{2} + \left(\frac{2}{3} - \frac{1}{3}\right)$

3.5 More Fraction Problems

Focus on...
- solving problems with fractions
- using appropriate strategies and calculation methods
- explaining the problem solving process

Literacy Connections

Reading Problems
Read the problem. Write it in your own words.

What information are you given?

What information do you need?

What is the problem asking you to do?

Count the leftover sandwich pieces on the tray.
How many sandwiches are left?

Discover the Math

What different strategies can you use to solve problems containing fractions?

Example 1: Leftover Sandwiches

For a party, Janine's mother serves a tray with sandwiches cut in quarters. After the party, there are 9 pieces left. How many sandwiches is this?

Solution

Find how many sandwiches are left.

Count the sandwich pieces left on the plate. How many whole sandwiches is this?

104 MHR • Chapter 3

Do It!

Strategies
Make a picture or diagram

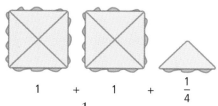

There are $2\frac{1}{4}$ sandwiches left.

Look Back

Multiply.
$9 \times \frac{1}{4} = \frac{9}{4}$

$\frac{9}{4}$ is another way of saying $2\frac{1}{4}$.

Example 2: Leftover Muffins and Oranges

The coach prepared a snack for her team. She cut muffins in half and oranges in eighths. After the game, there were five pieces of muffin and 13 pieces of orange left. How many muffins and how many oranges is this?

Solution

Place pieces together to make whole muffins and whole oranges.

Strategies
Look for a pattern

1 muffin = 2 pieces   4 muffins = 4 pieces   3 muffins = 6 pieces

There are more than 2 whole muffins. This is $2\frac{1}{2}$.

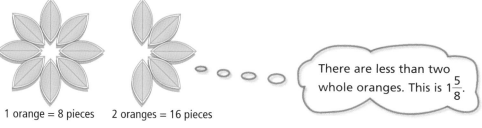

1 orange = 8 pieces   2 oranges = 16 pieces

There are less than two whole oranges. This is $1\frac{5}{8}$.

There are $2\frac{1}{2}$ muffins and $1\frac{5}{8}$ oranges left.

Key Ideas

- When solving a problem, read what is being asked. Write it in your own words.
- Decide what strategy to use. Use a diagram or manipulatives if you need to.
- Check your answer by using a different strategy to solve the problem.

Communicate the Ideas

1. Describe one strategy you could use to solve this problem.

   The head server in a restaurant cuts pies into 6 equal pieces. At the end of the day, there are 8 pieces left. How much pie is left?

2. Solve the pie problem using a different strategy. Show your solution.

3. In a group or as a class, discuss the various strategies you used for questions 1 and 2. Which strategies are most efficient? Explain.

4. How many whole sandwiches are there? How do you know?

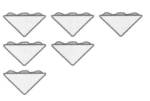

Check Your Understanding

Practise

5. What fraction does each set of diagrams represent?

   a)

   b)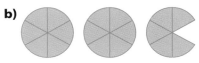

   c)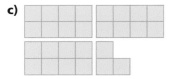

6. Write a fraction for each situation.

   a) 5 red pens out of a package of 12 pens
   b) 3 baseballs out of a box of 8 balls
   c) 2 green T-shirts in a stack of 5 T-shirts

Apply

For help with questions 7 to 9, refer to the Example.

7. A plate contains sandwiches cut into quarters. After lunch, there are 5 pieces left. How many sandwiches is this?

8. On pizza day, several 8-slice pizzas are ordered. At the end of lunch, there are 13 slices left. How many pizzas is this?

9. Several paper plates are cut in half for an arts and crafts class. After the class, there are 7 pieces left. How many paper plates is this?

For help with questions 10 and 11, refer to Example 2.

10. A large snack plate contains sandwiches cut into quarters and oranges cut into eighths. At the end of snack time, there are 14 pieces of sandwich and 12 pieces of orange left. How many sandwiches and how many oranges is this?

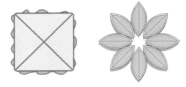

11. The school chef cut up muffins and pizzas to give to his helpers. He cut muffins in half and pizzas in sixths. After serving his helpers, the chef had three muffin pieces and eight pizza slices left. How many muffins and how many pizzas is this?

12. A fruit plate contains orange wedges. Each wedge is $\frac{1}{8}$ of a whole orange. There are enough orange wedges left on the plate to make $2\frac{1}{4}$ whole oranges. How many wedges are left on the plate?

13. Sixteen marbles are placed in a bag. Five of the marbles are blue, 3 are purple, 2 are white, and 6 are green.

    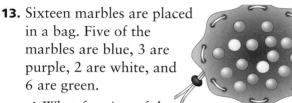

    a) What fraction of the marbles is each colour?
    b) What fraction of the marbles is green or blue?
    c) What fraction of the marbles is not purple?

14. Aleta offered her 5 friends watermelon wedges one hot summer afternoon. She cut two slices of watermelon into equal-sized wedges. She and her friends each ate one wedge. What fraction of a whole slice was each wedge?

15. There are 25 students in Selena's class. Her teacher divides the class into 5 teams for a game.
    a) What fraction of the class is on each team?
    b) The next day, four students are absent. What fraction could the teacher divide the class into now to have an equal number of students on each team? Describe your solution.

16. The length of a rectangular carpet is 15 m. The width is $\frac{1}{3}$ of its length.
    a) Sketch a diagram of the carpet and label the given information.
    b) Find the area of the carpet.
    c) Find the perimeter of the carpet.

17. Malak's birthday cake is cut into 12 pieces. His family eats $\frac{1}{3}$ of the cake for dessert.
    a) What fraction of the cake is left over? Describe your solution.
    b) The next day Malak shares the leftover cake with friends. He and his friends each have one piece and the cake is all gone. How many friends did he share with?
    c) If Malak cut the leftover pieces of cake in half, how many friends could he share with? Describe your solution.

Extend

18. a) The sum of each row and column must be 1. Find the missing values.

$\frac{1}{3}$	$\frac{1}{4}$	
$\frac{7}{12}$		$\frac{1}{6}$
	$\frac{1}{2}$	$\frac{5}{12}$

    b) Create a similar fraction square in which the sums equal 2.

19. Ms. Getfit is a math teacher who also teaches physical education. To get her students thinking about math in their physical education classes, Ms. Getfit uses a new format for a relay race. There are 8 teams competing and each team has 3 people. The first person on each team runs $\frac{1}{4}$ of a lap of the track. The second person runs $\frac{1}{6}$ of a lap of the track, and the third person runs $\frac{1}{3}$ of a lap of the track. How many laps of the track are run in total by all 8 teams combined? Describe your solution.

CHAPTER 3 Review

Key Words

Match each term with its meaning. In your notebook, write each term with its correct meaning.

1. The ▬▬▬ of 4 are 4, 8, 12, 16, ….
2. $\frac{6}{4}$ is an example.
3. $1\frac{1}{2}$ is an example.
4. $\frac{1}{2}, \frac{2}{4}, \frac{3}{6},$ and $\frac{4}{8}$ are examples.
5. 2 is this part of the fraction $\frac{2}{3}$.
6. 4 is this part of the fraction $\frac{3}{4}$.
7. $\frac{1}{2}$ and $\frac{2}{3}$ have 6 as a ▬▬▬ ▬▬▬.

A numerator
B equivalent fractions
C denominator
D multiples
E divisor
F improper fraction
G common denominator
H mixed number

3.1 Add Fractions Using Manipulatives, pages 86–89

8. Write an addition sentence to represent the fraction of each hexagon that is covered.

   a) 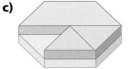   b)

   c) 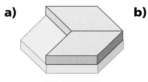   d)

9. Use pattern blocks or diagrams to model each addition.
   a) $\frac{1}{2} + \frac{1}{3}$
   b) $\frac{2}{3} + \frac{1}{6}$
   c) $\frac{1}{6} + \frac{1}{2}$

10. Draw a diagram to show the answer to each addition in question 9.

11. Suppose 1 hexagon = 1 whole.
    a) Cover a hexagon with 1 trapezoid, 1 rhombus, and 1 triangle. Draw a diagram to represent this situation.
    b) Write an addition statement to represent the diagram. Evaluate.

12. Use pattern blocks or diagrams to show each sum.
    a) $\frac{1}{3} + \frac{1}{3}$
    b) $\frac{1}{6} + \frac{1}{6} + \frac{1}{6} + \frac{1}{6} + \frac{1}{6}$

13. What multiplication statement does each part in question 12 represent?

3.2 Subtract Fractions Using Manipulatives, pages 90–93

14. Write a subtraction sentence to represent each diagram.

    a)    b)

15. Draw a diagram to show the answer to each subtraction in question 14.

16. Write a subtraction sentence to represent each diagram.

a) 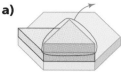   b)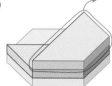

17. Draw a diagram to show the answer to each subtraction in question 16.

18. Use pattern blocks or diagrams to model each subtraction.

a) $\frac{2}{3} - \frac{1}{3}$ b) $\frac{1}{2} - \frac{1}{6}$
c) $\frac{5}{6} - \frac{2}{3}$

3.3 Find Common Denominators, pages 94–97

19. State a common denominator for each pair of diagrams.

a)   b)

20. Use paper folding or a diagram to find a common denominator for each pair of fractions.

a) $\frac{1}{3}$ and $\frac{5}{6}$ b) $\frac{3}{4}$ and $\frac{1}{6}$

21. Use multiples to find a common denominator for each pair of fractions.

a) $\frac{1}{4}$ and $\frac{2}{3}$ b) $\frac{2}{5}$ and $\frac{1}{2}$

22. Find two common denominators between 10 and 20 for $\frac{1}{2}$ and $\frac{1}{3}$.

3.4 Add and Subtract Fractions Using a Common Denominator, pages 98–103

23. Write and evaluate the addition or subtraction sentence represented by each diagram.

a)    b)

24. Draw a diagram to represent each addition or subtraction.

a) $\frac{5}{6} - \frac{3}{8}$ b) $\frac{3}{4} + \frac{3}{5}$

25. Evaluate.

a) $\frac{9}{10} - \frac{2}{3}$ b) $\frac{1}{4} + \frac{5}{6}$

3.5 More Fraction Problems, pages 104–107

26. A plate contains sandwiches cut into thirds. After lunch, there are 7 pieces left over. How many sandwiches is this?

27. Twenty-four cars are parked in a parking lot. Six of the cars are red, 4 are blue, 9 are white, and 5 are grey.
    a) What fraction of the cars is each colour?
    b) What fraction of the cars is red or blue?
    c) What fraction of the cars is not white?

28. A snack plate contains muffins cut half and apples cut into sixths. At the end of snack time, there are 5 muffin pieces and 11 pieces of apple left. How many muffins and how many apples is this?

CHAPTER 3 Practice Test

| Strand Questions | NSN 1–14 | MEA | GSS | PA | DMP |

Multiple Choice

For questions 1 to 4, select the correct answer.

1. What fraction of the hexagon is covered?

   A $\frac{1}{3}$ B $\frac{1}{2}$
   C $\frac{2}{3}$ D $\frac{5}{6}$

   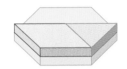

2. What fraction of the hexagon is covered?

   A $\frac{2}{6}$ B $\frac{4}{6}$
   C $\frac{1}{6}$ D $\frac{5}{6}$

3. Evaluate the subtraction represented by the diagram.

   A $\frac{5}{6}$ B $\frac{2}{6}$
   C $\frac{1}{2}$ D $\frac{1}{6}$

4. Suppose 1 hexagon = 1 whole. The diagram models

   A $\frac{5}{6} - \frac{1}{3}$ B $\frac{1}{3} + \frac{2}{6}$
   C $\frac{5}{6} + \frac{1}{3}$ D $\frac{1}{3} - \frac{2}{6}$

5. Suppose 1 hexagon = 1 whole.

   The diagram shows

   A $\frac{2}{3} + \frac{1}{6}$ B $\frac{1}{2} + \frac{1}{3}$
   C $\frac{1}{3} + \frac{2}{6}$ D $\frac{3}{6} + \frac{1}{3}$

Short Answer

6. Suppose 1 hexagon = 1 whole. The diagram shows 2 blue rhombi covering 1 yellow hexagon.

   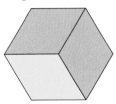

   You want to subtract $\frac{1}{2}$.

   a) Draw a diagram to show this subtraction.
   b) Write and evaluate the subtraction statement represented.

7. Rewrite each repeated addition as a multiplication statement and evaluate. Show your answer as an improper fraction and a mixed number, if necessary.

   a) $\frac{1}{5} + \frac{1}{5} + \frac{1}{5} + \frac{1}{5}$

   b) $\frac{2}{7} + \frac{2}{7} + \frac{2}{7} + \frac{2}{7} + \frac{2}{7}$

8. Find a common denominator for each pair of fractions.

   a) $\frac{1}{2}$ and $\frac{3}{4}$ b) $\frac{2}{3}$ and $\frac{2}{5}$

9. Find two common denominators between 10 and 30 for $\frac{1}{3}$ and $\frac{1}{4}$.

10. Evaluate.

a) $\dfrac{3}{8} + \dfrac{2}{3}$

b) $\dfrac{5}{6} - \dfrac{4}{5}$

11. a) Which is greater, $1 - \dfrac{3}{4}$ or $1 - \dfrac{3}{10}$?

b) Explain the strategy you used to solve part a).

12. Eric and Afsha are both trying to add $\dfrac{3}{5} + \dfrac{2}{3}$. Eric says the answer is $1\dfrac{4}{15}$. Afsha says the answer is $\dfrac{5}{8}$. Which student is correct? Show how you know.

13. Several apples are cut into slices for a class snack. Each slice is $\dfrac{1}{12}$ of a whole apple. After the snack, there are enough apple slices left over to make $1\dfrac{1}{3}$ whole apples. How many apple slices are left over?

Extended Response

14. Mia and Steven both found a common denominator for $\dfrac{1}{2} + \dfrac{2}{3} + \dfrac{3}{4}$. Mia said, "The common denominator is 12." Steven said, "The common denominator is 24."

a) Who is correct? Explain.

b) Use a common denominator to evaluate $\dfrac{1}{2} + \dfrac{2}{3} + \dfrac{3}{4}$

Chapter Problem Wrap-Up

1. You are asked to create a puzzle for the grade 6 class to help them understand fractions. Your puzzle is to be $\dfrac{1}{6}$ red, $\dfrac{1}{6}$ blue, and $\dfrac{1}{3}$ green. The rest of the puzzle is to be yellow.

a) Design a puzzle that fits this description. Use grid paper, coloured tiles, pattern blocks, centimetre cubes, or another material of your choice.

b) Show how you know your puzzle is $\dfrac{1}{3}$ green.

c) What fraction of your puzzle is yellow? Show how you know.

2. Create a puzzle that is $\dfrac{1}{4}$ red, $\dfrac{1}{4}$ blue, and $\dfrac{1}{3}$ green. The rest is yellow. Describe as many fraction relations as you can about this puzzle. Use pictures, words, and numbers in your report.

Data Management and Probability

- Collect and organize data on tally charts.
- Develop concepts of probability.
- Identify favourable outcomes, and state probabilities.
- List outcomes using tree diagrams, modelling, and lists.
- Use and apply probability, including in sports and games.

Key Words

tally chart
frequency table
probability
outcome
favourable outcome
random
tree diagram
simulation

CHAPTER 4

Probability and Number Sense

Auto racing is a thrilling sport. Drivers with good winning records are in hot demand. That is because these drivers and their teams have a good chance of winning again.

What makes each race exciting is the fact that each driver has an equal chance of winning. The driver's skill and the speed of the team that checks and fuels the car, however, increase the chances of winning.

By the end of this chapter, you will be able to develop an exciting race-car game. You will be able to check the probability of each player winning. You will also be able to make the game fair for at least two players. Have fun developing and playing the games in this chapter!

Chapter Problem

Some board games are based on auto racing. You can use spinners or number cubes to provide moves or chances.

What makes a game fair? What can you do to give each player an equal probability of winning a game?

Get Ready

Calculate Means

The **mean** or average of a set of numerical data is the sum of the data, divided by the number of data items. For example, look at this set of student heights.

mean = (165 + 178 + 156 + 158 + 153 + 165) ÷ 6
     = 162.5

The mean height for this group of students is 162.5 cm.

Carrie	165 cm
Arturo	178 cm
Jamilla	156 cm
Robert	158 cm
Wendy	153 cm
Jésus	165 cm

1. Find the mean of each set of data.
   a) 2, 4, 2, 2, 4, 4
   b) 150 cm, 170 cm, 158 cm, 166 cm, 184 cm
   c) 54 kg, 38 kg, 49 kg, 61 kg, 55 kg, 64 kg
   d) 13 mm, 17 mm, 12 mm, 16 mm, 18 mm
   e) 12.3 m, 11.9 m, 12.7 m, 13.0 m, 12.5 m, 11.9 m, 12.1 m, 12.8 m
   f) 35 jellybeans, 32 jellybeans, 37 jellybeans, 35 jellybeans, 34 jellybeans

2. Four different groups performed 50 coin tosses each. Their results were as follows:

   Group 1   23 heads, 27 tails
   Group 2   25 heads, 25 tails
   Group 3   21 heads, 29 tails
   Group 4   28 heads, 22 tails

   Calculate the mean number of heads.

Equivalent Fractions

These fraction strips show that the fractions $\frac{4}{6}$ and $\frac{2}{3}$ are equivalent.

The fractions $\frac{3}{5}$ and $\frac{5}{8}$ are not equivalent.

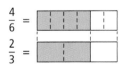

3. Draw fraction strips for each pair of fractions. State which pairs are equivalent.
   a) $\frac{2}{4}$ and $\frac{1}{2}$
   b) $\frac{2}{3}$ and $\frac{1}{2}$
   c) $\frac{2}{5}$ and $\frac{3}{8}$
   d) $\frac{7}{9}$ and $\frac{42}{54}$

4. a) Maria has 15 T-shirts, 6 of which are white. Express the number of white T-shirts as a fraction of the total number.
   b) Find an equivalent fraction for your fraction in part a). Hint: Use 5 as the denominator.

Compare and Order Fractions

You can compare fractions by showing them with a common denominator.

For example, Devon and Monica are comparing the amount of chocolate bar they have left.

Making Connections

You worked with fractions in Chapter 3.

 I have $\frac{1}{4}$ left. I have $\frac{5}{12}$ left.

To find a common denominator, list the multiples of 4 and 12.

The multiples of 4 are 4, 8, (12,) 16, 20, ...

The multiples of 12 are (12,) 24, ...

$\frac{1 \times 3}{4 \times 3} = \frac{3}{12}$

$\frac{5}{12}$

Twelve is a common denominator.

The fraction $\frac{5}{12}$ is greater than $\frac{3}{12}$. Monica has more chocolate left.

5. Write each pair of fractions with a common denominator. Which fraction is greater?

a) $\frac{1}{3}$ and $\frac{3}{8}$ b) $\frac{2}{5}$ and $\frac{1}{2}$ c) $\frac{4}{6}$ and $\frac{7}{10}$

6. Order these fractions from least to greatest.

$\frac{7}{10}$ $\frac{1}{6}$ $\frac{11}{15}$ $\frac{2}{3}$ $\frac{3}{5}$

7. Look at this spinner.

a) Express each colour as a fraction of the whole spinner.

b) Write the colours in the order of their fractions, from least to greatest.

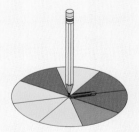

Convert Fractions to Decimals

To change a fraction to a decimal, divide the denominator into the numerator.

$\frac{7}{10}$ = ⓒ 7 ÷ 10 = 0.7 $\frac{1}{4}$ = ⓒ 1 ÷ 4 = 0.25

8. Show each fraction as a decimal.

a) $\frac{3}{10}$ b) $\frac{6}{10}$ c) $\frac{23}{100}$

9. Show each fraction as a decimal.

a) $\frac{1}{2}$ b) $\frac{3}{8}$ c) $\frac{2}{5}$

4.1 Introducing Probability

Focus on...
- probability
- outcomes

At the beginning of a soccer game, a referee tosses a coin. The winning captain chooses which end of the field to take.

How do the team captains know that they each have an equal chance of winning the toss? How can you express this chance as a number?

Discover the Math

Materials
- coin

tally chart
- used to record experimental results or data
- counts the data

frequency table
- used to show the total numbers of occurrences in an experiment or survey

probability
- the chance that something will happen
- often expressed as a proper fraction, or a decimal between 0 and 1

How can you answer questions about chance?

1. Trudy, the Snowcats captain, calls the toss at the start of a soccer match. Last night, Trudy tossed a coin 100 times and got 60 heads, so she decides to call heads. Is Trudy making a winning decision? Explain.

2. Test Trudy's decision by tossing a coin for about 5 min.

   a) Before you start, estimate the number of heads and the number of tails you will record.

   b) Record your results in a **tally chart** and **frequency table**, like this one.

	Tally	Frequency
Heads	|||||	
Tails	|||	
Total Trials		

3. a) Estimate the **probability** of getting heads, based on your results. State your estimate as a fraction, then convert it to a decimal.

   b) Based on your results, comment on Trudy's decision.

4. Combine results from the whole class. Use the decimal estimates to calculate the mean of the probabilities. What do you notice?

5. **Reflect** Can you think of another way to determine the probability? Use your ideas to help Trudy see the coin toss in a different way.

Example: Calculate Probability by Outcomes

A jar of 30 jellybeans has 7 red, 6 black, 4 yellow, 5 orange, and 8 green jellybeans.

a) How many possible **outcomes** are there when you pick a jellybean from the jar?
b) If you are hoping for a red jellybean, how many **favourable outcomes** are there?
c) What is the probability of picking a red jellybean, if you are not looking as you pick? Write the probability as a fraction and a decimal, rounded to the nearest hundredth.
d) What is the probability of picking a black jellybean at **random**, as a fraction and a decimal?

outcome
- one possible result of a probability experiment

favourable outcome
- an outcome that counts for the probability being calculated

random
- a choice or pick in which each outcome is equally likely
- if you do not look as you pick, you are choosing at random

Solution

a) There are 30 jellybeans in the jar. So, there are 30 possible outcomes.

b) There are 7 red jellybeans, so there are 7 favourable outcomes.

c) Probability(red jellybean) = $\dfrac{\text{favourable outcomes}}{\text{all outcomes}}$

$= \dfrac{\text{}}{\text{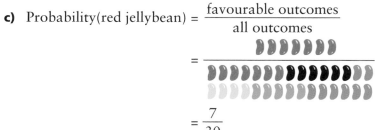}}$

$= \dfrac{7}{30}$

$\doteq 0.23$

The probability of a red jellybean is $\dfrac{7}{30}$, or approximately 0.23.

d) There are 6 black jellybeans.

Probability(black jellybean) = $\dfrac{\text{favourable outcomes}}{\text{all outcomes}}$

$= \dfrac{\text{number of black jellybeans}}{\text{total number of jellybeans}}$

$= \dfrac{6}{30}$

$= 0.2$

The probability of a black jellybean is $\dfrac{1}{5}$ or 0.2.

Strategies
How else might you show the fraction $\dfrac{6}{30}$?

The probability $\dfrac{6}{30}$ is equivalent to $\dfrac{1}{5}$. That's a 1 in 5 chance.

Key Ideas

- Probabilities can be estimated from repeated trials of an experiment.
- Probabilities can also be calculated.
- Probabilities can be shown as a fraction or as a decimal.

$$\frac{7}{30} = 7 \div 30$$
$$\doteq 0.23 \qquad \boxed{C}\,7\,\boxed{\div}\,30\,\boxed{=}\ 0.233333333$$

$$\text{Probability(red jellybean)} = \frac{\text{favourable outcomes}}{\text{all outcomes}}$$
$$= \frac{\text{red jellybeans}}{\text{all jellybeans}}$$
$$= \frac{7}{30}$$

- The outcomes of an experiment are the possible results. For example, in a coin toss, the two outcomes are heads and tails. The probability of getting heads is $\frac{1}{2}$ or 0.5.

Communicate the Ideas

1. Kyle says, "Let's use a coin toss. I call heads. I have one chance in two of winning." Devon says, "I call tails. My chance of winning is 0.5." Both Kyle and Devon are correct. Explain.

2. In some sports, a coin toss is not used. Discuss the different ways a sporting event can be started, using probability to make choices.

3. You are asked to pick a letter at random from the bag.
   a) What chance do you have of picking the letter A?
   b) What chance do you have of picking the letter S?
   c) Justify your answers.
   d) Try it. Use a tally chart to show your first 50 picks. Which letter did you pick the most often?

Check Your Understanding

Practise

4. Copy and complete this tally chart and frequency table.

	Tally	Frequency
Heads	|||| |||| |||| ||||		
Tails	|||| |||| |||| |||| |||| ||	
	Total Trials	

5. Copy and complete this tally chart and frequency table.

	Tally	Frequency
Red	|||| |||| |||	
Blue	|||| |||| ||||		
Yellow	|||| ||||		
	Total Trials	

6. In each situation, state the total number of outcomes, and the number of favourable outcomes.

   a) Helen likes roll-neck sweaters. She picks a sweater at random.

   b) Four winning tickets will be drawn.

   c) Christina wants an unchipped glass. She picks a glass at random.

7. For each part of question 6, state the probability as a fraction.

For help with questions 8 to 10, refer to the Example.

8. In a jar of 25 jellybeans, there are
   - 13 zippy zingers
   - 3 stomach stirrers
   - 2 tongue twisters
   - 7 face freezers

   a) How many possible outcomes are there?
   b) How many favourable outcomes are there for each flavour?
   c) What is the probability of picking a zippy zinger at random? Write the probability as a fraction and as a decimal.
   d) Find the probabilities of picking each of the other flavours at random. Show the probabilities as fractions and as decimals.

Apply

9. Suppose the numbers of each jellybean flavour in question 8 were doubled.

   a) Predict what will happen to the probability of each flavour.
   b) Determine the probability of picking each flavour at random. Do your findings confirm your prediction? Explain.

10. a) For the jar of jellybeans in question 8, what is the probability of picking a jellybean of *any* flavour at random? Explain.
    b) What is the probability of picking a knee knocker jellybean at random? Explain.

11. Karina's sock drawer has 3 pairs of grey socks, 4 pairs of white socks, and 6 pairs of tan socks. Each pair is rolled together. Every morning Karina picks a pair of socks at random.

    a) What is the probability that Karina will pick out a pair of white socks?
    b) Which colour of socks is Karina most likely to pick? Which colour of socks is she least likely to pick? Justify your answer.
    c) Explain why the probabilities change, depending on the colour.

12. Danae and Tom are about to start a board game. The game comes with a four-section spinner. Explain how to use the spinner to decide who goes first.

    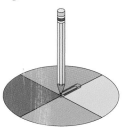

4.1 Introducing Probability • MHR 119

13. What's wrong? Cheryl repeatedly spun a spinner with four equal sections and got these results:

Colour   Total
orange   3
blue   4
yellow   0
red   2

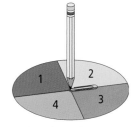

Is something wrong with Cheryl's spinner? How could she check?

14. Jonah's sock drawer is filled with unmatched socks. Each morning he pulls out two socks at random. On Monday morning, there are 2 green socks, 3 purple socks, and 1 orange sock in Jonah's drawer.

a) Use slips of paper in a paper bag to model Jonah's sock drawer. Record colours.

b) Pick two slips of paper at random. Note whether they match or not. Hint: Do not put the first slip of paper back before picking the second. Then, put both back.

c) Copy this tally chart and frequency table.

	Tally	Frequency
Matched		
Unmatched		
Total Trials		

Perform step b) repeatedly for 5 min. Record your results in the tally chart.

d) Use your results to estimate the probability that Jonah will get a pair of matched socks.

15. a) Veronica ate 2 red, 2 yellow, 1 orange, and 5 green jellybeans from a jar. She left behind 5 red, 6 black, 2 yellow, 4 orange, and 3 green jellybeans. Find the probability of picking each jellybean colour
  • before Veronica started eating
  • after Veronica had eaten her 10 jellybeans

b) Do you think Veronica picked her jellybeans at random? Explain.

16. A number cube was rolled 100 times. These results were recorded.

Number	Total
1	16
2	20
3	13
4	7
5	27
6	17

a) Use these results to estimate the probability of rolling a 5.

b) Do you think the cube is fair? Explain. How could you check?

c) If the same cube were rolled 6000 times, would you be surprised if the number 5 turned up 980 times? 320 times? Explain.

Extend

17. Create a probability experiment that involves picking an item out of a container. Use at least 50 items.

a) In repeated trials of the experiment, decide whether you should replace, or not replace, the item after each pick. Explain your decision.

b) Carry out repeated trials for 2 min. Then, estimate the probabilities of picking each type of item.

c) Count possible outcomes to determine the actual probabilities.

d) Compare your results for parts b) and c). Think about the items you used. Are all outcomes equally likely? Explain.

4.2 Organize Outcomes

Focus on...
- probability in games
- modelling outcomes
- favourable outcomes

Board games have been around for thousands of years. In many board games, you must roll a number cube or spin a spinner to move.

Suppose you are three board spaces away from "home." You need to roll a 3 or more. What is the probability that you will get home on this move?

Discover the Math

Materials
- white cardboard
- compasses or circular object to trace around
- ruler
- pencil crayons
- 2 paper clips
- 2 pencils

Optional:
- BLM 4.2A Spinner Templates

How can you organize outcomes to help with calculations?

1. Create a spinner with two equal-sized sections, as in the photograph. Colour the sections red and blue.

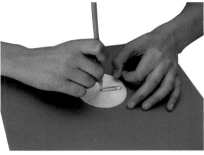

2. Create a spinner with four equal sections, coloured red, blue, yellow, and purple.

3. Create a way of organizing the combined outcomes when you spin both spinners.

4. Predict the probability of spinning red and yellow in the same spin.

5. Spin the pair of spinners repeatedly for about 5 min. Record your results in an organized list. Use your results to estimate the probability of spinning red and yellow.

6. **Reflect** How well does your organizer present your results? How could you improve it?

Strategies
Make an organized list

Example 1: Use a List or Modelling

Create an organizer, with probabilities, for
a) a spinner with equal red and blue sections
b) a spinner with equal red, blue, yellow, and purple sections

Literacy Connections

Reading Percents
The expression "50-50 chance" means a probability of 50%. You can read 50% as $\frac{50}{100}$. What simple fraction is equivalent to this?

Solution

a) *Method 1: Make a List*

spinning red   probability(red) = $\frac{\text{red}}{\text{red or blue}}$

= $\frac{1}{2}$

spinning blue   probability(blue) = $\frac{\text{blue}}{\text{red or blue}}$

= $\frac{1}{2}$

Strategies
Make a model

Method 2: Test the Spinner
Make and test the spinner. You should find that the two outcomes are equally likely.

Probability(red) = $\frac{1}{2}$

Probability(blue) = $\frac{1}{2}$

There are two outcomes. Each one has a probability of $\frac{1}{2}$. That's a 50% chance.

b) spinning red $\frac{1}{4}$ spinning blue $\frac{1}{4}$

spinning yellow $\frac{1}{4}$ spinning purple $\frac{1}{4}$

There are four outcomes. Each one has a probability of $\frac{1}{4}$.

Example 2: Use a Tree Diagram

a) Create a **tree diagram** to show the possible outcomes from spinners A and B.
b) What is the probability of spinning a 1 and a 4?
c) What is the probability of spinning a 1 and a 2?
d) What is the probability of spinning a total of 4?

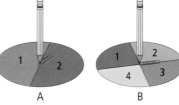

A   B

Solution

a) Spinner A   Spinner B   Outcome

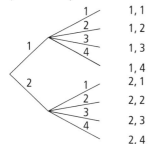

1, 1
1, 2
1, 3
1, 4
2, 1
2, 2
2, 3
2, 4

tree diagram
- diagram that shows outcomes as sets of branches
- useful for organizing combined outcomes

122   MHR • Chapter 4

b) There is only one favourable outcome, (1, 4).

Probability(1 and 4) = $\dfrac{\text{favourable outcomes}}{\text{all outcomes}}$

= $\dfrac{1}{8}$

c) (1, 2) and (2, 1) are the favourable outcomes.

Probability(1 and 2) = $\dfrac{\text{favourable outcomes}}{\text{all outcomes}}$

= $\dfrac{2}{8}$

Another way of saying this is $\dfrac{1}{4}$.

I changed the combined outcomes from pairs to totals. Two outcomes show a sum of 4.

d)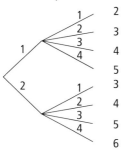

Probability(total of 4) = $\dfrac{\text{favourable outcomes}}{\text{all outcomes}}$

= $\dfrac{2}{8}$

Literacy Connections

Reading Tree Diagrams

Read tree diagrams from left to right.
- The branches on the left of the tree show the outcomes for one spinner.
- The branches on the right show the outcomes for the other spinner.
- At the far right of the diagram, read off the combined outcomes.

Key Ideas

- Tree diagrams and lists help to organize the outcomes of a probability experiment.
- Tree diagrams and lists can be used to determine probabilities. With this spinner, the probability of spinning red and then red in two spins is $\dfrac{1}{4}$.

 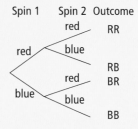

- Modelling with physical objects can help with creating an organized list or tree diagram.

Communicate the Ideas

1. How can a tree diagram help you organize the possible outcomes from this spinner and cube?

2. Draw a spinner for each set of probabilities.

  a) spinning red $\dfrac{1}{3}$ spinning blue $\dfrac{1}{3}$ spinning green $\dfrac{1}{3}$

  b) spinning yellow $\dfrac{1}{5}$ spinning purple $\dfrac{4}{5}$

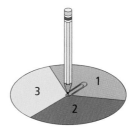 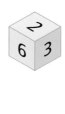

4.2 Organize Outcomes • MHR 123

Check Your Understanding

Practise

For help with questions 3 and 4, refer to Example 1.

3. Create an organized list of outcomes, with probabilities, for each spinner.
   a)   b)

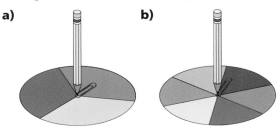

4. Create an organized list, with probabilities, for each spinner.
   a)   b)

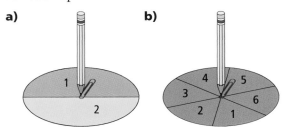

For help with questions 5 to 10, refer to Example 2.

5. Two spinners are numbered as shown.

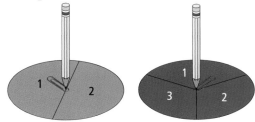

   a) Create a tree diagram for spinning both spinners.
   b) What is the probability of spinning a 1 and a 3?
   c) What is the probability of spinning a 1 and a 2?

6. a) What is the probability of spinning a 1 and a 3 on these spinners?

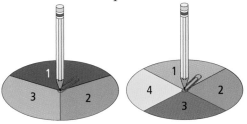

   b) What is the probability of spinning a 2 and a 4?
   c) What is the probability of spinning a total of 4?

Apply

7. a) Choose numbers for your own pair of spinners.
   b) Draw a tree diagram, showing the total scores.
   c) For each total score, determine the probability.

8. A number cube is labelled 2, 3, 3, 4, 5, 6.
   a) What is the probability of rolling a 4?
   b) What is the probability of rolling a 3?
   c) What is the probability of rolling a number less than 3?

9. Chantal needs to roll 3 or more on a number cube to win the game. Determine the probability that Chantal wins. Justify your answer.

10. a) What is the probability of spinning green?
    b) What is the probability of spinning yellow?
    c) Draw a diagram of a red-blue spinner with probability(red) = $\frac{3}{5}$.

124 MHR • Chapter 4

Chapter Problem

11. A car rally game uses two spinners.

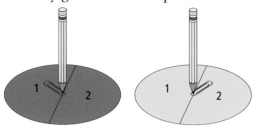

a) What are the possible sums when you add the results from these two spinners?

b) What is the probability of getting each sum?

12. Frances asked Davin to explain the probability lesson she missed. Davin decided to use a spinner to explain predicted probability of outcomes and tree diagrams.

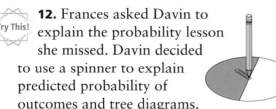

a) Draw a tree diagram to show the outcomes and predicted probability of spinning the spinner twice.

b) Frances wanted to test Davin's theory so she spun the spinner twice and got two reds. She argued with Davin that the results on her spin were different than what he explained. Write Davin's response.

Extend

13. Roll two number cubes. If the sum of the numbers showing is an even number, then Player 1 wins. If the sum is an odd number, Player 2 wins.

a) Play the game 50 times. Record who wins in a tally chart.

b) Estimate the probability of each player winning.

c) Explain whether one player is more or less likely to win. Justify your reasoning.

14. a) Create a probability game. Explain what materials are needed. List the rules clearly.

b) Explain whether some events in the game are more likely than others. If not, explain why not.

c) Create an organizer for your game. You can use a tree diagram, an organized list, or your own type of organizer.

d) Report on the probabilities involved in your game. If some probabilities cannot be determined from your organizer, try experimenting to get estimates.

Making Connections

What does math have to do with pop flavours?

Do you still drink the same beverages as you did in kindergarten? Do you like the same flavours?

The Ontario Beverage Company makes seven different flavours of pop. Over the past two years, they have been keeping track of how much of each flavour has been sold.

1. Based on this information, estimate the probability that a customer will buy Oliphaunt Orange.

2. Which flavours should the company discontinue? Explain, in terms of probability.

Drink Flavour	Sales (1000s)
Koala Cola	13
Lizard Lime	5
Lemur Lemon	8
Gorilla Grape	3
Roary Root Beer	9
Oliphaunt Orange	10
Jumping Ginger	2

4.3 Use Outcomes to Predict Probabilities

Focus on...
- listing outcomes
- predicting probabilities

Examine this set of numbered cards.

What is the probability of drawing a card numbered 3?

Discover the Math

How can you work with outcomes to predict probabilities?

Example 1: Probability in Number Cards

Crazy Eights can be played with a set of 40 cards, as pictured on this page. Determine the probability of each draw.
a) the 8 of ♠
b) any one of the Crazy Eights
c) any red card

Making Connections

You will find the rules to *Crazy Eights* on page 130.

Solution

a) There is only one 8 of ♠, so the probability of picking that card is $\frac{1}{40}$.

b) There are four 8s altogether, so the probability of picking an 8 is $\frac{4}{40}$ or $\frac{1}{10}$.

c) *Method 1: Count Outcomes*
There are 20 red cards, so the probability of picking a red card is $\frac{20}{40}$ or $\frac{1}{2}$.

Method 2: Use Proportions
Exactly half of the set is red, so the probability of picking a red card is $\frac{1}{2}$.

Strategies
Solve a simpler problem

Example 2: Outcomes of Multiple Coin Tosses

Madison and Robyn want to share a bag of potato chips. They cannot agree on the flavour. They decide to toss a coin and play "the best of three wins."
- If there are at least two heads out of three tosses, Madison decides on the flavour.
- Otherwise, Robyn decides.

What is the probability that Madison gets to decide?

Solution
Method 1: Use a Tree Diagram

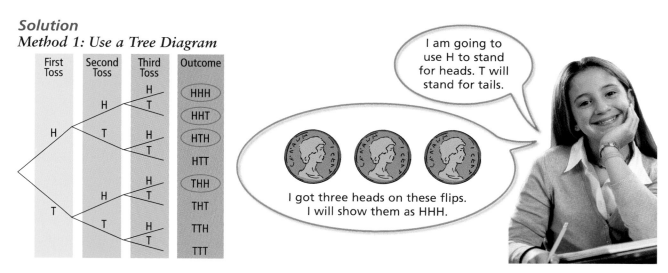

Four favourable outcomes show two heads. So, the probability that Madison gets to decide is $\frac{4}{8}$ or $\frac{1}{2}$.

Method 2: Create a List

HHH   TTT   HHT   TTH   HTH   THT   HTT   THH

The favourable outcomes for Madison are HHH, HHT, HTH, and THH.

So,

Probability(Madison decides)
$= \frac{\text{favourable outcomes}}{\text{all outcomes}}$
$= \frac{4}{8}$
$= \frac{1}{2}$

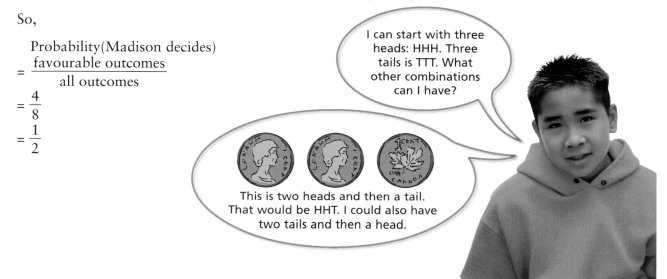

Key Ideas

- Outcome organizers such as tree diagrams and lists help to predict probabilities.
- When choosing a method to solve a probability problem, think about what might work best for the problem.

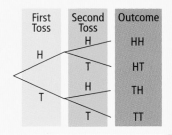

Communicate the Ideas

1. John e-mailed a friend about the probability lesson he missed. John explained that the teacher used a tree diagram to determine all the possible outcomes of tossing a coin three times. What could John tell his friend about the probability of winning a best of three?

2. For each situation, describe an appropriate method to illustrate the possible outcomes.
   a) rolling odds or evens with a number cube
   b) drawing a Crazy Eight from a 40-card set

3. Look at this tree diagram. What method would you use to find each probability, and why?
   a) exactly two heads
   b) at least one head and at least one tail

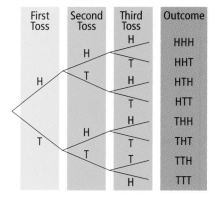

Check Your Understanding

Practise

For help with questions 4 to 6, refer to Example 1.

4. Determine the probability of each draw from a 40-card set. (See the picture on page 126.)
   a) the 3 of ♦
   b) the 7 of ♣
   c) any black card
   d) any 9
   e) any card except a Crazy Eight

5. Determine the probability of each draw from a 40-card set. (See the picture on page 126.)
   a) the 4 of ♠
   b) the 7 of ♥
   c) any red Crazy Eight
   d) a 7 or 8
   e) a 3 or 4

6. Look at this 20-card set. Which probabilities in question 4 change? Which stay the same? Explain.

For help with questions 7 to 10, refer to Example 2.

7. For each situation, list all the possible outcomes.

   a) You roll a standard number cube.

   b) You roll a cube labelled A, A, B, B, B, and C.

   c) You draw a jellybean from a bag with the colour selection shown.

8. For each part of question 7, choose one outcome. Calculate the probability.

9. You flip a coin and spin a three-section spinner. List the possible outcomes.

Apply

10. What is the probability of drawing the 8, the 9, or the 10 of ♠ from this set? Explain.

11. For each situation,
    • state the probability
    • justify your response
    a) rolling an odd number on a number cube
    b) rolling a 2 or a 5 on a number cube

12. What is the probability of picking a vowel from a bag containing the letters of the word MILLION? Explain.

13. A letter is drawn repeatedly and at random from the word MATHEMATICAL. You can choose one letter. Every time your choice is drawn, you get a point. Letters are replaced after each draw. Which letter should you choose? Explain.

14. Madeleine and Adeena disagree about probabilities with a standard number cube. Madeleine knows that rolling an odd number is more likely than rolling any individual number. Explain how Madeleine should convince Adeena.

15. Mya decides to use an 8-sided die when playing her favourite board game.

    a) How do the probabilities differ from using a standard 6-sided number cube?
    b) Another 8-sided die is numbered 1, 1, 2, 3, 4, 4, 4 and 5. Which number is most likely to be rolled? Explain.

Chapter Problem

16. A car rally game uses two number cards and a spinner. Pick a card at random. Then, spin the spinner. Subtract.

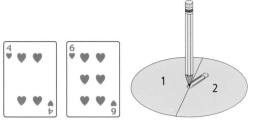

   a) What are the possible differences (card number − spinner number)?
   b) Try it for 3 min. List your results.
   c) What is the probability of getting a 3? Explain.

 17. Create your own probability game that uses simple probability. You can use number cubes, spinners, coloured chips, or other materials. Find and explain the probabilities for your contest.

Extend

18. Rafina is choosing from a menu. She can select one item from each category.

noodle: Shanghai or udon
protein: chicken, beef, or tofu
sauce: spicy Thai or peanut
fruit drink: mango or watermelon

   a) Draw the tree diagram displaying Rafina's possible choices.
   b) Is the probability of each protein choice really the same? Justify your reasoning.
   c) How is the number of choices reduced if Rafina decides on chicken? Explain, in terms of your tree diagram.

19. Two friends are using a coin to play "best of three wins." What is the probability that the winner is chosen after the second toss?

Making Connections

What does math have to do with card games?

Here is one version of the rules for *Crazy Eights*.

- Shuffle the cards. Deal all the cards out evenly to the players.
- The player to the dealer's right begins by laying down a card.
- Players take turns to play cards. Play always moves to the right.
- Usually, you must follow by suit or number. For example, if the 3 of hearts has just been played, you must play another 3 or another heart.
- You can play a Crazy Eight on any turn. You don't have to follow the same suit or number. If you do so, you immediately play a second card. The next player must now follow your second card.
- If you cannot play any card, you miss a turn.
- The winner is the first person to play his or her final card.

1. Play *Crazy Eights* with two or three friends.

2. Discuss what you learned about winning strategies.

3. How does understanding probability help you in *Crazy Eights*?

4.4 Extension: Simulations

Focus on...
- simulations
- probability experiments
- simulations to make best probability estimates

Companies often use "instant-win" promotions to entice customers. How good a deal is this sort of promotion?

Discover the Math

Materials

Any one of
- number cube
- six-section spinner
- slips of paper and a container

Optional:
- BLM 4.4A Run a Simulation

simulation
- a probability experiment used to model a real situation

How can you simulate probability problems?

1. The bottle cap for a new brand of pop has a letter inside it. If you collect all the letters that spell Y-O-U-W-I-N, you win a game console. How many bottles of pop do you think Jasmine needs to buy to win the game console?

2. Select a tool to simulate buying the pop. Describe a method for running your **simulation** and recording your results.

3. Use your simulation method to test how many bottles Jasmine buys. Compare your simulation results with your estimate from step 1. Were they close? Explain why or why not.

4. **a)** Compare your results with some other classmates, or as a class.
   b) Use the combined results to calculate the mean number of bottles Jasmine would have to buy.

5. **Reflect** Consider the simulations you compared in step 4.
   a) Explain what the simulations tell you about the contest.
   b) Why was it useful to compare results?

Example: Simulate With a Spinner

Jasmine used a spinner to simulate the promotion. She spun the spinner and recorded her results in a tally chart. She continued to spin until she had landed on all six letters.

a) How many bottles of pop did Jasmine have to "buy"?
b) Which letter appears to have been last? Explain your reasoning.

Letter	Tally
Y	IIII I
O	IIII
U	I
W	II
I	IIII
N	IIII

Solution

a) Add up the tallies. Jasmine bought 22 bottles of pop.

b) The letter U only has 1 tally. It must have been the last letter.

Key Ideas

- A simulation is an experiment that can be used to model a real situation involving probabilities.
- There are many different ways to simulate a situation.

  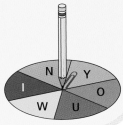

Communicate the Ideas

1. Will simulation outcomes be the same for each student? Explain.

2. Jasmine used a spinner to simulate buying bottles of pop with the letters for Y-O-U-W-I-N. What other methods might she have used?

Check Your Understanding

Practise

For help with questions 3 and 4, refer to the Example.

3. This tally chart shows the results of a simulation.

   a) How many rolls were needed to get all six letters?

   b) Which letter was last? How do you know?

Letter	Tally
S	II
C	IIII
O	I
R	III
E	II

4. This tally chart shows the results of a number cube simulation.

   a) How many rolls were needed to get all six numbers?

   b) Which numbers could have been last? Explain your reasoning.

Number	Tally
1	III
2	II
3	I
4	IIII
5	I
6	IIII

5. Describe an item that could be used to simulate each situation. Explain why each item is appropriate.

   a) A student does not know the answer to a True or False question.
   b) You are picking 1 pizza topping at random from 8 choices.
   c) You can win a phone by collecting the letters on bottle caps to spell P-H-O-N-E.

Apply

6. a) Create a simulation item for question 5c). Using your item, simulate the contest and record your results.
   b) Explain the results of your simulation.
   c) Explain how the results of your simulation compare to those of your classmates.

7. Your favourite chocolate bar has a letter on the inside wrapper. You can win if you collect B, A, and R. There are 10 different letters on the inside wrappers.

   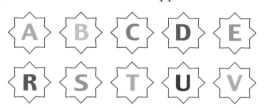

   a) Develop a simulation item. Simulate this contest.
   b) Run the simulation 10 times. For each simulation, how many bars were needed to get B, A, and R?
   c) Did you need to buy the same number

   of bars in each simulation? Explain.

8. At a street intersection, 100 vehicles, on average, pass through in an hour. 50% are passenger cars, 25% are vans, and 25% are trucks.

   a) Design a simulation item to simulate the situation, and conduct the simulation.
   b) Predict what the next 10 vehicles will be.

9. Describe a situation that can be simulated using each item.

   a) five marbles in a bag, each of a different colour
   b) a spinner divided in half, and one of the halves is divided into two quarters
   c) a cube labelled A, A, B, B, C, C

Try This!

10. Palo has been practising his aim at darts. Out of 10 throws, he hits the bulls-eye 3 times.

    a) Estimate the probability of Palo hitting a bulls-eye.
    b) Estimate the probability of Palo not hitting a bulls-eye.
    c) Describe an item that can be used to simulate the situation. Hint: How could you use a spinner?
    d) Predict whether Palo's next five throws will include any bulls-eyes.
    e) Conduct the simulation and compare with your prediction.

Extend

11. The pop company decides to change the contest. To win a game console, customers must collect the letters to spell W-I-N-N-E-R.

    a) Describe two different ways the contest could be run. What is the probability of drawing an N, with each way?
    b) Explain how a number cube can be used to simulate the contest. Which way in part a) does the simulation fit? Explain.
    c) For each way in part a), create an imaginary tally chart that shows the result of a simulation. Explain your method.

4.5 Apply Probability in Sports and Games

Focus on...
- applying probability
- probability in sports and games

Games are a fun way to practise your knowledge and skills.

What can you tell about the game being played in this photograph? What do you need to know to win?

Discover the Math

Materials
- coloured tiles, counters, or markers
- paper bag

Strategies

Act it out

How can you use probability to help you win?

The game of *Match or No Match* is played in pairs. Here is how you play.
- Place two green tiles and one blue tile in a bag.
- Before starting, decide who will be *Match* and who will be *No Match*.
- Player 1 draws a tile from the bag. Then, Player 2 draws a tile from the bag.
- The *Match* player gets a point if the tiles match. The *No Match* player gets a point if the tiles do not match.
- Replace the tiles. Repeat.

1. Create a tally chart to act as a score sheet.

2. Before playing the game, write two explanations:
   a) why you think you might win
   b) why you think you might lose

3. Play for about 5 min. The winner is the person with the most points. Who wins?

4. **Reflect** How could you improve your chances of winning?

Example 1: Use Tree Diagrams to List Game Outcomes

Kaia and Elana play *Match or No Match*. Kaia is *Match* and she thinks that she is less likely to win because of the possible outcomes. She says the game is not fair. Is Kaia correct?

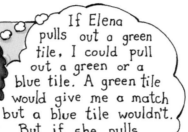

Literacy Connections

"Fair" games provide an equal chance for each player to win. If there are two players, there is an equal probability that each player will win. We say that each has a 50% chance of winning. This probability can be shown as the fraction $\frac{1}{2}$.

In a fair game, three players should each have a one in three chance of winning. This can be shown as $\frac{1}{3}$.

Solution

Understand Find the probability of *Match*. That will tell whether Kaia is correct or incorrect.

Plan Use a tree diagram to represent the game and determine the probability of *Match*.

Do It!

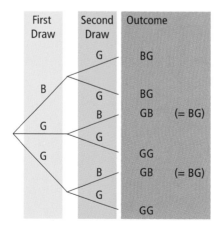

There are two favourable outcomes, GG and GG. So, the probability of *Match* is $\frac{2}{6}$ or $\frac{1}{3}$. Kaia is right. The game is not fair.

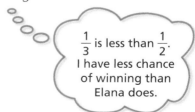

Look Back
- There are more combinations for *No Match* than for *Match*.
- It does not matter that there are two matching tiles and only one unmatched tile. *Match* is less likely.

4.5 Apply Probability in Sports and Games • MHR   135

Example 2: Probability in Baseball

Baseball star Jumpin' Jack Flazio's batting average is .300.
a) Estimate the probability of Flazio getting a hit.
b) Estimate the probability of Flazio not getting a hit.
c) How many hits, on average, will Flazio get in 40 at-bats?

Solution

a) A batting average of 0.300 means that Flazio has had 300 hits in 1000 at-bats. So, the estimated probability of scoring a hit is $\frac{300}{1000}$, or $\frac{3}{10}$.

b) Out of every 1000 at-bats, Flazio is expected *not* to get a hit 700 times. The estimated probability of not getting a hit is $\frac{700}{1000}$ or $\frac{7}{10}$.

c) *Method 1: Use Fraction Strips*

$\frac{3}{10} = \frac{12}{40}$

Flazio will get 12 hits on average.

Method 2: Use Equivalent Fractions

$$\frac{3}{10} = \frac{3 \times 4}{10 \times 4}$$
$$= \frac{12}{40}$$

Flazio will average 12 hits in 40 at-bats.

Literacy Connections

Reading Sports Data

Batting averages are really fractions out of a thousand. For example, ".342" means $\frac{342}{1000}$, or 0.342. Out of 1000 times at bat, the batter has scored a hit 342 times.

Key Ideas

- Knowledge of probability can help in sports and games. Strategies that give more favourable outcomes improve your chances of winning.
- Batting averages in baseball and goals scored in hockey or soccer are an indication of an athlete's performance.

Batting average .350

Another way of showing this is $\frac{350}{1000}$. Out of every 1000 times at bat, this player had 350 hits. He did not get a hit 750 times.

Communicate the Ideas

1. Why is it sometimes helpful to predict probabilities in a game?

2. **a)** Elana thinks the game of *Match or No Match* is fair. Do you agree? Explain.
   b) Suggest ways of changing the game to make it fair for two players.

Check Your Understanding

Practise

3. Use a tree diagram to show the chances of randomly picking matching marbles.

4. Use a tree diagram to show the probability of randomly grabbing two different types of fruit.

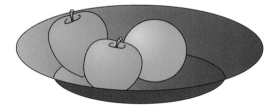

5. Use a tree diagram to show the probability of randomly getting two different coins.

6. **a)** Christian scores 2 goals for every 10 shots on net. What is his chance of *missing* a goal?
   b) Ramona wins 6 races of every 15 she enters. What is her chance of *losing* a race?

Apply

For help with questions 7 and 8, refer to Example 1.

7. In *Heads and Tails*, two players each toss a coin. If either or both coins come up heads, Player 1 gets 1 point. If the result is two tails, Player 2 gets 2 points.
   a) Draw a tree diagram for the score in one round of the game.
   b) Is *Heads and Tails* a fair game? Explain.

8. In *Match or No Match*, imagine adding another blue tile to the bag. You now have two tiles of each colour. How does this affect the probability of matching?

For help with questions 9 and 10, refer to Example 2.

9. Imant's batting average at baseball is .250 or 0.250.
   a) Estimate the probability of Imant getting a hit.
   b) Estimate the probability of Imant *not* getting a hit, as a fraction and as a decimal.
   c) How many hits, on average, will Imant get in 24 at-bats?

10. Raquel averages 17 hits in every 50 at-bats.
   a) Estimate Raquel's probability of getting a hit.
   b) Write Raquel's batting average with three decimal places. How many hits would Raquel average out of 1000?

11. Lydia entered an egg-and-spoon contest. In 10 runs, she dropped the egg 4 times.
   a) Estimate the probability of Lydia dropping the egg.
   b) What is the estimated probability that Lydia will *not* drop the egg, the next time she races?
   c) Add your answers for parts a) and b) together. Explain your result.

12. A hockey goalie has made 9 saves in 10 shots.
   a) The coach has to guess at the probability of the goalie not letting a shot in. What is the best estimate?
   b) Out of 100 shots on net, how many goals is the goalie likely to let in?
   c) If there were 20 shots on net, how many goals would be expected?

13. Design your own coloured-tiles game. Determine the winning probabilities for each player. Is your game fair? Does it matter who draws first?

14. At Mount St. Louis Ski Resort, 3 ski lifts are open. Five trails are open heading down the mountain.
   a) How many different ways can a skier go up and down the mountain? Use an organizer to justify your answer.
   b) What is the probability that a skier, choosing at random, will choose any trail in a)? Justify your answer.

15. In MONOPOLY®, a player must remain in jail unless she or he rolls a double in three rolls. Players use two number cubes.
   a) What is the predicted probability of rolling doubles in one roll?
   b) What is the predicted probability of not rolling doubles in one roll?
   c) In terms of the probability, explain why a "get out of jail" card would be useful to have.

Chapter Problem

16. A car rally game uses two spinners. Cars move the sum of each spin. They move along a track 10 spaces long.

Is this game fair? Explain.

17. a) In a board game, you can move any number of spaces up to and including the value you roll on a number cube. What is the probability that you will win on your next roll, if you are
   • four spaces away from the finish line?
   • one space away from the finish line?
   • eight spaces away from the finish line?
   b) Is this case, explain what a probability of 0 means. What does a probability of 1 mean?

Extend

18. From 35 yards, Didier successfully kicks one field goal in two. What is the probability that Didier scores

   a) five field goals in a row?
   b) four out of five field goals?

19. James and Karina are playing "best of five" with a coin. James calls tails. After two tosses, Karina has scored two heads.

   a) What is the probability that the third toss will be tails?
   b) Is this probability affected by the results of the first two tosses?
   c) What is the probability that Karina will win the best of five, given her two heads so far?

20. Tien Gow, meaning "Heaven Nine," is a Cantonese domino game for four players. It is played with a set of 32 Chinese dominoes. Each player starts with eight randomly chosen dominoes.
   - At each move, you must place a new domino touching one that has already been played.
   - For each pair of touching dominoes, the halves that touch must have equal numbers of spots.

   Which are the best dominoes to hold, and why?

Making Connections

Sharing Birthdays

When is your birthday? Does someone in your class share your birthday? Do two classmates have the same birthday?

Amazingly, the probability of two people in a class of 24 sharing a birthday is more than 50%. This is true even if you don't count twins.

1. Find out whether your class has two people (not twins) with a shared birthday. You can count your teacher.

2. Survey some other classes to find out if class members also have shared birthdays. What did you find?

CHAPTER 4 Review

Key Words

For questions 1 to 3, copy the statement and fill in the blanks. Use some of these words:

tally chart
probability
favourable outcomes
tree diagram
frequency table
outcomes
random
simulation

1. The ▉▉▉▉▉▉▉▉ of getting heads in a coin toss can be estimated. You would use a ▉▉▉▉ ▉▉▉▉▉ and ▉▉▉▉▉▉▉▉ ▉▉▉▉▉ to record your trials.

2. To determine a probability, count the total ▉▉▉▉▉▉▉▉ and the ▉▉▉▉▉▉▉▉ ▉▉▉▉▉▉▉▉.

3. A ▉▉▉▉ ▉▉▉▉▉▉▉ can be used to organize outcomes.

4. Rearrange the circled letters in questions 1 to 3 to make a key word. Define this word.

4.1 Introducing Probability, pages 116–120

Use this visual for questions 5 to 7.

5. Josephine randomly draws a marble from the bag.
   a) Is she as likely to draw a blue marble as a white marble?
   b) What is the probability of drawing a yellow marble?

6. What is the probability of drawing
   a) an orange or a white marble?
   b) any marble other than blue?

7. a) Which colour marble is Josephine most likely to pick? least likely to pick? Why?
   b) Explain why the probabilities change, depending on the colour.

8. Work with a classmate. Write the numbers 1 to 10 on separate cards, shuffle them, and place them face down in a row. Take turns turning over a card. If the number on the card is even, Player 1 wins. If it is odd, Player 2 wins. Turn the card back over and shuffle again.
   a) Play the game for 5 min. Record who wins in a tally chart and frequency table.
   b) Estimate the probability of each player winning.
   c) Explain whether one player was more or less likely to win. Justify your reasoning.

4.2 Organize Outcomes, pages 121–125

9. Connie is having trouble deciding which sweatshirt to buy. She has a choice of
   - 2 styles: buttoned or pullover
   - 3 colours: blue, black, or red
   a) Draw a tree diagram showing all her possible choices.
   b) How many different choices are there?
   c) Decide and explain whether the probability of selecting each sweatshirt combination is the same.

10. For each situation, choose a method for organizing the outcomes. Justify your choice.

  a) spinning a two-section spinner marked 1 and 2, rolling a number cube, and adding the values

  b) tossing a coin three times

4.3 Use Outcomes to Predict Probability, pages 126–130

11. For each question,
  - state the probability
  - explain your reasoning

  a) rolling a number less than 4 on a number cube

  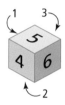

  b) choosing the letter Z in a random draw from letters in the phrase PIZZA BIZARRO

12. Two spinners are spun at the same time.

  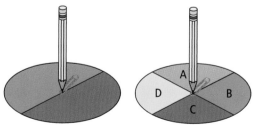

  a) Draw a tree diagram for the possible outcomes.
  b) List all the possible outcomes.
  c) Which combinations are most likely? Justify your reasoning.

13. A 10-sided die with letters A through J is rolled 60 times. How many times would you expect the outcome E? Why?

4.4 Extension: Simulations, pages 131–133

14. In each box of Yona's favourite cereal, there is a toy figure promoting the movie *Chasm of Doom*. Yona wants to collect all 5 figures.

  a) Predict how many boxes Yona will have to buy. Justify your answer.
  b) Explain what item you will use to simulate buying the boxes.
  c) Perform the simulation. Record your results in a tally chart.
  d) Justify why you repeated the simulation the number of times you did.
  e) Compare your prediction and your simulation results.

4.5 Apply Probability in Sports and Games, pages 134–139

15. Janet made up a game from a 10 by 10 game sheet with 100 empty squares. She placed the letters A, E, I, O, and U randomly on the board. The letters included 11 As, 7 Es, 12 Is, 4 Os, and 8 Us. She left 9 squares blank. She coloured the remaining squares.

  a) If Janet closes her eyes and touches a square, what is the probability that she will touch
   - a blank square?
   - a vowel?
   - neither a blank nor a vowel?

  b) Which type of square is she most likely to choose? Justify your response.

Chapter 4 Practice Test

Strand Questions	NSN 5, 6, 11	MEA	GSS	PA	DMP 1–13

Multiple Choice

For questions 1 to 6, select the best answer.

1. What is the probability of guessing a wrong answer on a multiple choice test, when there are 4 options?

    A $\frac{3}{4}$ **B** $\frac{1}{4}$ **C** 0 **D** 1

2. A cube numbered 1, 1, 2, 3, 4, 5 is rolled. The probability of rolling less than 3 is

    A $\frac{1}{3}$ **B** $\frac{2}{3}$ **C** $\frac{4}{6}$ **D** $\frac{1}{2}$

3. When you roll a number cube, which outcome is most likely?

    A rolling a 3
    B rolling a number greater than 2
    C rolling an even number
    D rolling a sum of 7

4. This spinner is spun 50 times. Approximately how many times would you expect the outcome E?

    A 4 **B** 7
    C 10 **D** 13

    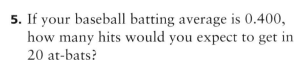

5. If your baseball batting average is 0.400, how many hits would you expect to get in 20 at-bats?

    A 4 **B** 12
    C 8 **D** 6

Short Answer

6. Each letter of the word PROBABILITY is written on a separate card and placed face down. You choose a card.
    a) What is the probability of choosing the letter I?
    b) What is the probability of choosing a consonant?
    c) What is the probability of choosing one of the first 5 letters of the alphabet?

7. Su Yeon spins the following spinner and flips the coin.

    a) Draw a tree diagram showing the possible outcomes.
    b) List all the possible outcomes.
    c) What is the probability that Su Yeon spins yellow and the coin lands heads up?
    d) What is the probability that she spins red or blue and the coin lands tails up?

8. a) A coin is tossed. State the probability of heads, as a fraction and as a decimal.
    b) Compare this value for the probability to an estimate based on repeatedly tossing a coin. How are the values related? How are they different?

9. Rebecca picks a card at random and rolls the number cube.

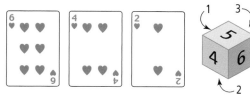

a) Draw a tree diagram.
b) List all the possible outcomes.

Extended Answer

10. Jason orders a submarine sandwich for lunch every Friday. He has a choice of
- bread: white or whole wheat
- filler: turkey, salami, or cheese

a) How many different ways can Jason order his submarine? Use a tree diagram to organize your answer.
b) Is the probability of each submarine choice the same? Justify your reasoning.

11. Vladimir walked into class and was surprised to find a quiz at his desk. Having missed the last week of classes, Vladimir had no knowledge of the quiz content. There were four true-or-false questions that he had to guess at.

a) Based on Vladimir's lack of content knowledge, predict how many true-or-false answers he will get correct. Justify your answer.
b) Using a spinner, simulate Vladimir's performance on the true-or-false questions. Record your results in a tally chart.
c) Estimate the probability of each score, based on your results.
d) Why might your simulation give different results than your prediction in part a)?

Chapter Problem Wrap-Up

Dana started making a car rally game. Can you help her finish it?

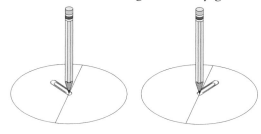
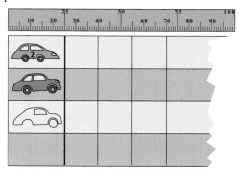

- The game should be fair for at least two people.
- Include a set of rules.

1. Show why your game is fair.
2. Test the game. Modify it if you need to.

Making Connections

What does math have to do with nutrition?

Dietitians help people eat better. They study what a person eats every day and suggest diet changes. For example, they may suggest ways for the person to lower cholesterol. One way is to reduce the daily fat intake to a fraction of the amount eaten.

1. The average person consumes 2000 calories a day. Marta consumes 3000 calories a day. Marta's nutritionist suggests that she reduce her daily caloric intake to the average. By what fraction does Marla need to reduce her caloric intake?

2. Rick consumes 1800 calories a day. One-third of these are in the form of fat. How many calories does Rick eat as fat?

Making Connections

What are the chances?

In the game *Up and Down*, one player moves counters up the board and the other moves down. The object is to move all your counters off the opposite end of the board.

- Begin with all 15 counters stacked on your first space.
- Take turns to roll two number cubes.
- If you roll 3 and 5, for example, you can move one counter 3 spaces and another 5 spaces, or you can move a single counter 8 spaces.
- If you roll doubles, for example, 3 and 3, you get to make *four* moves of 3.
- Spaces with two or more counters of the *same* colour are "protected."
- If you have a single counter on any space, and your opponent has a roll that will reach that space, he or she can "hit" your counter and take it off the board.
- You must bring a counter that has been hit back on before making any other move.
- You can move a counter off if you roll high enough to move it beyond your last space.
- The winner is the first player to move all 15 counters off.

Materials
- 30 counters, half one colour, half another
- 2 sheets of plain paper
- ruler and pencil

Optional
- BLM 3/4 Task A Up and Down

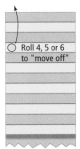

Develop a Fair Game

1. Copy and cut out the spinner and the shapes shown. Use them to make a game that is fair for two players. How will you use the shapes in the spinner? Will you use the outside shapes? If so, how?

2. Use the materials to make a game that is fair for two players.

3. Develop rules and instructions for your game.

4. Use your knowledge of probability to justify your game plan. Explain how each player is equally likely to win.

5. Can you develop more than one fair game? Show your results.

Materials
- plain paper
- ruler
- pencil crayons
- scissors

Optional:
- BLM 3/4 Task B Develop a Fair Game

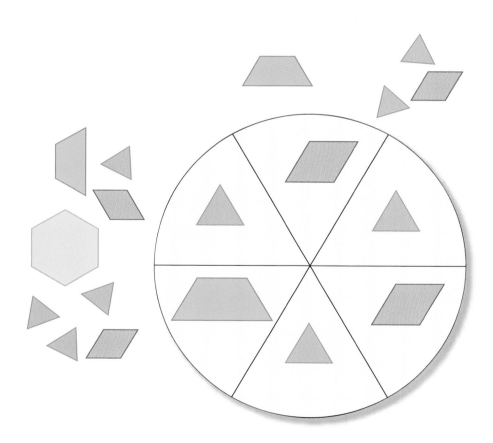

Task: Develop a Fair Game • MHR 145

Chapters 1–4 Review

Chapter 1 Measurement and Number Sense

1. Find the perimeter of each shape.

   a)

   b)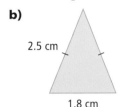

2. Find the area of each shape.

   a)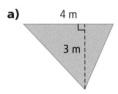

   b)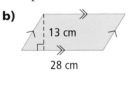

3. Evaluate each expression.

   a) $3 + 4 \times 5$
   b) $2 \times 6 - 8 \div 2$
   c) $5 + 6 \div 3 \times 2$
   d) $(7 + 4 - 2) \times (5 - 2)$
   e) $3 + 12 \times (16 \div 4 \div 4)$

4. A park is in the shape of a trapezoid, as shown.

   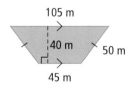

   a) Find the length of fence needed to surround the park.
   b) Find the area of the park.

5. A trapezoid has a perimeter of 18 cm and an area of 18 cm². Draw the trapezoid on grid paper.

Chapter 2 Two-Dimensional Geometry

6. A 5-m ladder leans with its lower end 1 m from the wall. What type of triangle is formed?

   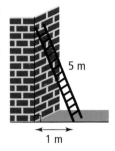

7. Sketch and name the geometric shape(s) with the following properties.

   a) three sides; each with a different length
   b) three sides; two with the same length
   c) four sides of equal length
   d) four sides, two opposite pairs have the same length

8. Ms. Jung picked up six erasers left behind by her students. She noticed that they were different shapes and sizes. She sketched the shapes she saw.

   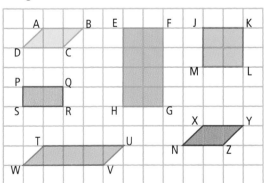

   a) What geometric shapes can be seen?
   b) Which of the shapes are congruent? Explain.
   c) Which of the shapes are similar? Explain.

146 MHR • Chapter 4

Chapter 3 Fraction Operations

9. Write an addition sentence to represent the fraction of each yellow hexagon that is covered.

   a)   b)

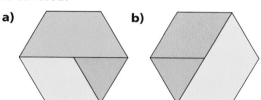

10. Use pattern blocks or diagrams to model each subtraction.

    a) $\dfrac{1}{2} - \dfrac{1}{3}$ b) $\dfrac{2}{3} - \dfrac{1}{6}$

11. Find a common denominator for each pair of fractions.

    a) $\dfrac{3}{5}$ and $\dfrac{1}{2}$ b) $\dfrac{1}{3}$ and $\dfrac{3}{4}$

12. Evaluate.

    a) $\dfrac{1}{2} + \dfrac{2}{5}$ b) $\dfrac{3}{4} - \dfrac{2}{3}$

13. Paula spends her Mondays as follows.
    8 h sleeping
    6 h at school
    3 h at gymnastics
    2 h doing homework
    The rest of the time she is eating or relaxing.
    a) What fraction of the day does Paula spend sleeping?
    b) What fraction of the day is she at school?
    c) How many hours does she spend eating or relaxing? What fraction of the day is this?

Chapter 4 Probability and Number Sense

14. A number cube and a coin are tossed simultaneously.
    a) Draw a tree diagram showing all the possible outcomes.
    b) What is the probability of a 1 and tails?
    c) What is the probability of a 2 or a 3 and heads?
    d) What is the probability of an odd number and tails?
    e) What is the probability of a number less than 4 and heads?

15. Two spinners are spun. Find the probability of spinning each of the following.

    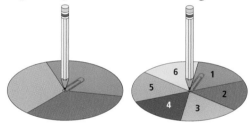

    a) red and the number 4
    b) red or blue and the number 4
    c) green and an even number
    d) blue or green and a number less than 4

16. A domino set has 28 tiles. Seven of the tiles are doubles. Doubles have the same number of dots on each half. All the dominoes are placed face down on the table. Find the probability of each outcome.
    a) a domino that is a double
    b) a domino that is not a double

Number Sense and Numeration
- Compare and order fractions, decimals, and percents.
- Solve problems by converting between fractions, decimals, and percents.
- Use estimation to justify or assess the reasonableness of calculations.
- Explain the process of problem solving.

Measurement
- Estimate and calculate areas of trapezoids and irregular two-dimensional shapes.

Data Management and Probability
- Display data on bar graphs, pictographs, and circle graphs.
- Use and apply a knowledge of probability in sports and games.

Key Words
statistic

repeating decimal

percent

CHAPTER 5

Fractions, Decimals, and Percents

In 2002, Canada scored a remarkable double. Both the women's and men's hockey teams won gold at the Olympics.

Action and skill play a big part in sports. But the difference between a very good athlete and a great one often comes down to numbers.

By the end of this chapter, you will have a better understanding of the numbers used to report on athletic performances. You will be able to discuss and compare sports statistics using fractions, percents, and decimals.

Chapter Problem

A hockey goalie's performance can be measured in two ways:
- the goals-against average
- the save percentage

What does each statistic (or stat) tell about the goalie? How might such information help you compare two goalies?

Get Ready

Write Fractions as Decimals

You can use base 10 blocks to represent fractions and decimals. For example, let a hundreds flat represent 1. The fractions $\frac{1}{10}$ and $\frac{17}{100}$ can be represented as in the diagram.

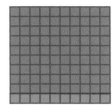

The fraction $\frac{1}{10}$ is the same as the decimal 0.1.

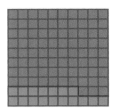

$\frac{17}{100}$ is the same as 0.17.

1. Use base 10 blocks or grid paper to represent each fraction. Then, write each fraction as a decimal.

a) $\frac{3}{10}$ b) $\frac{7}{10}$ c) $\frac{1}{100}$ d) $\frac{23}{100}$

2. Write each fraction as a decimal.

a) $\frac{6}{10}$ b) $\frac{32}{100}$ c) $\frac{97}{100}$ d) $\frac{5}{10}$

Compare and Order Decimals

To order decimals, look at place values. For example, to order 0.74, 0.7, and 0.732, you can use a place value chart.

Ones	Tenths	Hundredths	Thousandths
0	7	4	
0	7	0	
0	7	3	2

The order of the decimals, from least to greatest, is 0.7, 0.732, 0.74.

This can also be written as 0.7 < 0.732 < 0.74.

I added a zero placeholder to 0.7 to compare the hundredths place.

All three numbers have the same tenths digit. So, look at the hundredths digit. 0.74 is the greatest. 0.70 = 0.7 is the least.

3. Order each set of decimals from least to greatest.

a) 0.25, 0.225, 0.2
b) 1.34, 1.334, 1.43

4. Order each set of decimals from greatest to least. Use the symbol ">" to stand for "is greater than."

a) 0.082, 0.0802, 0.08
b) 5.45, 5.545, 5.454

Representing Percents

You can use a hundred grid to show a percent. For example, 25% means 25 out of 100. To show 25%, colour 25 squares on a hundred chart.

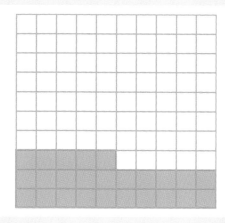

5. Show each percent on a hundred grid.
   a) 50%   b) 75%

6. Show each percent on a hundred grid.
   a) 100%   b) 60%

Decimals and Percents

To write a decimal as a percent, multiply the decimal by 100%. Remember to add the % sign. For example:

0.71 = 0.71 × 100%
     = 71%

The decimal point moves two places to the right.

0.2 = 20%

Add a zero because you move two decimal places.

To write a percent as a decimal, express the percent as a fraction with denominator 100 and drop the % sign. For example:

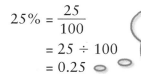

The decimal point moves two places to the left.

Add a zero placeholder because there are two decimal places.

7. Write each decimal as a percent.
   a) 0.43   b) 0.6
   c) 0.05   d) 0.02

8. Write each percent as a decimal.
   a) 75%   b) 90%
   c) 3%   d) 45%

5.1 Fractions and Decimals

Focus on...
- comparing and ordering fractions and decimals
- writing fractions as decimals
- writing decimals as fractions

Suppose your hockey team won 38 out of 50 games last year. How could you show this statistic as a fraction? a decimal?

What other sports statistics use fractions? Which use decimals? How are they used?

Discover the Math

Materials
- grid paper
or
- counters

How can you relate fractions and decimals?

1. Look at the first three columns in this table. Who do you think is having the best season? Explain why you think so.

Goalie	Total Goals Against	Games Played	Goals-Against Average
Akina	36	12	
Julie	60	15	
Katrien	35	14	

statistic
- a value calculated from a set of data

2. Goalies can be rated by their goals-against average. This **statistic** compares the number of goals scored against a goalie to the number of games played.

$$\text{Goals-against average} = \frac{\text{total goals against}}{\text{total games played}}$$

   a) Copy and complete the table. Write the goals-against average for each goalie as a fraction and as a decimal.
   b) Do you want to change your answer from step 1? Explain.

Strategies
Choose a formula

3. **a)** How does the goals-against average help you predict a goalie's performance?
   b) Is it better for a goalie to have a lower goals-against average or a higher one? Explain why.

4. **Reflect** The goals-against average is usually stated as a decimal.
   a) How is the decimal form different from the fraction form?
   b) How are they related?
   c) Which form is more useful? Justify your choice.

Example 1: Compare and Order Fractions

In a pie-eating contest, Evan ate $2\frac{3}{4}$ pies, Tia ate $2\frac{7}{10}$ pies, and Gustav ate $2\frac{4}{5}$ pies. The person who ate the most pies won the contest. Order the contestants from first place to third place.

Solution
Method 1: Model the Amounts

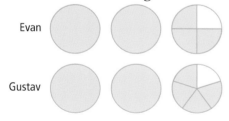

I made fraction strips to compare the amounts.

The fraction strips show that Gustav ate the most pie. Tia ate the least pie. The contestants, from first place to third place, are Gustav, Evan, and Tia.

Method 2: Draw Diagrams

Evan    Tia

Gustav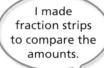

Strategies
Make a picture or diagram

The diagrams show that Gustav ate the most pie. Tia ate the least pie. The contestants, from first place to third place, are Gustav, Evan, and Tia.

Method 3: Convert to Decimals

Evan:
$2\frac{3}{4} = 2 + 3 \div 4$
$= 2 + 0.75$
$= 2.75$

Tia:
$2\frac{7}{10} = 2 + 7 \div 10$
$= 2 + 0.7$
$= 2.7$

Gustav:
$2\frac{4}{5} = 2 + 4 \div 5$
$= 2 + 0.8$
$= 2.8$

$2.8 > 2.75 > 2.7$
The contestants, from first place to third place, are Gustav, Evan, and Tia.

Strategies
What other strategy could you use to compare and order these fractions?

Example 2: Repeating Decimals

Use a calculator to write each fraction as a **repeating decimal**. Then, order the fractions from least to greatest.

a) $\frac{1}{3}$ b) $\frac{2}{3}$ c) $\frac{1}{6}$

repeating decimal
- a decimal with a digit or group of digits that repeats forever
- write the repeating digits with a bar: $0.333… = 0.\overline{3}$

Solution

a) $\frac{1}{3} = 1 \div 3$ ⓒ 1 ÷ 3 = 0.333333333

$= 0.333…$
$= 0.\overline{3}$ **Place a bar above the repeating digit.**

b) $\frac{2}{3} = 2 \div 3$ ⓒ 2 ÷ 3 = 0.666666667

$= 0.666…$
$= 0.\overline{6}$ **The calculator displays the final 7 because it rounds up. It would show more 6's if it had a larger display.**

c) $\frac{1}{6} = 1 \div 6$ ⓒ 1 ÷ 6 = 0.166666667

$= 0.1666…$
$= 0.1\overline{6}$ **The digit 1 does not repeat. Place the bar above the 6.**

$0.1\overline{6} < 0.\overline{3} < 0.\overline{6}$

So, $\frac{1}{6} < \frac{1}{3} < \frac{2}{3}$.

Literacy Connections

Reading Repeating Decimals

$0.\overline{3}$ means that the 3 repeats. Another way of showing this is $0.333…$.

$0.1\overline{6}$ means that the 6 repeats. Another way of showing this is $0.1666…$.

Example 3: Convert Decimals to Fractions

a) What fraction of a dollar is $0.75?
b) What fraction of a dollar is $0.44?

Solution

a) $0.75 = \frac{75}{100}$

$0.75 is $\frac{75}{100}$ of a dollar.

Another way of saying this is $\frac{3}{4}$ of a dollar.

b) $0.44 = \frac{44}{100}$

$0.44 is $\frac{44}{100}$ of a dollar.

Another way of saying this is $\frac{11}{25}$ of a dollar.

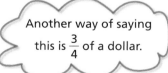

Making Connections

If you need help with equivalent fractions, go to Chapter 3, page 85.

Key Ideas

- To write a fraction as a decimal, divide the numerator by the denominator. For example,
$$\frac{1}{2} = 1 \div 2$$
$$= 0.5$$

- To express a decimal as a fraction, think of place value.
For example, $0.7 = \frac{7}{10}$ and $0.37 = \frac{37}{100}$.

- Compare and order fractions using diagrams, manipulatives, number lines, or common denominators, or by writing them as decimals.

- Compare and order decimals using place value.

- Some fractions produce repeating decimals. For example, $\frac{5}{6} = 0.8333....$ Another way of writing this is $0.8\overline{3}$.

$\frac{1}{4} = 0.25$ $\frac{1}{2} = 0.5$ $\frac{3}{4} = 0.75$

Communicate the Ideas

1. Write three fractions. Describe two methods you can use to order these fractions.

2. Write each repeating decimal correctly. Explain what error has been made.
   a)  0.444... = 0.4
   b) 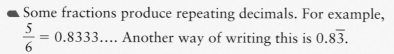 2.1666... = 2.$\overline{16}$

3. a) How could you compare the numbers in a mixed list of fractions and decimals?
   b) What other strategy might be useful? Explain.
   c) Use one strategy to compare $\frac{4}{5}$, 0.6, 0.9, and $\frac{3}{4}$. Order the numbers from greatest to least.

4. a) Express 0.7 and 0.51 as fractions.
   b) How many numerals are after the decimal of each number? How many zeros are in the denominator of each fraction? Explain how the answers to these questions are related.

Check Your Understanding

Practise

5. Show each fraction or mixed number using a diagram.
   a) $\frac{1}{2}$
   b) $\frac{3}{4}$
   c) $1\frac{2}{5}$
   d) $3\frac{3}{8}$

6. Write each fraction or mixed number in question 5 as a decimal.

7. Order these decimals from greatest to least.
   a) 1.3, 1.301, 0.34, 0.3
   b) 0.29, 0.489, 0.06, 0.2

For help with questions 8 and 9, refer to Example 1.

8. In a pie-eating contest, Robert ate $2\frac{3}{4}$ pies, Dora ate $2\frac{3}{10}$ pies, and Gianetta ate $2\frac{2}{5}$ pies.
   a) Who won?
   b) Order the contestants from first place to third place.

9. Order these fractions from least to greatest.
   a) $\frac{4}{9}, \frac{1}{2}, \frac{1}{3}$
   b) $1\frac{2}{3}, 1\frac{5}{6}, 1\frac{3}{4}, 1\frac{3}{8}$

For help with questions 10 and 11, refer to Example 2.

10. Write these decimals using repeating notation.
    a) 0.3333333333
    b) 0.555555556
    c) 2.166666667
    d) 0.183333333

11. Use a calculator to write each fraction or mixed number as a repeating decimal.
    a) $\frac{1}{6}$
    b) $1\frac{1}{3}$
    c) $3\frac{2}{3}$
    d) $2\frac{5}{6}$

For help with questions 12 and 13, refer to Example 3.

12. What fraction of a dollar is $0.50?

13. What fraction of a kilometre is 250 m?

14. Express each decimal as a fraction.
    a) 0.2
    b) 0.3
    c) 0.6
    d) 0.4
    e) 0.25
    f) 0.85
    g) 0.64
    h) 0.78

Apply

15. a) Order these numbers from least to greatest.
    $\frac{17}{20}$ 1.04 $\frac{21}{27}$
    0.87 $\frac{23}{25}$ $\frac{6}{5}$

    b) Order these numbers from greatest to least. Use two different methods.
    12.84 $12\frac{5}{6}$ $\frac{319}{25}$

Chapter Problem

16. Several teams used different methods to show goals-against averages. You are the coach of an all-star team. You want to choose the best goalie to start your first game. Which one would you choose?

    Andrea: $\frac{44}{20}$ Mustafa: $\frac{32}{8}$
    Delilah: $\frac{27}{8}$ Thomas: 2.7

    a) What strategies might you use to answer this question?
    b) Choose one strategy and solve the problem.

17. Carlo's last three quiz scores in math class were $\frac{8}{8}, \frac{9}{12}$, and $\frac{14}{16}$.
   a) Arrange these quiz scores from greatest to least.
   b) Explain your method.

18. Trina's three cats all love Meow Munch cat food. One day, the supermarket ran out of Meow Munch, and Trina had to buy Kitty Kibble instead. That evening,
   - Fuzzball ate $\frac{2}{3}$ of his dinner
   - Princess ate $\frac{3}{4}$ of her meal
   - Pounce ate $\frac{3}{5}$ of his supper

   a) Show visually the fraction of food that each cat ate.
   b) Who liked Kitty Kibble the most? Who disliked Kitty Kibble the most? Explain.
   c) Can you be sure of your answers in part b)? Explain.

19. Some of these gymnastics scores out of 10 were given as mixed numbers, and some as decimals.
   $6.7, 6\frac{2}{5}, 6\frac{1}{3}, 6\frac{3}{8}, 6.05, 6\frac{9}{20}$
   a) Order the scores from greatest to least.
   b) What strategy did you use to solve this question? What other strategies might you use? Describe one.

20. Create your own problem involving fractions and decimals. Trade problems with a partner and solve your partner's problem.

21. Express each decimal as a fraction. Simply the fraction.
   a) 0.75   **b)** 0.8
   c) 0.50   **d)** 0.65

22. A Student Council gives a "School Spirit Award" to the class with the best attendance at a dance. Four classes were surveyed.

Teacher	Number of Students	Number to Attend
Mr. Tamaki	30	21
Mrs. Galante	32	24
Ms. Serafini	26	19
Mr. Hetfield	28	20

   a) Without calculating, predict which class should win. Explain your thinking.
   b) Which class actually wins the School Spirit Award? Justify your decision.

Making Connections

In Chapter 10, you will learn more about conducting and analysing surveys.

Extend

23. Use a calculator to explore $\frac{4}{7}$ as a decimal.
   a) Do you think this fraction produces a repeating decimal? Why or why not?
   b) What additional information would help you decide?
   c) Try to find fractions with similar decimal forms to $\frac{4}{7}$. Write a brief report on what you discover.

24. The table shows United Nations estimates for Earth's population. What is the current growth rate of Earth's population? Use fractions and/or decimals to answer this question in at least two different ways.

Year	Earth's Population
2000	6.06 billion
2001	6.14 billion
2002	6.22 billion
2003	6.30 billion

5.2 Calculate Percents

Focus on...
- writing fractions and decimals as percents
- representing percents

Each year in Ontario, a few grade 7 and grade 8 students are chosen to assist Members of Provincial Parliament (MPPs). These students are called pages. They run errands and deliver messages.

To qualify as a page, you need an overall average mark of at least 80%. How could you check whether you qualify?

Literacy Connections

What is a page?
Page has more than one meaning. Right now, you are reading the page of a book. "Page" can also mean a young person who runs errands and carries messages in a hotel, theatre, or parliament.

Discover the Math

How can you calculate percents from fractions or decimals?

1. Maia wants to become a page. On her last four math quizzes, she got $\frac{19}{25}$, $\frac{13}{20}$, $\frac{18}{20}$, and $\frac{8}{10}$. She needs an average of 80%.

   a) Look at Maia's quiz scores. Find a common denominator and write Maia's scores as equivalent fractions.

   b) Show each equivalent fraction on a hundred grid or using base 10 blocks.

   c) Explain why using equivalent fractions makes it easy to compare the scores.

2. a) Write each of Maia's quiz scores as a **percent**.

   b) Find the mean of Maia's percent scores.

   c) Does Maia need to improve her scores? Explain.

3. **Reflect** Describe how you wrote Maia's scores as percents. What other method could you have used? Describe it.

Materials

- grid paper
or
- base 10 blocks (hundreds, tens, and ones)

percent
- out of 100
- 50% means $\frac{50}{100}$ or 0.5

Making Connections

If you need help with calculating means, go to Chapter 4, page 114.

Example: Write Fractions and Decimals as Percents

On a Student Council, it takes 60% of the votes to make a major decision. The students on the Council have voted on whether to hold a dance. The results are shown in the table.

Grade Level	Votes in Favour	Number on Council
Grade 7	4	6
Grade 8	7	9

a) Did the Student Council decide to hold the dance?
b) What is the minimum number of votes needed to make a decision?

Solution

a) There were 4 + 7 = 11 votes in favour. The total number of votes was 6 + 9 = 15.

$\frac{11}{15} = 0.7333...$
$= 0.7\overline{3}$
$= 0.7\overline{3} \times 100\%$
$\doteq 73\%$

First, I divided 11 by 15 to write this fraction as a decimal.

Then, I multiplied by 100 to find the percent.

73% is greater than 60%. The council decided to hold the dance.

b) Try 10 votes out of 15:
$\frac{10}{15} = 0.\overline{6}$
$= 0.\overline{6} \times 100\%$
$\doteq 67\%$

Half of 15 is 7.5. 60% is more than half. I'll start by trying 10.

So, 10 is more than 60%.

Try 9 votes out of 15:
$\frac{9}{15} = 0.6$
$= 0.6 \times 100\%$
$= 60\%$

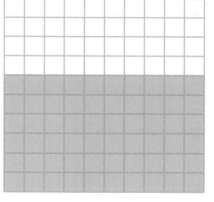

The minimum number of votes to make a decision is 9 out of 15.

Key Ideas

- To write a fraction as a percent, write the fraction as a decimal. Then, multiply by 100%. For example:

  $\frac{5}{8} = 0.625$
  $= 0.625 \times 100\%$
  $= 62.5\%$

- You can represent percents visually.

$\frac{1}{4} = 0.25$ or 25%

5.2 Calculate Percents • MHR 159

Communicate the Ideas

1. Describe two methods to write $\frac{16}{25}$ as a percent.

2. Represent 75% visually.

3. What percent is green? Explain.

Check Your Understanding

Practise

For help with questions 4 to 9, refer to the Example.

4. Write each fraction as a decimal.
   a) $\frac{7}{10}$ b) $\frac{17}{20}$ c) $\frac{4}{5}$

5. Show each fraction in question 4 as a percent.

6. Write each fraction as a decimal, and then as a percent.
   a) $\frac{3}{20}$ b) $\frac{27}{50}$ c) $\frac{19}{25}$

7. Represent each percent in question 5 visually.

Apply

8. On a Student Council, it takes 55% of the votes to make a major decision. The results on a decision to set up a study support group are shown in the table.

Grade Level	Votes in Favour	Number on Council
Grade 7	3	8
Grade 8	7	12

   a) Did the Student Council decide to set up the study support group?
   b) What is the minimum number of votes needed to make a decision?

9. A Student Council is made up as follows:
   - Grade 6   4 students
   - Grade 7   6 students
   - Grade 8   6 students

   A majority of 60% is needed to make major decisions. What is the minimum number of votes needed?

10. Leah got 32 out of 45 on her geography test. What was her score as a percent? Round your answer to the nearest percent.

11. a) David got 45 out of 60 on a test. What percent did he get?
    b) What other method could you use to solve this question? Explain.

12. In a recent taste test, 177 out of 250 people said they preferred the taste of Fizzo to Splash.
    a) What percent of the people surveyed preferred Fizzo?
    b) What percent preferred Splash?
    c) The makers of Fizzo announce that "7 out of 10 people prefer Fizzo to Splash." Is this fair? Explain.

13. Amir got 65% on his science quiz. Amir said to a friend, "I got 65% out of 100." Did Amir tell his friend the correct score? Explain.

14. Rocco and Biff are koala bears. They are trying to eat eucalyptus leaves that are almost out of reach.

- Rocco successfully reaches his target leaf on 12 out of 15 tries.
- Biff succeeds on 16 of 20 tries.

a) Predict which is the more acrobatic bear. Explain why you think so.
b) Write each bear's score as a fraction.
c) Write each bear's score as a percent.
d) Can you determine which is the more acrobatic bear? Explain.

15. Suppose a friend borrowed your notes for this lesson. Later, the friend sent you an e-mail.

> Have I got this right? To change a decimal to a percent, I move the decimal point 2 places to the left and add the % sign?

a) Does your friend understand the procedure? Explain your thinking.
b) Write an e-mail reply to your friend. Make sure your friend understands why your explanation makes sense.

16. On her next four math quizzes, Maia got $\frac{16}{25}$, $\frac{14}{20}$, $\frac{18}{20}$, and $\frac{8}{10}$. She wants to be a page in the Provincial Legislature. To be accepted, she needs an average of 80%.

a) Write Maia's scores as percents. What is her mean score?
b) Maia's next quiz is out of 25. What minimum score should she try to get on the quiz?
c) Use pictures, numbers, and words to explain how you answered part b).

Extend

17. When no party wins more than 50% of the seats in an election, a minority government is formed.

a) There are 300 seats in Canada's federal Parliament. Give an example of how a minority government could be formed in a federal election. Show the number of seats for each of these parties: Liberal, Conservative, NDP, Bloc Québecois.
b) Who would likely be the governing party in your example? What percent of the seats do they hold?
c) What is the total percent held by non-governing parties?
d) Explain why a minority government has less power than a majority government.

> **Did You Know?**
>
> Would you like to learn more about how our government works? You can, by becoming a legislature page. Find out how to do this. Go to www.mcgrawhill.ca/links/math7 and follow the links.

18. Approximately 70% of Earth's surface is covered by water. The total surface area of Earth is about 510 000 km².

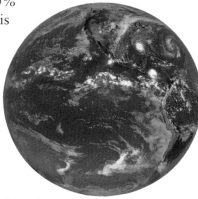

a) What total area of Earth is covered by water?
b) What is the total land area on Earth?
c) What effect do you think global warming could have on these values? Explain your answer. Hint: Research any information you need.

5.3 Fractions, Decimals, and Percents

Focus on...
- presenting numerical data
- converting between fractions, decimals, and percents
- estimating

Why are election results often reported as percents?

Discover the Math

Materials
- pencil crayons

How do percents make data easier to understand?

1. At Sherwood Junior High School, three students ran for president of the Student Council. Each student picked a different colour for signs and buttons. The results of the election were as follows. Who won the election? How can you tell?

Candidate	Colour	Percent of Votes
Justine	red	21%
Bram	green	27%
Lucia	blue	52%

2. Design a bar graph for the school newspaper to represent the election results.

3. Write a paragraph about the results of the election. Report the results as fractions. Choose approximate fractions with small denominators. Hint: Use your estimation skills.

4. **Reflect** What are the advantages of reporting the results as percents? as approximate fractions? What other reporting methods could be useful? Explain.

Example: Write Percents as Fractions and Decimals

Ari is studying tree species in the Canadian Shield. Ari's research in one woodlot is shown in the circle graph.
a) What fraction of the forest is made up of each tree species?
b) To plan for harvesting, Ari needs the decimal equivalent of each percent. Convert each percent to decimal form.
c) The forest has approximately 2000 trees. Approximately how many are pine?

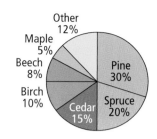

Solution

a)

Type of Tree	Percent	Fraction
Pine	30%	$\frac{30}{100}$
Spruce	20%	$\frac{20}{100}$
Cedar	15%	$\frac{15}{100}$
Birch	10%	$\frac{10}{100}$
Beech	8%	$\frac{8}{100}$
Maple	5%	$\frac{5}{100}$
Other	12%	$\frac{12}{100}$

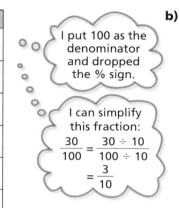

I put 100 as the denominator and dropped the % sign.

I can simplify this fraction:
$\frac{30}{100} = \frac{30 \div 10}{100 \div 10} = \frac{3}{10}$

b)

Type of Tree	Percent	Decimal
Pine	30%	$30 \div 100 = 0.3$
Spruce	20%	$20 \div 100 = 0.2$
Cedar	15%	$15 \div 100 = 0.15$
Birch	10%	$10 \div 100 = 0.1$
Beech	8%	$8 \div 100 = 0.08$
Maple	5%	$5 \div 100 = 0.05$
Other	12%	$12 \div 100 = 0.12$

c) Number of pine trees = 0.3 × total number of trees
= 0.3 × 2000
= 600

Approximately 600 of the trees are pine.

Key Ideas

- Fractions, decimals, and percents are useful forms of data for different types of presentations.
- To express a percent as a fraction, write the value over 100 and drop the % sign.
  For example, $31\% = \frac{31}{100}$.

Lucia got just over $\frac{1}{2}$ the vote.

Communicate the Ideas

1. What fraction is yellow? Explain how to write the fraction as a decimal and as a percent.

2. What approximate fraction is 24%? Explain.

Check Your Understanding

Practise

For help with questions 3 to 5, refer to the Example.

3. Write each percent as a fraction. Simplify the fraction.
   a) 25%   b) 50%   c) 10%
   d) 20%   e) 75%   f) 100%

4. Write each percent as a fraction. Can you simplify the fraction?
   a) 40%   b) 55%   c) 72%
   d) 4%   e) 5%

5. Write each percent as a decimal.
   a) 32%   b) 64%   c) 70%
   d) 83%   e) 5%

6. Write each percent as a fraction.
   a) 53%   b) 77%
   c) 9%   d) 26%

7. Choose approximate fractions with small denominators to show the fractions from question 6 in a different way.

8. Write each fraction as a decimal. Then, write the decimal as a percent.
   a) $\frac{13}{20}$ b) $\frac{14}{25}$ c) $\frac{3}{8}$
   d) $\frac{9}{16}$ e) $\frac{2}{15}$

9. Copy and complete the table.

Fraction	Percent	Decimal
		0.15
	65%	
$\frac{1}{50}$		

Apply

10. Dilip and Françoise tried a quiz with 20 multiple-choice questions. Their scores were
    Dilip 95%   Françoise 80%
    a) To skip the homework on this topic, students had to get at least 17 out of 20 correct answers. Who had to do the homework?
    b) What strategy did you use to solve this problem? What other strategy might you use?

11. A 25-question multiple-choice quiz has these passing grades:
    A at least 80%
    B at least 68%
    C at least 56%
    How many correct questions do you need for each passing grade?

12. What's wrong? Clara wrote 0.6 as a fraction this way.
    $$0.6 = 6\%$$
    $$= \frac{6}{100}$$
    $$= \frac{3}{50}$$
    Correct Clara's mistake. Explain your answer.

13. A music store had these sales for the month of January. Write the percent sales
    a) as fractions (for stock-taking)
    b) as decimals (for tax calculations)

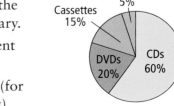

14. Wally bought 8 tomato plants, 3 cucumber plants, and 5 pepper plants for his vegetable garden. He wants to allow equal space for each plant.

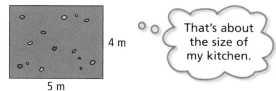

a) What fraction of Wally's garden should he use for each type of plant?
b) What percent of the garden does each type of plant take?
c) Sketch a diagram to show how Wally could organize his plants.
d) What area will each type of plant need?

15. A newspaper report claims that "63% of children in the developing world do not have a safe water supply."

a) Rewrite this statement, using an approximate fraction.
b) You want to check how accurate the percent figure is. What information would you need to check? What relationship should you look for in the data? Explain.

16. A sales breakdown for the Ontario Beverage Company shows 22% cans, 67% bottles, and 9% party packs. Report the sales breakdown using approximate fractions with small denominators.

17. The Barkerville Dog Pound adopts out dogs and cats. One year, they found homes for 800 animals. 53% of these were cats.

a) Write the percent of cats as a decimal.
b) How many cats were adopted out?
c) How many dogs were adopted out?

 18. Marie competes in chess tournaments. In her first year of competition, she won 9 games, lost 12, and tied 4.

a) Find the percents of wins, losses, and ties.
b) Marie plans to play 25 games a year in the next three years of competition. How many of these games do you expect Marie to win? Explain your answer.
c) Think about your answer to part b). What factors might change your answer? How could you investigate these factors?

Extend

19. Refer to question 18.

a) Create a single statistic that describes Marie's performance. Give your statistic a name. Calculate Marie's statistic for her first-year record.
b) Suppose Marie's record in her second year is 10 wins, 7 losses, and 8 ties. How will her statistic change?

20. Earth's mass is 0.0031 times the mass of Jupiter. This can be written as Earth = 0.0031 Jupiters. Similarly,

Mercury = 0.0002 Jupiters
Venus = 0.0026 Jupiters
Mars = 0.0003 Jupiters
Saturn = 0.2989 Jupiters
Uranus = 0.0457 Jupiters
Neptune = 0.0539 Jupiters
Pluto = 0.000 007 Jupiters

a) What is Earth's mass as a percent of Jupiter's mass?
b) Write the other planets' masses as percents of Jupiter's.
c) Determine each planet's mass, as a percent of the *total* planetary mass in our solar system. Illustrate your answer visually.

5.4 Apply Fractions, Decimals, and Percents

Focus on...
- fraction, decimal, and percent problems
- multi-step problems
- estimation

Your basketball team is in a tight playoff game. You are awarded a free throw. Scoring on this free throw could make the difference between winning and losing.

How can you use statistics to help you decide which player to send on the court?

Discover the Math

What strategies might you use to solve problems like these?

Example 1: Compare Sports Data

Pemba, Heather, and Roberto are the best free-throw players on your basketball team.

Player	Free-Throw Attempts	Good/No Good
Pemba	25	✓✓✓✗✗✓✓✗✗✗✓✓✗✓✗✗✗✗✓✓✗✗✓✓✗
Heather	25	✗✓✗✓✗✓✗✗✗✓✓✓✓✓✗✗✗✓✓✓✓✗✗✓✓
Roberto	20	✗✗✗✓✓✗✗✓✓✗✗✗✗✓✓✓✗✓✓✓✗

a) Based on these records, who should you send on the court?
b) What other factors might you consider in making your decision?

166 MHR • Chapter 5

Solution

Understand

a) Assume the player with the best record is the person most likely to make a free throw. You need to identify the best free-throw record.

> **Strategies**
> Make an assumption

Plan

1. Show each player's record as a fraction: $\dfrac{\text{successful free throws}}{\text{number of attempts}}$

2. Express the fractions as decimals or percents.
3. Compare to decide who has the best chance of scoring on a free throw.

Do It!

1. Show each player's record as a fraction.

   Pemba: $\dfrac{\text{successful free throws}}{\text{number of attempts}} = \dfrac{12}{25}$

   Heather: $\dfrac{14}{25}$ Roberto: $\dfrac{9}{20}$

2. Express the fractions in a different way.

   Method 1: Write Fractions as Decimals

   Pemba's record: $\dfrac{12}{25} = 12 \div 25$
   $= 0.48$

   Heather's record: $\dfrac{14}{25} = 14 \div 25$
   $= 0.56$

   Roberto's record: $\dfrac{9}{20} = 9 \div 20$
   $= 0.45$

   Method 2: Write Fractions as Percents

   Pemba's record: $\dfrac{12}{25} = 12 \div 25$
   $= 0.48 \times 100\%$
   $= 48\%$

   Heather's record: $\dfrac{14}{25} = 14 \div 25$
   $= 0.56 \times 100\%$
   $= 56\%$

   Roberto's record: $\dfrac{9}{20} = 9 \div 20$
   $= 0.45 \times 100\%$
   $= 45\%$

3. Compare the free-throw records.

   Method 1: As Decimals   *Method 2: As Percents*
   $0.56 > 0.48 > 0.45$ $56\% > 48\% > 45\%$

Look Back

Heather scored on more than half of her attempts. Less than half of Pemba's and Roberto's attempts were successful. Heather should take the free throw.

> **Did You Know?**
> The statistic in Example 1 is called the free-throw percentage. It is used in several leagues, including the National Basketball Association (NBA).

b) Other factors to consider are
- current injuries
- most recent performances
- the ability to perform under pressure

Example 2: Calculate Percent of Area

A community park has a swimming pool.
a) Estimate the percent of the area of the park taken up by the pool.
b) Calculate the percent of the area taken up by the pool. Round your answer to the nearest percent.

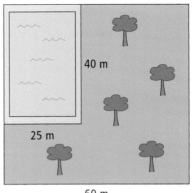

Solution

a) The pool appears to take up approximately a quarter of the park. This is approximately 25%.

b) Calculate the areas.

	Shape	Area Formula	Area (m²)
Swimming Pool	Rectangle	$A = l \times w$	$A = l \times w$ $A = 40 \times 25$ $A = 1000$
Park	Square	$A = s^2$	$A = s^2$ $A = 60 \times 60$ $A = 3600$

$$\frac{\text{area of pool}}{\text{area of park}} = \frac{1000}{3600}$$
$$= 0.2\overline{7}$$

Write this decimal as a percent.
$$0.2\overline{7} = 0.2\overline{7} \times 100\%$$
$$= 27.\overline{7}\%$$
$$\doteq 28\%$$

Round to the nearest percent.

The pool takes up approximately 28% of the park.

Strategies
What strategy is used here? What other strategy might be used?

Making Connections
If you need help with calculating areas, go to Chapter 1, page 11.

Key Ideas

- In problems involving fractions, decimals, and percents, choose which form of the number to work with.
- Estimation is a good tool to check your conversion work. For example, the pool in this park is less than $\frac{1}{4}$ of the area of the park.

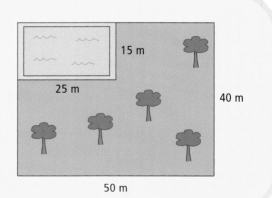

Communicate the Ideas

1. Pemba attempted 25 free throws. She made 12 of them and missed 13. Show her free-throw results as a fraction. Explain why a fraction is a useful way to show these statistics.

2. What other methods might you use to show Pemba's free-throw results? Describe how to do this.

3. Jasmine's apartment has a living room, a bedroom, a kitchen, and a bathroom. How can Jasmine find the percent of her apartment that is used for each room? Describe different strategies.

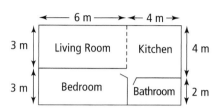

Check Your Understanding

Practise

4. Write each fraction as a decimal.
   a) $\dfrac{3}{16}$
   b) $\dfrac{13}{20}$
   c) $\dfrac{5}{6}$
   d) $\dfrac{17}{25}$

5. Write each fraction in question 4 as a percent.

6. Write each percent as a fraction. Simplify the fraction.
   a) 40%
   b) 55%
   c) 28%
   d) 66%

For help with questions 7 and 8, refer to Example 1.

7. Tamika was selling chocolate bars for a youth group. Two out of every five people she asked bought a chocolate bar. Show this result as
   a) a fraction
   b) a decimal
   c) a percent

8. In a survey, 7 out of every 20 people preferred hot chocolate to any flavour of juice. Show this result as a fraction, a decimal, and a percent.

Apply

For help with question 9, refer to Example 2.

9. A community park has a swimming pool.

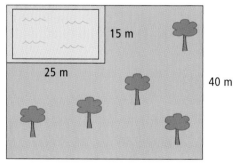

   a) Estimate the percent of the area taken up by the pool.
   b) Calculate the percent of the area taken up by the pool. Round your answer to the nearest percent.

10. During hockey penalty shots, Paco can score a goal 9 times out of 10. Frederica can score a goal 13 times out of 15.

   a) Show each player's statistic as a fraction, a decimal, and a percent. Use a table or some other method to organize your answer.

   b) Who is the better scorer? Explain why.

11. During a snowy January, Snowbelt City buses were on schedule 60% of the time. What fraction of the time were the buses *not* on schedule?

12. a) Rico played only 18 out of 32 hockey games last season due to an ankle injury. What percent of the season did Rico miss?

   b) What strategy did you use to solve part a)? What other strategy might be useful?

13. The make-up of Earth's atmosphere is shown in the circle graph. What approximate fraction of air is made up of

   a) nitrogen?   **b)** oxygen?
   c) other gases?

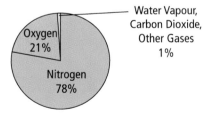

14. Kendra's CD collection has 12 dance CDs, 15 alternative CDs, and 18 rock CDs.

   a) How large is Kendra's collection?

   b) Estimate the percent of each type of CD in Kendra's collection.

   c) Calculate the percent of each type of CD.

   d) What fraction of the collection does each type of CD represent?

   e) Describe another method you could have used in part d).

15. The Ontario Beverage Company makes seven different flavours of pop. Sales for the past two weeks are shown in the table.

Drink Flavour	Sales (Thousands)
Koala Cola	13
Lizard Lime	5
Lemur Lemon	8
Gorilla Grape	3
Roary Root Beer	9
Oliphaunt Orange	10
Jumping Ginger	2

Write a brief report on drink sales. Include
- total sales of drinks
- a visual showing percent sales of each flavour

Literacy Connections

Visuals in Reports
Bar graphs and circle graphs are good ways to show sales data. Visuals and few words make reports easy to read.

Try This!

16. Santino got $\frac{26}{40}$ on a test. He needs to get 80% to be on the honour roll.

   a) Will he make it? Explain.

   b) If not, how many marks does he need to get on the make-up test?

Extend

17. The producers of a 90-min action film decide that between 50% and 65% of the film should be action scenes. The film editor has 70 min of good action footage. Should the director ask the editor to cut any of this footage? If so, how much? Use pictures, numbers, and words to explain your answer.

18. In a board game, you are trapped in "detention." You have two chances to escape, based on rolling a pair of number cubes:
- roll doubles (the same number on both cubes)

or

- roll a total of at least 9

There is a catch: You have to decide in advance to try for one way or the other. Should you go for doubles, or a total of 9 or more? Justify your decision, using calculations.

19. A jacket is marked down 30% from the regular price of $150.
  a) Find the sale price.
  b) The sale price is not the final price! On most clothing in Ontario, there is a provincial sales tax (PST) of 8% and a Goods and Services Tax (GST) of 7%. Find the final price of the jacket.

Making Connections

Planet Earth in Percents

Good atlases and almanacs contain interesting data. Sometimes you have to work with data to create percents.

Research some information about planet Earth. Use fractions and percents to report your findings.

Literacy Connections

Keeping Track of Sources
File cards can help you research. Use one file card for each source.

Galbraith, Don. Sciencepower 7.
McGraw-Hill Ryerson, 1999.
p. 338: Earth's crust is between $\frac{5}{6376}$ and $\frac{35}{6406}$ of its radius.

$1216 + 2270 + 2885 + 5$ — 5 km to 35 km
$1216 + 2270 + 2885 + 35$

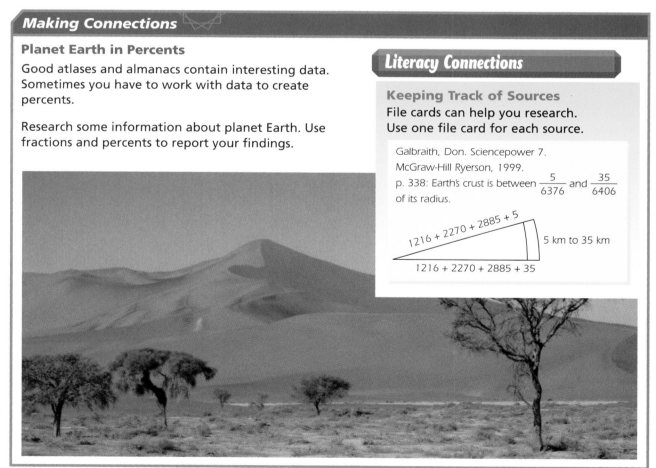

Chapter 5 Review

Key Terms

For questions 1 to 3, copy the statement and fill in the blanks. Use some of these words:

fraction   percent
repeating decimal   statistic

1. A ■■■(■)■■■■■■ is a value calculated from a set of data.

2. You can write a ■■■(■)■■■■(■) decimal using a bar or a dot.

3. Another way to say "out of a hundred" is "(■)■■■■■■."

4. Unscramble the circled letters in questions 1 to 3 to answer this question:

   Who called the MPP to the committee meeting?

5.1 Fractions and Decimals, pages 152–157

5. Show each fraction using a diagram.
   a) $\frac{3}{5}$ b) $\frac{2}{3}$ c) $\frac{5}{8}$

6. A car has just come into Otto's Auto Body for emergency repairs. Otto checks to see which mechanic is closest to finishing his or her current job.

   Nico: "I'm $\frac{1}{3}$ of the way finished, boss."

   Jacques: "I'm about halfway done, Otto."

   Leah: "I still have about $\frac{3}{4}$ of the job to do, boss."

   a) Use a fraction to show how much of his or her work each mechanic has done.
   b) Who should be assigned the new job? Justify your answer.

7. Write each fraction as a decimal.
   a) $\frac{7}{12}$ b) $\frac{5}{9}$ c) $\frac{17}{31}$

8. Order the fractions in question 7 from least to greatest.

5.2 Calculate Percents, pages 158–161

9. Write each decimal as a percent.
   a) 0.65   b) 0.4
   c) 0.237   d) 0.008

10. Write each percent as a decimal.
    a) 26%   b) 94.3%
    c) 7%   d) 2%

11. Write each fraction as a decimal.
    a) $\frac{33}{100}$ b) $\frac{3}{50}$
    c) $\frac{7}{9}$ d) $\frac{11}{90}$

12. Sheila decides to put her best test into her portfolio. Her three top test scores are $\frac{28}{35}$, $\frac{33}{39}$, and $\frac{31}{37}$.
    a) Write each test score as a percent.
    b) Which test should Sheila put into her portfolio? Explain your decision.

13. On average, Chantal gets a hit 3 out of every 10 times at bat.
    a) What percent of the time does Chantal get a hit?
    b) Express this batting average as a decimal.
    c) How many hits would you expect Chantal to get in 180 at-bats? Explain how you found your answer.

5.3 Fractions, Decimals, and Percents, pages 162–165

14. Write each percent as a fraction. Simplify the fraction.
   a) 20%   b) 72%
   c) 8%   d) 98%

15. Copy and complete the table.

Percent	Decimal	Fraction
	0.88	
36%		
		$\frac{1}{5}$
42%		
4%		

16. A lasagna recipe calls for three types of cheese.

Cheese	Amount (Makes 8 Servings)
Mozzarella	400 g
Ricotta	350 g
Parmesan	250 g
Total	1000 g

   a) Show each cheese type as a fraction of the total.
   b) You want to calculate the amounts of cheese for three servings. Which form should you use for the amount of each cheese: fraction, decimal, or percent? Explain your choice.
   c) Find the amount of each cheese in three servings of lasagna.

5.4 Apply Fractions, Decimals, and Percents, pages 166–171

17. Professor Bish has a collection of reference books. He estimates that he has these percents of each type of book.

Type of Book	Percent
Engineering	35%
Mathematics	30%
Science	25%
Other	10%

Professor Bish has three equal-sized shelves that can easily fit all of his books.

   a) How do you think Professor Bish should organize his books? Illustrate with a diagram.
   b) Professor Bish has 180 books in total. How many would be placed on each shelf? Discuss any assumptions you must make.

18. Professor Bish organizes his office desk as shown.

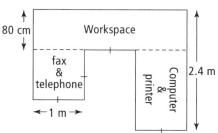

Professor Bish feels that he needs at least 50% of his desk for workspace. Should he ask the university for a larger desk? Support your answer with calculations.

Chapter 5 Practice Test

Strand Questions	NSN 1–11	MEA 11	GSS 11	PA	DMP

Multiple Choice

For questions 1 to 6, choose the best answer.

1. In a bag of coloured candy, 25% of the candies are green. If the bag contains 48 candies, how many are green?
   - **A** 12
   - **B** 15
   - **C** 24
   - **D** 25

2. The decimal equivalent of $1\frac{1}{6}$ is
   - **A** 1.16
   - **B** $0.1\overline{6}$
   - **C** $1.1\overline{6}$
   - **D** $1.\overline{6}$

3. The fraction equivalent of 4% is
   - **A** $\frac{1}{4}$
   - **B** $\frac{2}{5}$
   - **C** $\frac{1}{25}$
   - **D** $\frac{1}{400}$

4. Vincent scored 17 out of 20 on his geography test. What percent did he achieve?
   - **A** 75%
   - **B** 85%
   - **C** 90%
   - **D** 95%

5. In a school survey, 75% of students were wearing blue jeans. What fraction of students surveyed does this represent?
   - **A** $\frac{3}{4}$
   - **B** $\frac{3}{5}$
   - **C** $\frac{75}{10}$
   - **D** $\frac{4}{5}$

6. The number 3.75 is greater than which fraction or mixed number?
   - **A** $\frac{19}{5}$
   - **B** $3\frac{2}{3}$
   - **C** $\frac{16}{3}$
   - **D** $3\frac{4}{5}$

Short Answer

7. Copy and complete the table.

Fraction	Decimal	Percent
		28%
$\frac{3}{8}$		
	0.45	
		5%
$\frac{24}{240}$		

8. Three types of recordable CD-ROMs are checked for faults.

Brand	Number Tested	Number Passed
Electro-Zip	20	19
Ultraback	10	7
A-Retrieve	30	23

   a) What fraction of each brand passed the test?
   b) Which brand of CD-ROM seems the most reliable? the least reliable? Justify your answers.

9. The table shows Hoshi's performance during a unit in science class.

Evaluation Item	Result
Lab	55%
Quiz	half of multiple-choice questions answered correctly
Test	29 out of 50

   a) Arrange Hoshi's evaluation items from strongest performance to weakest.
   b) Describe the method you used.

174 MHR • Chapter 5

Extended Response

10. The table shows estimated percents for various tree species in a forest.

Type of Tree	Percent
Maple	30%
Oak	25%
Pine	20%
Birch	15%
Other	10%

a) What fraction of the forest is made up of each tree species?
b) Write each percent as a decimal.
c) The forest has approximately 1500 trees. Would you expect more or fewer than 500 maples? Explain.

11. At a fairground game, you can throw a dart at the target to win a prize.

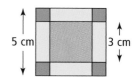

a) Which do you think is easiest to hit: red, yellow, or blue? Explain your choice.
b) Every dart that lands in one of the coloured regions is a winner. What fraction of the total area is each colour?
c) Rank the colours in order from greatest to least. Which colour should get the best prize, and why?

Chapter Problem Wrap-Up

A hockey coach is choosing a new goalie for her team. She has 20 games left in the season and only three players to choose from.

1. Use the information in the table to recommend which player the coach should choose. Provide as many reasons as you can for your recommendation.

2. What can the coach hope for from her goalie in the next 20 games? Explain.

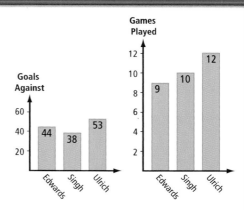

Goalie	Save Percentage
Edwards	0.88
Singh	0.96
Ulrich	0.92

Patterning and Algebra
- Find and describe patterns in sequences of numbers and shapes.
- Extend a pattern, complete a table, and write words to explain the pattern.
- Use patterns to make predictions.
- Use patterning strategies in problem solving.
- Interpret a variable as a symbol that stands for a number.
- Evaluate simple algebraic expressions.

Data Management and Probability
- Organize, analyse, interpret, and present data mathematically and on graphs.

Key Words
natural numbers
variable expression
pattern rule
table of values
relationship
algebraic equation

CHAPTER 6

Patterning

Dominoes can be set up to fall in a chain reaction called a domino topple. The world record for the most dominoes toppled in one attempt is 3 847 295. For record-breaking domino topples, the dominoes are often set up in unusual and interesting designs.

To prevent accidental toppling of the entire chain during set-up, stoppers are placed between the dominoes at regular intervals. The stoppers are removed just before the first domino is toppled.

How can math help in the topple-safe design of a large domino chain?

Chapter Problem

Describe all the patterns you can see in this domino layout.

Get Ready

Explore Patterns

In math, many patterns can be predicted with complete accuracy. For example, to extend this pattern of rectangles,
- turn each rectangle through 90°
- add a square to the right
- the square's sides are the length of the rectangle
- put the rectangle and the square together to form a new rectangle
- repeat

This instruction predicts the entire pattern.

Examine the sequence of rectangle widths: 1, 1, 2, 3, 5, 8, 13, …. Each number in the sequence is the sum of the two previous numbers. For example,
1 + 1 = 2
1 + 2 = 3
2 + 3 = 5

By following this rule, you can predict the next two numbers in this sequence.
8 + 13 = 21
13 + 21 = 34

Did You Know?

The sequence 1, 1, 2, 3, 5, 8, 13, … is called a Fibonacci sequence. Leonardo Fibonacci discovered it over 800 years ago, while investigating how quickly rabbits breed. Fibonacci sequences describe many patterns in nature. For example, the spiral of blades on a pine cone follows a Fibonacci sequence.

1. Describe the pattern in each sequence. Then, continue each pattern for three more items.

   a) , …
   b) 2, 4, 6, 8, …
   c) 100, 95, 90, 85, …
   d) 3, 6, 12, 24, …
   e) △, △, △△, △△, …
   f) $\frac{1}{1}, \frac{1}{2}, \frac{1}{4}, \frac{1}{8},$ …
   g) ab, abc, abcd, ab, abc, abcd, …
   h) ☹, ☺☹, ☺☺☹, …

2. Extend each number pattern.
   a) Start with 8. Repeatedly add 7. List the next six numbers.
   b) Start with 2. Repeatedly add 15. List the next three numbers.
   c) Start with 100. Repeatedly subtract 12. List the next four numbers.
   d) Start with 4. Repeatedly multiply by 3. List the next five numbers.
   e) Start with 128. Repeatedly divide by 2. List the next three numbers.
   f) Start with 2, 5. Repeatedly add the previous two numbers. List the next five numbers.

Use Graphing Skills

(5, 1) is an **ordered pair**. To plot (5, 1) on a **coordinate grid**,
- put your pencil at (0, 0)
- move 5 units to the right for the **x-coordinate**
- then, move up 1 unit for the **y-coordinate**
- mark a dot and label the coordinates

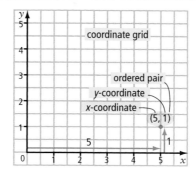

These points are plotted on the coordinate grid to the right.
Q(0, 3), R(9, 0), S(4, 7)

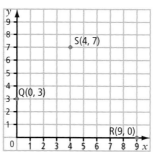

3. a) Draw a coordinate grid with numbers from 0 to 6 on each axis.
   b) Plot these ordered pairs on your coordinate grid.
   A(1, 0), B(2, 1), C(0, 2), D(2, 2), E(3, 3), F(4, 2), G(6, 2), H(4, 1), I(5, 0), J(3, 1)
   c) Join the points in alphabetical order. Finish by joining J back to A.
   d) Identify the shape.

4. For each ordered pair, copy and complete this statement: "Move ■ units to the right. Then, move ■ units up."
   a) (4, 2)   **b)** (7, 3)   **c)** (3, 1)   **d)** (1, 3)

5. The organizer of a fair plots the fair's layout on a coordinate grid. Write an ordered pair for each location.

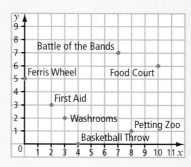

6.1 Investigate and Describe Patterns

Focus on...
- patterns with numbers
- patterns with shapes
- describing patterns

The numbers on a calendar make some interesting mathematical patterns. If the last day of March is a Monday, what are the dates of the Fridays in April? How did you find these dates? What other patterns do you see?

Discover the Math

Materials

Optional:
- BLM 6.1A Calendar Patterns
- calendar page showing one month

What patterns can you find and describe in a calendar?

Look at this calendar page.
1. a) Can you find any patterns in the red square $\begin{array}{|c|c|}\hline 5 & 6 \\ \hline 12 & 13 \\ \hline\end{array}$ of the calendar?
   b) Shift the square one position to the right. See if you get the same results.
   c) Describe in words the pattern or patterns that you found.

2. Find other patterns in the calendar.

3. **Reflect** Describe in words all the patterns that you found in step 2. Are some patterns similar to others? Try grouping your patterns.

180 MHR • Chapter 6

Example 1: Describe the Number Pattern

a) Describe the pattern 1, 4, 9, 16, 25, … in as many ways as you can. Use pictures, words, and numbers.
b) Find the next three numbers in the pattern.

Solution

a) *Method 1: Use Words and Multiplication*
$1 \times 1 = 1$ $2 \times 2 = 4$ $3 \times 3 = 9$
$4 \times 4 = 16$ $5 \times 5 = 25$ …

Multiply 1 by itself. Then, multiply 2 by itself, 3 by itself, and so on. The answers form the pattern.

Method 2: Use Pictures

1 square   4 squares   9 squares   16 squares

The total number of squares at each stage creates the pattern 1, 4, 9, 16, 25, ….

Method 3: Draw a Number Line

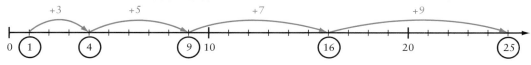

The pattern jumps to the right in steps of ordered odd numbers.

b) $6 \times 6 = 36$ $7 \times 7 = 49$ $8 \times 8 = 64$
The next three numbers are 36, 49, and 64.

> **Strategies**
> What other methods can you use to find the next three numbers?

Example 2: Describe the Geometric Pattern

Study these figures. What geometric pattern do they show?

Solution

Method 1: Describe How to Create the Pattern
- Begin with a square.
- For the next square, halve the side length.
- Draw this smaller square in the upper-right corner of the previous square.
- Repeat this pattern over and over.

Method 2: Describe What the Pattern Looks Like

This is a pattern of squares within squares. Each new square has $\frac{1}{4}$ of the area of the previous square, and is in the upper-right corner.

Key Ideas

- In a pattern, you can predict what comes next.
- Some patterns are based on number operations. Other patterns are based on geometric shapes.
- To describe a pattern, identify the first item. Then, describe how the numbers or shapes that follow are generated. Relate each new item to previous items, or to the counting numbers 1, 2, 3, ...

Communicate the Ideas

1. Michel and Fareeha described the pattern 2, 4, 6, 8, 10, ...

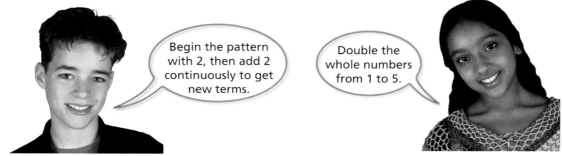

   Begin the pattern with 2, then add 2 continuously to get new terms.

   Double the whole numbers from 1 to 5.

   Decide whether Michel, Fareeha, or both are correct. Explain.

2. Create your own pattern. Is there more than one way to describe the pattern? Explain.

Check Your Understanding

Practise

For help with questions 3 to 8, refer to Example 1.

3. a) Describe this pattern. Use pictures, words, and numbers.
   5, 10, 15, 20, 25, ...
   b) Find the next three numbers in the pattern.

4. a) Describe this pattern in as many ways as you can. Use pictures, words, and numbers.
   1, 7, 13, 19, 25, ...
   b) What are the next three numbers in the pattern?

5. a) Describe the pattern 0, 6, 14, 24, 36, 50, ... using numbers.
   b) Predict the next two numbers in the pattern.

6. Describe each pattern. State the next three items.
   a) 1, 2, 2, 3, 3, 3, …
   b) 2, 4, 6, 8, 10, …
   c) 100, 95, 90, 85, …
   d) abc, bcd, cde, …
   e) 256, 128, 64, 32, …
   f) vxz, uwy, tvx, …

7. Make a number pattern and a letter pattern. Trade with a friend. Solve each other's patterns.

8. a) Describe this pattern using numbers.

• •• ••• ••••
• •• ••• ••••

   b) List the next three numbers of dots.

For help with questions 9 to 12, refer to Example 2.

9. Study these figures. Describe the geometric pattern.

10. Study these figures. Describe the geometric pattern.

11. Describe the geometric pattern.

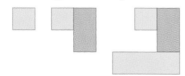

12. Make a geometric pattern. Describe your pattern.

Apply

13. Copy and complete each pattern by replacing each ■ with the appropriate value.
   a) 5, ■, 15, ■, … (addition pattern)
   b) 3, ■, 12, ■, … (multiplication pattern)
   c) ■, 16, ■, 12, … (subtraction pattern)
   d) ■, 200, ■, 50, … (division pattern)

14. The old Chinese calendar used three 10-day weeks.

1	2	3	4	5	6	7	8	9	10
11	12	13	14	15	16	17	18	19	20
21	22	23	24	25	26	27	28	29	30

   a) Choose any four numbers that form a square.
   b) Multiply each pair of numbers in the diagonals.
   c) Repeat this with another square of four numbers.
   d) Describe what you notice.

Use this calendar page to answer questions 15 and 16.

JANUARY

S	M	T	W	T	F	S
			1	2	3	4
5	6	7	8	9	10	11
12	13	14	15	16	17	18
19	20	21	22	23	24	25
19	20	21	22	23	24	25
26	27	28	29	30	31	

15. a) Add the date of the first Wednesday to the date of the second Wednesday. Repeat for the first and second Thursdays, Fridays, and Saturdays.
   b) Describe the pattern.
   c) Use your description to find the next three numbers in the pattern.
   d) Predict whether or not similar patterns will occur in later weeks. Explain your reasoning.

6.1 Investigate and Describe Patterns • MHR **183**

16. a) Multiply the date of the first Wednesday by the date of the second Wednesday. Repeat for the first and second Thursdays, Fridays, and Saturdays.
b) Describe the pattern.
c) Use your description to find the next three numbers in the pattern.

17. Create a pattern using simple geometric figures. Explain your pattern.

18. A pattern begins with the fraction $\frac{1}{2}$.
Then, 1 is added to both the numerator and the denominator.
a) State the next five fractions in the pattern.
b) As the pattern continues, do the fractions get bigger or smaller? Explain your reasoning.

> **Making Connections**
> In question 18, use what you learned about fractions in Chapter 3 to develop and describe a pattern.

19. Study this growing pattern.

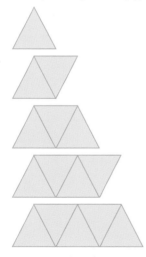

a) Describe the pattern.
b) Why do you think it is called a growing pattern?
c) Draw the next two stages in the pattern.

20. a) Use toothpicks to make your own growing pattern. Describe your pattern in words.
b) Ask a partner to draw the next two stages in your pattern.

21. Study this shrinking pattern.

a) Describe the pattern.
b) Draw the next two stages in the pattern.

 22. a) Use toothpicks to make a shrinking pattern. Describe your pattern in words.
b) Ask a partner to draw and describe your pattern.

Extend

23. Alex gave Sasha the pattern 2, 4, … , and asked him to write the next three numbers. Does Alex's question have a single answer? Explain your thinking.

24. Regularized Islamic years have 6 months with 30 days, alternating with 6 months of 29 days.
a) How long is an Islamic year?
b) How many Islamic years will it take to "gain" one year, compared to the Western calendar? Use a pattern to explain your answer.

> **Did You Know?**
> In the original Islamic calendar, each month begins when the first, thin crescent of the new moon is seen.
>
> Research the history of calendars. Go to www.mcgrawhill.ca/links/math7 and follow the links to get started.

6.2 Organize, Extend, and Make Predictions

Focus on...
- describing patterns using numbers
- describing patterns using variables
- extending patterns
- making predictions

This is a house screen painting from Port Simpson, British Columbia. Identify the patterns on this painting. How many times does each pattern repeat?

Discover the Math

Materials
- toothpicks
or
- plain paper or triangle dot paper

Optional
- BLM 6.2A Patterns With Toothpicks

How can you organize and predict patterns?

1. Use toothpicks to build the first four stages of this triangle pattern. Record all the patterns you can see.

2. a) How many toothpicks would you need for a triangle diagram with a base of 5 toothpicks?
   b) Predict the number of toothpicks in a triangle diagram with a base of 6 toothpicks.
   c) Check your predictions by modelling with toothpicks. How close were your predictions to your models? Explain.

3. **Reflect** How could you organize each stage of the pattern? How would this help you to extend the pattern?

Example 1: Organize and Extend a Pattern

a) Determine the perimeter (the number of toothpicks) of each of the first four triangles.
b) Show how each perimeter relates to the set of **natural numbers**.
c) Find the next three perimeters in the pattern.

natural numbers
- the numbers 1, 2, 3, ...
- also called positive integers

Solution

a) The first four perimeters use 3, 6, 9, and 12 toothpicks.

b)
Natural Number (Base Length)	Perimeter	Operation
1	3	3 × 1
2	6	3 × 2
3	9	3 × 3
4	12	3 × 4

The natural numbers 1, 2, 3, ... are the base lengths. I multiply each number by 3 to get the perimeter.

c) The next three perimeters are
  3 × 5 = 15   3 × 6 = 18   3 × 7 = 21

Strategies
How could you show this pattern using addition?

Example 2: Describe and Extend a Pattern Using Variables

a) Show the relationship between the number of cubes in each stack and the number of vertical faces.
b) Use a **variable** to write a **variable expression** for the pattern.
c) Determine the number of vertical faces in a stack of 100 cubes.

variable
- a letter that can stand for any number

variable expression
- numbers and variables, combined by operations

Solution

a) *Method 1: Use a Table*

Number of Cubes	Number of Vertical Faces	Expression
1	4	4 × 1
2	8	4 × 2
3	12	4 × 3

Method 2: Use Equations
4 × 1 = 4
4 × 2 = 8
4 × 3 = 12

Strategies
What other strategy could you use to show this relationship?

b) Use the variable n to stand for the number of cubes in a stack. Multiply the number of cubes, n, by 4. The variable expression for the number of vertical faces is $4 \times n$, or $4n$ for short.

c) $4n = 4 \times 100$ **Substitute 100 for n.**
   $= 400$
There will be 400 vertical faces in a stack of 100 cubes.

Literacy Connections

Reading Expressions
A number beside a variable means you multiply the variable by the number. So, $4n = 4 \times n$.

Key Ideas

- Any letter can be used as a variable to represent a number.
- A variable can be used in an expression that shows how a pattern works. For example, $3 \times n$ or $3n$ shows how the perimeters of toothpick triangles grow.

Communicate the Ideas

1. Look at these growing patterns:
2, 4, 6, 8, …   2, 4, 8, 16, …
Explain how to find the next number in the growing patterns. What different techniques can you use?

2. You tell two friends a story.
- These friends each tell another two friends and then they each tell another two friends.
- Assume no two friends are the same.

Explain how you can find out how many different people know the story after 5 rounds of telling it. Use pictures and words.

Check Your Understanding

Practise

For help with questions 3 and 4, refer to Example 1.

3. a) Determine the perimeter of each of the first four squares.

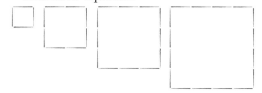

b) Show how each perimeter relates to the set of natural numbers.

4. a) Determine the perimeter of each of the first four hexagons.

b) Describe how each perimeter relates to the natural numbers.

5. Describe each number pattern. Relate each item to the set of natural numbers.
a) 5, 10, 15, 20, …   **b)** 1, 4, 9, 16, 25, …
c) 1, 3, 5, 7, …   **d)** $\frac{1}{2}, \frac{1}{3}, \frac{1}{4}, \frac{1}{5}, \ldots$

6. For each pattern in question 5, predict the next three numbers.

For help with questions 7 to 9, refer to Example 2.

7. a) Show the relationship between the number of cubes in these blocks and the number of exposed faces.

   b) Write a variable expression for the pattern.
   c) Use your variable expression to determine the number of exposed faces in a block of 100 cubes.

8. In a room with 1 person, there are 8 large joints: 2 knees, 2 hips, 2 elbows, and 2 shoulders. How many large joints are there in a room with
   a) 2 people?   **b)** 3 people?
   c) 4 people?   **d)** x people?
   e) 73 people?

9. Sam always eats 3 more candies than his friend. How many candies will Sam eat if his friend eats
   a) 5 candies?   **b)** 12 candies?
   c) 100 candies?   **d)** k candies?

Apply

10. One trillium has 3 petals.

   a) How many petals do 2, 3, and 4 trilliums have?
   b) How many petals do 100 trilliums have?
   c) If there are 3000 petals, how many trilliums are there?
   d) Explain the strategy you used to solve these questions.

11. a) Study these three toothpick shapes. Describe the pattern of the number of toothpicks in each shape.

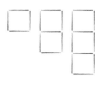

   b) Predict how many toothpicks would be in the fourth diagram. Give a reason for your prediction.
   c) Draw the fourth diagram. If you need to change your prediction, explain why.
   d) Predict how many toothpicks would be in the fifth diagram.
   e) Consider the method you used to make pattern predictions in parts b) and d). Describe another method for predicting growing patterns.

12. A pattern of zigzags is made from line segments.

   a) Draw the next two stages in the pattern.
   b) Describe the relationship between the number of line segments and the number of acute angles.
   c) Choose a variable to represent either line segments or angles. Develop an expression to show the relationship.

Making Connections

You identified acute angles in Chapter 2.

Chapter Problem

13. a) Describe the pattern in this chain of dominoes.

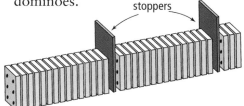

   b) If the pattern continues, how many stoppers are needed for a 40-domino chain? a 100-domino chain?

14. Try this magic math game. Copy the table to record your answers.

Directions	How the Numbers Change			Expression
Think of a number.				n
Add 7.				
Multiply by 3.				
Subtract 8.				
Subtract 3 × your original number.				

a) Think of a number. Add 7. Triple your answer. Subtract 8. Now subtract three times your original number.
b) Try this two more times with different numbers.
c) Using n as a variable, write an expression for each step.
d) Explain what you learned about this expression.

15. Create your own magic math question. Fill in a table to show how it works.

Did You Know?

Are you good at brain teasers? You might like to find out about Mensa, an international society that encourages and supports good thinkers and problem solvers. Go to www.mcgrawhill.ca/links/math7 and follow the links.

Extend

16. A solitaire game, sometimes called *The Tower of Hanoi*, uses different-sized coins or counters. For example, you could use a toonie, a loonie, a quarter, a nickel, a penny, and a dime.

- You have three tower bases.
- To begin with, the tower is on the left-hand base.
- You must rebuild the tower on another base.
- You can only move one counter at a time to any other base.
- You cannot place a counter on top of a smaller counter.

a) Try the game, with different tower heights. Discover and explain any patterns used in a successful game.
b) Investigate the minimum number of moves for each tower height to complete the game.

17. Sanjay is making hexagon trails. The table shows two different methods to find the perimeter for each trail length.

Steps	Perimeter	Pattern A	Pattern B
1	6	2 + 4	6
2	10	2 + 4 + 4	6 + 4
3	14	2 + 4 + 4 + 4	6 + 4 + 4

a) Choose either pattern A or Pattern B. Use your chosen pattern to find the perimeter of a trail with 150 hexagons.
b) Determine the perimeter of the 10th item in this pattern.

6.3 Explore Patterns on a Grid or in a Table of Values

Focus on...
- relationships between numbers and variables
- describing patterns on a grid

Engineers can design long, steep ski jumps. They use mathematics to make the smooth curve by plotting ordered pairs on a grid.

How can you use a grid to represent where the skier is right now?

Discover the Math

Materials
- grid paper

How can you explore patterns on a grid?

1. Copy the coordinate grid below onto a piece of grid paper. Plot the ordered pair (1, 2) as shown.

2. a) From the point (1, 2), move 1 space right and 2 spaces up. Plot a second point.
   b) Repeat two more times, moving 1 space right and 2 spaces up each time.
   c) Describe the resulting pattern of points.

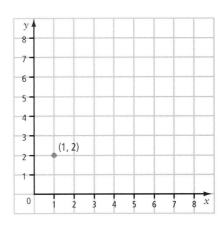

190 MHR • Chapter 6

3. Suppose the x-coordinate is 25. Predict what the y-coordinate will be.

4. a) State a **pattern rule** for determining the y-coordinate when you are given the x-coordinate.
   b) How would your pattern rule change if the initial point were (3, 1)? Explain.

pattern rule
- a simple statement that tells how to form or continue a pattern

5. **Reflect** How does displaying a pattern on a grid help you to describe and extend the pattern?

Example 1: Plot Points and Describe Patterns

a) Plot the points given in the **table of values**.

x	0	1	2	3	4	5
y	0	1	4	9	16	25

table of values
- table listing two sets of numbers that may be related

b) Describe the pattern of points.

Solution

a) [Graph showing points (0,0), (1,1), (2,4), (3,9), (4,16), (5,25)]

That's a ▬▬▬ line.

b) The pattern of points curves upward. The y-values grow very quickly.

Literacy Connections

Look for a Pattern
Use your finger to trace a line joining the dots. What kind of line have you traced?

Example 2: Plot Points and Examine Relationships

Describe the relationship between the base and the perimeter of the toothpick triangles.

relationship
- pattern formed between two sets of numbers
- often seen in a table of values
- can be plotted on a coordinate grid

Solution

The relationship connects base length and perimeter.

1. Make a table of values comparing the base and perimeter.
2. Describe the relationship seen in the table.

Do It!

1. Table of values:

Base, b	Perimeter, P	(b, P)
1	3	(1, 3)
2	6	(2, 6)
3	9	(3, 9)
4	12	(4, 12)

2. As the base increases by 1, the perimeter increases by 3.

1 × 3 = 3: That's (1, 3).
2 × 3 = 6: That's (2, 6).

Look Back

The fifth diagram should have 15 toothpicks.

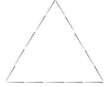

Key Ideas

- Patterns can be shown by listing ordered pairs in a table of values.
- A pattern rule is a description, in words, of a pattern. It is often used to predict the pattern. For example, in the pattern shown in the table, you multiply the base by 3 to get the perimeter.

Base, b	Perimeter, P	(b, P)
1	3	(1, 3)
2	6	(2, 6)
3	9	(3, 9)
4	12	(4, 12)

- A pattern between two sets of numbers is called a relationship.
- Plotting ordered pairs on a grid can help identify relationships.

Communicate the Ideas

1. Andrea used a table of values to show the number of shelves in a library shelving unit. Carlos used a coordinate grid to show the same information.

Number of Vertical Supports	2	3	4	5
Number of Shelves	5	10	15	20

   Which method do you find easier to use? Explain why.

   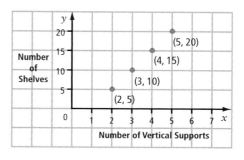

2. Make a table of values for the ordered pairs shown on the grid. Then, create a question that leads to this pattern.

   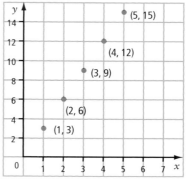

Check Your Understanding

Practise

For help with questions 3 to 5, refer to Example 1.

3. Plot the points given in each table of values. Then, describe the pattern of points.

   a)
x	0	1	2	3	4	5
y	0	1	2	3	4	5

   b)
x	0	1	2	3	4	5
y	15	12	9	6	3	0

4. Plot the points given in each table of values. Then, describe the pattern of points.

   a)
x	3	4	5	6	7	8	9
y	1	2	3	4	5	6	7

   b)
x	0	1	2	3	4	5	6
y	10	9	8	7	6	5	4

5. a) Plot the points given in the table of values.

x	0	1	2	3	4	5	6
y	0	7	12	15	16	15	12

   b) Describe the pattern of points. How is this pattern different from the patterns in questions 3 and 4?

For help with questions 6 and 7, refer to Example 2.

6. Describe the relationship between the base and the perimeter of these toothpick squares.

6.3 Explore Patterns on a Grid or in a Table of Values • MHR   193

7. Describe the relationship between the width and the length of these toothpick rectangles.

8. Describe the relationship between the width and the length of these toothpick rectangles.

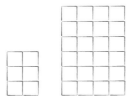

Apply

9. Give a pattern rule for the number of toothpicks in each stage of this pattern.

10. a) Find, and describe in words, a relationship between the width and the area of the squares in question 6.
    b) Repeat part a) for question 7. Hint: Use the shorter sides.
    c) Repeat part a) for question 8.

11. In the game of Battleship, ships sit on a 12 by 12 grid. Battleships can cover 4 points in a row, vertically, horizontally, or diagonally. A battleship sits with one endpoint at (5, 7). List all the points for each possible set of ordered pairs for this battleship. Hint: Use a picture or diagram.

12. When plotting these points on a grid, Ginny reversed the x- and y-coordinates. What happened to the resulting pattern?

x	0	1	2	3	4	5
y	0	1	4	9	16	25

13. a) Make a table of values for the expression $4x + 1$. Use x-values from 0 to 5.
    b) Plot the points on a coordinate grid.
    c) Describe the pattern.

Making Connections

In question 13, you gave the variable x six values: 0, 1, 2, 3, 4, and 5. Learning how to use variables is a useful skill. This knowledge can help you program computers. You will work with variables in Chapter 12.

Extend

14. Asumi threw a ball into the air. The table of values gives
    • the horizontal distance, d, in metres, from Asumi
    • the height, h, in metres, of the ball at the same instant

d (horizontal distance)	h (height)
0	1
1	16
2	21
3	16
4	1

   a) Plot the points on a grid.
   b) Describe the shape of the path.
   c) Explain the shape of the path at its highest point.

15. a) A pattern rule says that the y-coordinate is 5 more than the x-coordinate. Rewrite the rule to find the x-coordinate from the y-coordinate.
    b) Create another pattern rule in the form "y-coordinate = one operation on x-coordinate." Rewrite your pattern to reverse it.
    c) Create a method for reversing any simple pattern rule of the type in parts a) and b).

6.4 Express Simple Relationships

Focus on...
- relationships between numbers and variables
- solving problems with patterns
- evaluating algebraic expressions

Yukio charges $6 per hour for babysitting the first child, plus $1 per hour for each extra child. What is the relationship between hourly rate and the number of children?

Discover the Math

How can you express mathematical relationships?

Example 1: Describe an Algebraic Equation

The table of values shows the cost of a taxi ride from 0 km to 5 km in length.
a) Copy and complete the table.
b) Describe an **algebraic equation** for the cost of the taxi ride.

algebraic equation
- an equation or formula that describes a relationship
- uses numbers and variables

Distance, d (km)	Cost, C ($)	(d, C)
0	0	(0, 0)
1	3	
2	6	
3	9	
4	12	
5	15	

Solution

a)

Distance, d (km)	Cost, C ($)	(d, C)
0	0	(0, 0)
1	3	(1, 3)
2	6	(2, 6)
3	9	(3, 9)
4	12	(4, 14)
5	15	(5, 15)

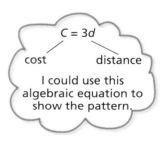

$C = 3d$
cost — distance
I could use this algebraic equation to show the pattern.

b) The cost, in dollars, is 3 times the distance, in kilometres.
The formula for the cost, in dollars, of a taxi ride is $C = 3d$.

Example 2: Describe a Relationship

The equation $y = 4x$ shows the relationship between the number of squares and the number of sides.

4 sides   8 sides

a) Create a table of values for $y = 4x$. Use four consecutive values for x.
b) Describe how the number of sides changes relative to the number of squares.

Solution

a) Choose four x-values, for example, 2, 3, 4, and 5 squares.

For $x = 2$,   For $x = 3$,
$y = 4(2)$ $y = 4(3)$
$y = 8$ $y = 12$

For $x = 4$,   For $x = 5$,
$y = 4(4)$ $y = 4(5)$
$y = 16$ $y = 20$

x	y	(x, y)
2	8	(2, 8)
3	12	(3, 12)
4	16	(4, 16)
5	20	(5, 20)

This is the same as relating the y-values to the x-values. To find the y-values, substitute the x-values in $y = 4x$.

Did You Know?

Emmy Noether was a great mathematician and teacher. She worked closely with her university students. Together they made many breakthroughs in algebra. In the 1930s, she fled to the United States to escape Nazi persecution.

b) As x, the number of squares, increases by 1, y, the number of sides, increases by 4.

You can also say, There are four times as many sides as there are squares.

Key Ideas

- You can use an algebraic equation to express a relationship.

  $C = 3d$

  cost of taxi ride   distance of taxi ride

- Ordered pairs of data can be analysed to identify a relationship.

  number of squares   number of sides

  (2, 8)

Communicate the Ideas

1. Develop an equation for the table of values to the right. Explain how you know you are correct.

Number of Wings, w	Number of Birds, b
2	1
4	2
6	3

2. Sara developed this table of values. She described the relationship: "My pay is eight times the number of hours I work." Then, she wrote this as

$$h = 8p$$
hours   pay

Is Sara correct? Explain.

Hours	Pay
1	8
2	16
3	24

Check Your Understanding

Practise

For help with questions 3 to 6, refer to Example 1.

3. List the set of ordered pairs for each table of values. State the pattern rule.

a)
x	y
1	5
2	10
3	15
4	20

b)
x	y
0	4
1	5
2	6
3	7

4. State the pattern rule for each relationship.

a)
x	y
5	1
6	2
7	3
8	4

b)
x	y
0	6
1	4
2	2
3	0

5. The table of values shows the cost of a bus ride across 5 zones.

Zone, z	Cost, C ($)	(z, C)
1	2	(1, 2)
2	3	
3	4	
4	5	
5	6	

a) Copy the table. Complete the third column.
b) Describe the pattern.

6. The table shows Yukio's babysitting rates.

Number Children, c	Hourly Rate, R ($)	(c, R)
1	6	(1, 6)
2	7	(2, 7)
3	8	(3, 8)

a) Copy the table, and add two extra rows for 4 and 5 children. Complete the second and third columns.
b) Describe the pattern.

For help with questions 7 and 8, refer to Example 1.

7. a) Create a table of values for $y = 2x$. Use any four consecutive values for x.
   b) Describe how the y-values change relative to the x-values.

8. For each equation, create a table of values. Describe the relationship.
   a) $y = 3x$
   b) $y = x + 2$
   c) $y = 10 - x$
   d) $y = 3x - 1$

Apply

9. a) Use the data in the table of values to write and solve a pattern problem involving money.
   b) Trade problems with a friend. Use different strategies to solve each other's problems.

x	y
1	15
2	30
3	45
4	60
5	75

10. For the equation $y = x + 1$:
   a) Complete a table of values for these values of x: 1, 2, 3, 4, and 5.
   b) Plot the ordered pairs on a grid.
   c) Explore how the y-values change as the x-values change. Explain what you see.

11. A box of Valentine candies is priced at $10.
   a) Complete a table of values for selling from 1 to 10 boxes of candy.
   b) Graph the information.
   c) Write a formula for the cost, C, to purchase n boxes of Valentine candies.
   d) Use your formula, or patterning, to predict the cost of 15 boxes of Valentine candies. Use your graph, or a diagram, to check your results.

Chapter Problem

12. A domino topple is being set up. Stoppers are used every 15 dominoes to prevent toppling during set-up.
   a) How many stoppers are needed for an n-domino chain? Explain your solution.
   b) How many stoppers are needed for a 1000-domino chain?
   c) An individual world record for domino toppling was set in 2003 by Ma Lihua, in Singapore. She used 303 621 dominoes. How many stoppers would she have needed?

13. For each set of ordered pairs (x, y), describe in words how y relates to x. Write an equation using x and y.
   a) (5, 4), (6, 5), (7, 6), (8, 7)
   b) (0, 0), (1, 6), (2, 12), (3, 18)
   c) (10, 6), (11, 7), (12, 8), (13, 9)
   d) (3, 0), (5, 2), (7, 4), (9, 6)

14. A catering company prepares meals for parties. They use the equation $M = p + 5$ to calculate the number of meals to make for a party. M stands for the number of meals. p stands for the number of people planning to go.
   a) Copy and complete the table of values for the values of p shown.

p	M = p + 5	(p, M)
30	$M = p + 5$ $M = 30 + 5$ $M = 35$	
40		
50		
60		
70		

   b) Describe how the values for M change relative to changes in p.
   c) Why would the catering company include "+ 5" in the equation?

Extend

15. a) Develop a table of values for the area of squares with sides 1 cm to 6 cm long.

b) Explain the relationship between the side length and the area.

16. For the last day of school, special T-shirts with the name of every student in Mr. Vaish's class were ordered. It costs $50 to set up for the printing and $5 for every shirt printed.

a) Make a table of values and a graph for this situation. Write an equation to represent it.

b) There are 23 students in the class. Which is the most efficient way, the graph or the equation, to predict the cost of T-shirts for the whole class? Justify your choice.

Making Connections

The Game of Bounce

The game of *Bounce* is for two or three players. Each player has two counters of the same colour.

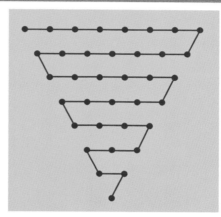

Materials
- BLM 6.4A Bounce Game Board
- two counters of the same colour per player
- number cube

1. Players take turns rolling the number cube and moving one of their counters forward by the number shown on the cube.

For example: This player started with a roll of 3.

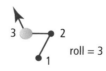

2. Suppose you move a counter onto a dot that already has one of your own counters on it. You can "bounce" one of your counters forward one line, as shown.

3. Suppose you move a counter onto a dot that has an opponent's counter on it. You can "bounce" the other counter *back* one line, as shown.

4. You must roll the exact number needed to land on the final dot. You can then take that counter off.

5. The winner is the first player to take both of her or his counters off.

CHAPTER 6 Review

Key Words

For questions 1 to 4, choose the letter representing the term that best matches each statement.

1. all the counting numbers (1, 2, 3, ...)
2. a table used to record x- and y-coordinates
3. a pattern between two sets of numbers
4. a description of a pattern

a) relationship
b) natural numbers
c) pattern rule
d) variable
e) table of values

5. Write each sentence in your notebook. Fill in each blank using appropriate words from the list.

   variable   algebraic equation
   ordered pair   variable expression

   a) $4x$ is an example of a/an ▨.
   b) (2, 7) is an example of a/an ▨.
   c) x is a/an ▨.
   d) $y = 4x$ is a/an ▨.

6.1 Investigate and Describe Patterns, pages 180–184

6. Describe each pattern in words. Predict the next two items.

   a)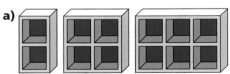

   b) 7, 9, 11, 13
   c) 243, 81, 27
   d) a, cc, eee, gggg
   e)
   f) 3, 7, 10, 17, 27, ...

7. Patterns that get smaller as they repeat forever are called fractals. Study this fractal pattern of triangles.

   a) Describe the rule used to create this pattern.
   b) Draw the next item in the pattern.

8. Invent a calendar for one month of your choice. Use your calendar to answer these questions.

   a) Choose any four numbers that form a square.
   b) Multiply each pair of numbers in the diagonals.
   c) Choose another four numbers that form a square. Repeat step b).
   d) Identify and describe any patterns that you notice.

6.2 Organize, Extend, and Make Predictions, pages 185–189

9. In a science experiment, Zena measured the growth of a bean plant.

Days After Germination	Height (cm)
1	1
2	3
3	5
4	7

   If this pattern continues, what will the height of the plant be

   a) after 5 days?
   b) after 9 days?
   c) after n days?
   d) after 18 days?

10. Study this sequence of stacked 1-cm squares.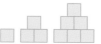
    a) Draw the next two stacks.
    b) Calculate the perimeter of each diagram. Organize the data. Hint: Use a table.
    c) How does the perimeter relate to the previous perimeter?
    d) How does the perimeter relate to the diagram number? Write an expression to show this.

6.3 Explore Patterns on a Grid or in a Table of Values, pages 190–194

11. Plot each set of points. Describe the pattern of points.

    a)
x	y
1	0
2	3
3	6
4	9
5	12
6	15

    b) (0, 21), (1, 13), (2, 9), (3, 7), (4, 6)

12. a) Describe the relationship between the width and the length of these toothpick rectangles.

    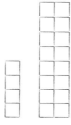

    b) Find, and describe in words, a relationship between the shorter side length and the area of the rectangles.

6.4 Express Simple Relationships, pages 195–199

13. a) List the ordered pairs in the table.
    b) Describe the pattern in words.

x	y
1	3
2	6
3	9
4	12

14. a) Copy this table of values. Use the equation $y = x + 3$ to complete the table for the values of x shown.

x	y	(x, y)
1		
2		
3		
4		

    b) List the ordered pairs.
    c) Describe how the y-values change relative to changes in x.

15. A new computer business predicted its sales as shown.

Month	Number of Computers Sold
1	13
2	26
3	39
4	52
5	65
6	78

    a) Show the information as ordered pairs.
    b) Describe the pattern rule.
    c) Using the pattern rule, predict the number of computers to be sold in the seventh month.

Chapter 6 Practice Test

| Strand Questions | NSN | MSN | GSS | PA 1–13 | DMP 4, 5, 6, 13 |

Multiple Choice

For questions 1 to 6, select the best answer.

1. The next two terms in the sequence 4, 7, 10, 13, … are
   - **A** 16, 19
   - **B** 17, 22
   - **C** 23, 36
   - **D** none of these

2. The next two terms in the sequence 7, 14, 28, 56, … are
   - **A** 70, 84
   - **B** 84, 106
   - **C** 84, 119
   - **D** 112, 224

3. To find the next term in the sequence 1, 2, 4, 7, 11, … ,
   - **A** add 8
   - **B** add 5
   - **C** add 11
   - **D** add 18

4. Match the table of values to the correct formula.

Number of Bees, b	Number of Legs, l
1	6
2	12
3	18

   - **A** $b = 6l$
   - **B** $b = l - 6$
   - **C** $l = 6b$
   - **D** $l = b \div 6$

5. The ordered pairs on the coordinate grid are
   - **A** (0, 0), (1, 3), (2, 6), (3, 9)
   - **B** (0, 1), (1, 3), (2, 5), (3, 7)
   - **C** (1, 0), (3, 1), (5, 2), (7, 3)
   - **D** (0, 1), (1, 2), (2, 4), (3, 6)

   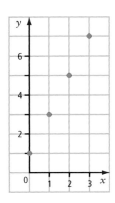

6. Which relationship describes the ordered pairs (0, 0), (2, 6), (4, 12), (6, 18)?
   - **A** $y = 2x$
   - **B** $6y = x$
   - **C** $y = 3x$
   - **D** $y = 6x$

Short Answer

7. **a)** Plot the data as ordered pairs on a coordinate grid.
   b) Describe the pattern in words.
   c) Write an algebraic equation for y in terms of x.
   d) Use your equation to find the value of y if $x = 15$.

x	y
2	4
3	5
4	6
5	7
6	8

8. Given the formula $y = 3x$, do the following:
   a) Create a table of values for $x = 0, 1, 2, 3,$ and 4.
   b) List the ordered pairs.
   c) Describe how the y-values relate to the x-values.

For questions 9 to 11, use this pattern of squares.

9. For a 2 by 2 square:
   a) How many 1 by 1 squares are there?
   b) How many 2 by 2 squares are there?
   c) What is the total number of squares of various sizes?

10. For a 3 by 3 square:
    a) How many 1 by 1 squares are there?
    b) How many 2 by 2 squares are there?
    c) How many 3 by 3 squares are there?
    d) What is the total number of squares of various sizes?

11. a) What pattern can you see in the answers to questions 9 and 10? Use your pattern to predict the total number of squares in a 4 by 4 square.
b) Draw a 4 by 4 square and check your prediction.

12. Kara is planning a garden path with this pattern. The table shows two different patterns to find the path's perimeter.

Steps	Perimeter	Pattern A	Pattern B
1	8	2 + 6	8
2	14	2 + 6 + 6	8 + 6
3	20	2 + 6 + 6 + 6	8 + 6 + 6
4	26	2 + 6 + 6 + 6 + 6	8 + 6 + 6 + 6

Choose either Pattern A or Pattern B. Use your chosen pattern to find the perimeter of a path with 8 steps.

Extended Response

13. After a "Reduce, Reuse, Recycle" campaign, the Environment Club measured the mass of paper collected each week for a month.

Week Number, n	Total Mass, M, of Paper (kg)
1	175
2	350
3	525
4	700

a) Plot the data as ordered pairs on a coordinate grid.
b) Describe the pattern in words.
c) Predict how much paper will be recycled in one school year. Hint: How many weeks are there in a school year? Will exam weeks generate more recycling than usual?

Chapter Problem Wrap-Up

A domino chain begins with 100 dominoes, then splits off into three rows of dominoes. One row has 200 dominoes, the second has 300 dominoes, and the third has 400 dominoes.

1. Stoppers are placed every 30 dominoes. How many stoppers are needed? Where would you place your stoppers?

2. Are there other ways to place the stoppers? Do they use the same number of stoppers? Explain.

Making Connections

What do patterns have to do with music?

In $\frac{4}{4}$ time, one measure of music has 4 beats.

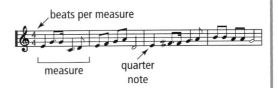

1. **a)** Calculate the total number of beats in 2, 3, and 4 measures.
   b) What is the total number of beats in 12 measures?
   c) What is the total number of beats in n measures?

In $\frac{3}{4}$ time, one measure of music has 3 beats. The nursery tune *Rock-a-Bye Baby* is in $\frac{3}{4}$ time.

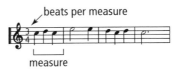

2. **a)** Calculate the total number of beats in 4 measures of $\frac{3}{4}$ time.
   b) How many beats are there in n measures of $\frac{3}{4}$ time?

Making Connections

What do you get when you mix mathematics and art?

Fractals!

Fractals are often used to model natural forms such as jagged shorelines and peaked mountains.

What patterns can you see in each fractal?

1.

2.

Fold Fractals

Fractals are mathematical patterns made by repeating the same patterns over and over again, at smaller and smaller sizes. Follow these steps to make a fractal sheet.

Materials
- cardboard or construction paper
- ruler
- scissors
- pencil

Optional:
- pencil crayons

1. a) Fold a piece of paper exactly in half, with the two short sides together.

b) Find and mark the $\frac{1}{4}$ and $\frac{3}{4}$ positions along the fold.

c) At both the $\frac{1}{4}$ and $\frac{3}{4}$ marks, make a vertical cut from the fold to halfway up the paper. Fold the midsection up to form an upside-down U-shape.

2. Repeat steps 1b) and c) using the smaller, double-layer rectangle formed by the cuts and folds. You should now have two sets of cuts.

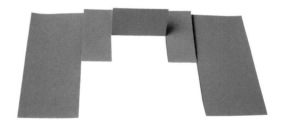

3. Unfold the paper. Pop out the fractals.

4. Investigate the shapes and patterns in your fractal sheet. Use your knowledge of shapes, patterns, fractions, and percents to report on what you see. What happens if you extend the pattern?

Number Sense and Numeration
- Compare, order, and represent decimals, multiples, factors, and square roots.
- Represent exponents as repeated multiplication.
- Represent perfect squares and their square roots in a variety of ways.
- Explain the process of problem solving using appropriate mathematical language.

Patterning and Algebra
- Identify and use a pattern.
- Explain and extend a pattern.

Key Words

square

cube

exponential form

base

exponent

square root

perfect square

power

CHAPTER 7

Exponents

A healthy lake or pond is home to many living things, including birds, fish, and plants. Small plants called algae live on the surface of the water. As they grow, they add oxygen to the water. They also provide food for fish and other animals.

Pollutants, such as fertilizers and sewage, can make the algae grow too quickly. This can make a large body of water look like green soup. When the huge numbers of algae die and decay, they use up much of the oxygen in the water. Then the water may not support life.

Large numbers like the number of algae in a polluted pond can be expressed using exponents. In this chapter, you will learn how exponents can be used to represent numbers.

Chapter Problem

How long will it take for algae to completely cover a polluted pond?

By the end of this chapter, you will be able to answer this question.

Get Ready

Factors and Multiples

The **factors** of 6 are 1, 2, 3, and 6.
The following pairs of these factors multiply to make 6.
$1 \times 6 = 6 \quad 2 \times 3 = 6 \quad 3 \times 2 = 6 \quad 6 \times 1 = 6$

The first four **multiples** of 5 are 5, 10, 15, and 20.
Each multiple is the product of 5 and a natural number.
$5 \times 1 = 5 \quad 5 \times 2 = 10 \quad 5 \times 3 = 15 \quad 5 \times 4 = 20$

1. List the factors of each number.
   a) 8   b) 17   c) 24

2. Which of the following numbers have 2 as a factor? How can you tell?
   100   301   456   294   279   193

3. List the first four multiples of each number.
   a) 4   b) 8   c) 6

4. List the first three multiples of each number.
   a) 10   b) 12   c) 20

True Statements

The symbol > means "is greater than." So, 3 > 2 is a true statement.
The symbol < means "is less than." So, 2 < 3 is a true statement.
The symbol = means "is equal to." So, 3 = 3 is a true statement.

5. Rewrite each statement. Replace ■ with >, <, or = to make a true statement.
   a) 13 ■ 14   b) 13.6 ■ 13.5
   c) 8 × 3 ■ 2 × 12

6. Find the missing number that makes a true statement.
   a) ■ + 7 = 12   b) 9 − ■ = 3
   c) 5 × ■ = 20   d) ■ ÷ 6 = 4

Unit Conversions

Here are some relationships between metric units of length.
Measurements are often converted from one unit to another.
$3 \text{ m} = 3 \times 100 \text{ cm}$
$\quad\quad\ = 300 \text{ cm}$

I am converting from metres to centimetres. I am converting to a smaller unit, so multiply.

$5000 \text{ m} = \dfrac{5000}{1000} \text{ km}$
$\quad\quad\quad\ = 5 \text{ km}$

I am converting to a larger unit, so divide.

10 mm = 1 cm
1000 mm = 1 m
100 cm = 1 m
1000 m = 1 km

7. Copy and complete each unit conversion.
  a) 2 km = ■ m
  b) 2000 mm = ■ m
  c) 30 cm = ■ mm

8. Convert.
  a) 200 m to kilometres
  b) 550 cm to metres
  c) 0.5 m to millimetres

Area of a Square

You can find the area of a square by multiplying the side length by itself.

$A = s \times s$
$A = 3 \times 3$
$A = 9$
The area is 9 m².

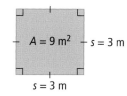

9. Find the area of the square.

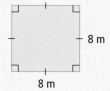

10. Find the area of the square.

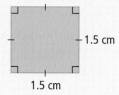

Volume of a Cube

You can find the volume of a cube by multiplying three edge lengths.

$V = l \times l \times l$
$V = 2 \times 2 \times 2$
$V = 8$
The volume is 8 cm³.

11. Find the volume of the cube.

12. Find the volume of the cube.

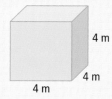

7.1 Understand Exponents

Focus on...
- squares
- cubes
- exponents
- exponential form

How are the designs of these square flags related?

Switzerland

International Red Cross

Discover the Math

Materials
- ruler
- centimetre grid paper or square tiles
- centimetre cubes

square number
- the product of two equal factors, for example, $3 \times 3 = 9$
- represents the area of a square

cubic number
- the product of three equal factors, for example, $2 \times 2 \times 2 = 8$
- represents the volume of a cube

How can you represent squares and cubes?

1. The diagrams represent the first three **square numbers** on a grid.
   a) Use centimetre grid paper or square tiles to continue the pattern.
   b) How is each diagram related to its number sentence?

$1 \times 1 = 1 \quad 2 \times 2 = 4 \quad 3 \times 3 = 9$

2. The diagrams show models that represent the first three **cubic numbers**.
   a) Use centimetre cubes or diagrams to represent the next cubic number.
   b) How is each diagram related to its number sentence?

$1 \times 1 \times 1 = 1 \quad 2 \times 2 \times 2 = 8 \quad 3 \times 3 \times 3 = 27$

3. Which part of a model of a cubic number represents a square number? Explain.

4. **Reflect** Describe a square number and a cubic number.

Example 1: Evaluate Squares

Find the area of a square with each side length.
a) 7
b) 3.1

Solution

a) $A = s^2$
$A = 7^2$
$A = 7 \times 7$
$A = 49$
The area is 49 square units.

Literacy Connections

Reading Squares
You can write expressions like 3×3 as squares.
$3 \times 3 = 3^2$
You can read 3^2 as "three squared."

210 MHR • Chapter 7

b) Use a calculator.
$A = s^2$
$A = 3.1^2$
$A = 9.61$

[c] 3.1 [x²] 9.61 or [c] 3.1 [×] 3.1 [=] 9.61
The area is 9.61 square units.

Example 2: Evaluate Cubes

Find the volume of a cube with each edge length.
a) 6   **b)** 15

Solution
a) $V = l^3$
$V = 6^3$
$V = 6 \times 6 \times 6$
$V = 36 \times 6$
$V = 216$
The volume is 216 cubic units.

Literacy Connections

Reading Cubes
You can write expressions like $4 \times 4 \times 4$ as cubes.

$4 \times 4 \times 4 = 4^3$

You can read 4^3 as "four cubed."

b) Use a calculator.
$V = l^3$
$V = 15^3$
$V = 3375$

Estimate: $15 \times 15 \doteq 10 \times 20 = 200$
$200 \times 15 = 3000$

[c] 15 [yˣ] 3 [=] 3375. or [c] 15 [×] 15 [×] 15 [=] 3375.
The volume is 3375 cubic units.

A number in the form 4^3 is in **exponential form**.
4^3 is the product of three 4s.
$4^3 = \underbrace{4 \times 4 \times 4}_{3 \text{ factors}}$

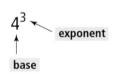

exponential form
- a shorter method for writing numbers expressed as repeated multiplication

$\underset{\text{repeated multiplication}}{4 \times 4 \times 4} = \underset{\text{exponential form}}{4^3}$

base
- the factor you multiply

exponent
- the number of factors you multiply

Example 3: Use Exponential Form

Write each repeated multiplication in exponential form.
a) 18×18 **b)** $11 \times 11 \times 11$

Solution
a) $18 \times 18 = 18^2$
There are two 18s, so the exponent is 2.

b) $11 \times 11 \times 11 = 11^3$
There are three 11s, so the exponent is 3.

Key Ideas

- Exponents represent repeated multiplication. For example, $7 \times 7 = 7^2$. The exponent is 2.

- A square is the product of two equal factors.

  $A = 5 \times 5$
  $A = 5^2$
  $A = 25$

- A cube is the product of three equal factors.

  $V = 2 \times 2 \times 2$
  $V = 2^3$
  $V = 8$

Communicate the Ideas

1. Exponents can be thought of as a form of shorthand. Explain why.
2. Use exponents to represent the area of this square.
3. Explain why 2^3 can be named "two cubed."

Check Your Understanding

Practise

For help with questions 4 and 5, refer to Example 1.

4. Find the area of each square.

   a)   b)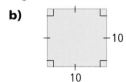

5. Find the area of a square with each side length.
   a) 6
   b) 12
   c) 11

For help with questions 6 and 7, refer to Example 2.

6. Find the volume of each cube.

   a)   b)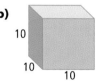

7. Find the volume of a cube with each edge length.
   a) 5
   b) 12
   c) 20

For help with questions 8 and 9, refer to Example 3.

8. Write each repeated multiplication in exponential form.
   a) 13×13
   b) $25 \times 25 \times 25$

9. Write as a repeated multiplication.
   a) 9^2
   b) 7^3
   c) 12^3

Apply

10. Rewrite each statement. Replace ■ with >, <, or = to make a true statement.
    a) 2^3 ■ 3^3
    b) 4^3 ■ 4^2
    c) 5^3 ■ 10^2
    d) 1^2 ■ 1^3

11. Evaluate using a calculator.
    a) 1.3^2
    b) 2.4^2
    c) 4.1^2
    d) 1.2^3
    e) 3.2^3
    f) 2.5^3

12. Does $2^3 = 2 \times 3$? Explain.

13. Write these expressions in order from the greatest value to the least value.
    $10 \times 10 \times 10 \quad 20^2 \quad 8^3 \quad 25 \times 25$

14. The pyramid of Menkaure in Egypt has a square base with a side length of 105 m. What is the area of the base?

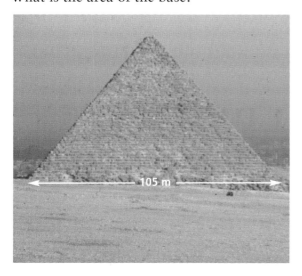

15. A square with a side length of 1 cm has an area of 1 cm². Explain why the unit of area includes the exponent 2.

16. The total length of all the edges on a cube is 36 cm. What is the volume of the cube?

17. The perimeter of this figure is 48 cm. What is its area?

18. a) Use diagrams to show why 16 and 4 are square numbers.
    b) The product of 8×2 is 16. The quotient of $8 \div 2$ is 4. Find three other pairs of numbers whose product and quotient are both square numbers.
    c) What is the relationship between the numbers in each pair?

Extend

19. Rama used his calculator to explore some patterns using exponents. Copy and complete the first three rows of each pattern. Then, write the next three rows. Describe each pattern in words.
    a) $1^3 = ■^2$
       $1^3 + 2^3 = ■^2$
       $1^3 + 2^3 + 3^3 = ■^2$
    b) $3^2 - 1^2 = ■^3$
       $6^2 - 3^2 = ■^3$
       $10^2 - 6^2 = ■^3$

20. The diagrams represent the first four triangular numbers.

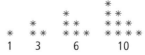

a) Describe a relationship between triangular numbers and square numbers.
b) Use diagrams to explain the relationship.

7.2 Represent and Evaluate Square Roots

Focus on...
- square roots
- perfect squares

A checkerboard is a square. The playing surface is divided into smaller squares. How many small squares are along each side? How many small squares are there altogether?

Notice that 8 is the side length of a square with an area of 64.
$8 \times 8 = 64$
The two factors, 8 and 8, are equal.
8 is the **square root** of 64.

square root (of a number)
- a factor that multiplies by itself to give that number

Materials
- ruler
- grid paper

Discover the Math

How can you represent and evaluate square roots?

1. Cheryl and Amar used grid paper to estimate the value of 3.5^2.

   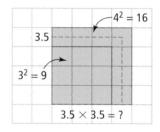

   a) Describe what each of them did.
   b) How does the diagram show that the square root of 12 is between 3 and 4?

I could count squares to show that 3.5^2 is about 12.

It must be between 9 and 16.

2. The diagram gave Amar an idea. He said, "I can use grid paper to show that the square root of 30 is between 5 and 6." Describe what he might have done.

3. **Reflect** Describe how you could use grid paper to
   a) estimate 4.5^2
   b) show that the square root of 42 is between 6 and 7

Example 1: Evaluate Square Roots

Find the side length of a square with the given area.
a) 81 cm² **b)** 11.56 m²

Solution
The side length is the square root of the area.
a) Find a factor that multiplies by itself to give 81.
$9 \times 9 = 81$
So, $\sqrt{81} = 9$
The side length is 9 cm.

b) Find a factor that multiplies by itself to give 11.56.
$3 \times 3 = 9$ **Too low.**
$4 \times 4 = 16$ **Too high.**
$\sqrt{11.56}$ must be between 3 and 4.
Use a calculator. ⓒ 11.56 √ 3.4
The side length is 3.4 m.

Example 2: Identify Perfect Squares

Decide if each number is a **perfect square**.
a) 121 **b)** 18

Solution
a) Try the natural number 10.
$10 \times 10 = 100$ **Too low.**
Try 11.
$11 \times 11 = 121$ **Correct!**
So, $\sqrt{121} = 11$
The square root of 121 is a natural number, 11.
So, 121 is a perfect square.

Strategies
Use systematic trial

b) Try the natural number 4.
$4 \times 4 = 16$ **Too low.**
Try 5.
$5 \times 5 = 25$ **Too high.**
Since 18 is between 16 and 25, the square root of 18 is between 4 and 5.
The square root of 18 is not a natural number.
So, 18 is not a perfect square.

Check using a calculator.
ⓒ 18 √ 4.242640687

Literacy Connections

Reading Square Roots
The $\sqrt{\ }$ symbol indicates the square root of a number.

Read $\sqrt{81} = 9$ as "the square root of 81 equals 9."

Technology Tip
- Key sequences vary. On some calculators, you need to enter ⓒ 11.56 √. On other calculators, you need to enter ⓒ √ 11.56 =.

perfect square
- a number whose square root is a natural number
- example is 4

Key Ideas

- The side length of a square represents the square root of a number.
- A perfect square is a number whose square root is a natural number.
- The $\sqrt{}$ symbol indicates the square root of a number.

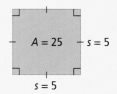

Communicate the Ideas

1. How does the diagram show the square root of 16?
2. How could you use grid paper, tiles, or blocks to show that $\sqrt{36} = 6$?
3. Decide if 49 is a perfect square. Show how you know.
4. Is it possible to find $\sqrt{6.25}$? Explain and justify your answer.

Check Your Understanding

Practise

5. State the side length of each square.

   a)   b)

For help with questions 6 and 7, refer to Example 1.

6. Find the side length of a square with the given area.
   a) 25 m^2 b) 49 cm^2
   c) 100 km^2 d) 9 m^2

7. Use a calculator to find the side length of a square with the given area.
   a) 1.69 cm^2 b) 12.25 m^2
   c) 0.04 mm^2 d) 1.96 cm^2

For help with question 8, refer to Example 2.

8. Decide if each number is a perfect square. Show how you know.
   a) 16   b) 24
   c) 58   d) 225

9. Evaluate.
   a) $\sqrt{64}$
   b) $\sqrt{144}$
   c) $\sqrt{400}$

10. Use a calculator to evaluate.
    a) $\sqrt{625}$
    b) $\sqrt{441}$
    c) $\sqrt{10\,000}$

11. Evaluate.
    a) $\sqrt{1.44}$ b) $\sqrt{2.25}$
    c) $\sqrt{5.76}$ d) $\sqrt{0.25}$

Apply

12. How does the game board show that $\sqrt{100} = 10$?

13. a) The square root of 81 is 9. How can you use your calculator to find this? Write your steps.
    b) Ask a classmate to check your steps.

14. On a baseball diamond, the first-base bag is a square with an area of about 1444 cm². What is the length of one side of the bag?

15. Find the perimeter of a square with each area.
    a) 9 m²
    b) 144 cm²

16. Each face of a cube has an area of 36 cm². What is the volume of the cube?

17. Which of the following square roots are between 6 and 7? Explain how you know.
    $\sqrt{52}$ $\sqrt{41}$ $\sqrt{35}$ $\sqrt{38}$ $\sqrt{45}$

18. The playing surface on a checkerboard has an area of 576 cm². What is the side length of each small square on the board?

19. A square picture has an area of 100 cm². It is centred on a square mat with an area of 324 cm².
    a) Draw a diagram to show the dimensions of the picture and mat.
    b) Identify one item that is approximately the size of the picture and one that is the size of the mat.
    c) How wide is the border around the picture? Justify your response.

Extend

20. Tamar is building a fence around her deck. The deck is a square with an area of 25 m². It costs $120 to build each metre of wood fencing. How much will Tamar spend?

21. Since $2^3 = 8$, 2 is known as the "cube root" of 8. Find the cube root of each of these numbers. Explain how you found your answers.
    a) 27   b) 125   c) 1 000 000

Making Connections

What does math have to do with recreation?

Some sports, such as judo, use squares on their playing surfaces. The diagram shows the square shapes on a judo mat.

Use your personal experiences or your research skills to find another sport that uses a square on its playing surface. Find the side length and the area of this square. Use the rules of the sport to describe the purpose of the square.

List other recreational activities that involve square surfaces. When might you need to know the perimeter or the area of each square?

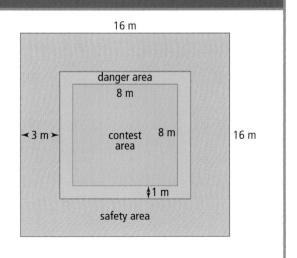

7.3 Understand the Use of Exponents

Focus on...
- exponential form
- powers

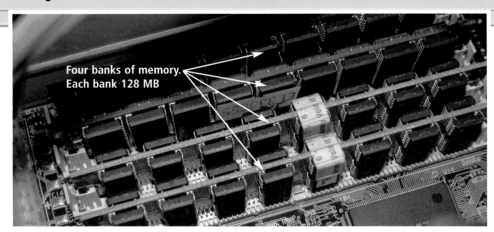

Computer memory can be measured in megabytes. The photograph shows some memory chips in a computer. How many megabytes of memory do the chips contain?

A megabyte can be defined as either a million bytes or as 2^{20} bytes. Do you think these two values are the same? To find out, you need to learn more about exponents.

Discover the Math

Materials
- scientific calculator
- sheet of paper

Optional
- BLM 7.3A Paper Folding Layers Worksheet

How do you write and evaluate numbers in exponential form?

1. Fold a sheet of paper in half. Then, fold it in half again. Count the layers of paper. Copy the table. Complete it as you continue to fold the paper. Notice that each number of layers is written in three ways in the table.

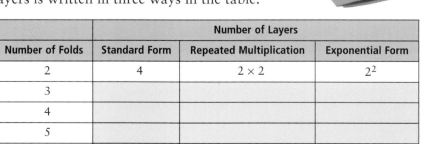

	Number of Layers		
Number of Folds	Standard Form	Repeated Multiplication	Exponential Form
2	4	2 × 2	2^2
3			
4			
5			
6			

2. a) Describe the pattern you see in the table.
   b) Predict the next two rows.

3. Use your calculator to evaluate 2^6. Describe the method you used.

4. Use your calculator to evaluate 2^{20}.

5. Compare the value of 2^{20} with 1 000 000.

6. Do you agree or disagree that 2^{20} bytes should be called a megabyte? Explain and justify your answer.

7. Reflect In step 1, you looked at three ways to represent the number of layers. Which way do you think is easiest to use and why? Are there times when either of the other two ways would have advantages? Explain.

power
- a number in exponential form
- includes a base and an exponent

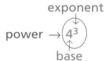

Example 1: Evaluate Powers
Evaluate the **power** 3^4.

Solution
Method 1: Use Paper and Pencil
$3^4 = 3 \times 3 \times 3 \times 3$
$ = 9 \times 3 \times 3$
$ = 27 \times 3$
$ = 81$

I see two sets of 3×3.
$3^4 = (3 \times 3) \times (3 \times 3)$
$ = 9 \times 9$
$ = 81$

Method 2: Use a Calculator
Calculators vary.
Try these methods on your calculator.
[C] 3 [y^x] 4 [=] 81.
or
[C] 3 [×] [=] [=] [=] 81.
or
[C] 3 [×] 3 [×] 3 [×] 3 [=] 81.

Literacy Connections

Reading Powers
You can name powers in words by describing the base and the exponent. These are some common ways to name 3^4:
- three to the fourth
- three to the exponent four
- the fourth power of three

Technology Tip
- On some calculators, the power key will appear as [y^x]. On others, the power key may appear as [a^x] or [x^y].

Example 2: Write Powers

Write each expression as a power. Then, name it in words.
a) $6 \times 6 \times 6 \times 6 \times 6$
b) $2 \times 2 \times 2 \times 2 \times 2 \times 2 \times 2$

Solution

a) $6 \times 6 \times 6 \times 6 \times 6 = 6^5$
The power is six to the fifth.

> You could also say six to the exponent five or the fifth power of six.

b) $2 \times 2 \times 2 \times 2 \times 2 \times 2 \times 2 = 2^7$
The power is two to the seventh.

> You could also say two to the exponent seven or the seventh power of two.

Powers can be described according to the value of the base.
For example, 2^2, 2^3, and 2^4 are powers of two.
They represent the repeated multiplication of 2 by itself.

Similarly, 3^2, 3^5, and 3^9 are powers of three.
They represent the repeated multiplication of 3 by itself.

Key Ideas

- Repeated multiplication can be represented using exponents.
  $7 \times 7 \times 7 \times 7 = 7^4$
- A number written in exponential form is called a power.
- A power includes a base and an exponent.

  power → 4^3 (exponent on the 3, base on the 4)

Communicate the Ideas

1. Describe how you would evaluate 6^4
   a) using a calculator
   b) using paper and pencil

2. Rachel said that 10^3 equals 30. How could you help Rachel correct her error?

3. Show the method you would use to write 625 as a power of 5.

4. David said that 3^4 is a power of 4. Is David correct? Explain.

Check Your Understanding

Practise

5. Identify the base and the exponent of each power.
   a) 2^4 b) 1^6 c) 4^3

For help with questions 6 and 7, refer to Example 1.

6. Evaluate.
   a) 4^1 b) 3^3 c) 5^4

7. Evaluate.
   a) 2^5 b) 1^7 c) 6^3

For help with questions 8 and 9, refer to Example 2.

8. Write each expression as a power.
   a) $5 \times 5 \times 5 \times 5 \times 5$
   b) $11 \times 11 \times 11 \times 11 \times 11 \times 11 \times 11$
   c) $100 \times 100 \times 100 \times 100 \times 100 \times 100$

9. Write each expression as a power.
   a) $4 \times 4 \times 4 \times 4 \times 4 \times 4$
   b) $9 \times 9 \times 9$
   c) $2 \times 2 \times 2 \times 2 \times 2 \times 2 \times 2 \times 2$

10. Name each power in words.
    a) 4^5 b) 8^3 c) 2^8

11. Express as a power.
    a) three to the sixth
    b) five to the exponent four
    c) seventh power of nine

12. Rewrite each statement. Replace ■ with >, <, or = to make a true statement.
    a) 3^3 ■ 2^5
    b) 6^4 ■ 10^3
    c) 8^1 ■ 2^3
    d) 5^4 ■ 4^5

13. Use a calculator to evaluate.
    a) 0.2^5
    b) 1.4^4
    c) 0.3^6

Apply

14. The metric system of measurement is based on powers of 10.
    a) The metric prefix *kilo-* means 1000. Write this number as a power of 10.
    b) The metric prefix *mega-* means 1 000 000. Write this number as a power of 10.

> **Making Connections**
>
> In grade 8, you will learn more about powers of 10 and about their importance in math and science.

15. Express as a power of 4.
    a) 64
    b) 256
    c) 4

16. Find the unknown number in each equation.
    a) 3^5 = ■
    b) $2^■$ = 64
    c) ■^{4} = 4096

17. Write the following expressions in order from the greatest value to the least value.
    2^4 17   19.5 $\sqrt{225}$ 1^{18} 18^1

18. Express the total number of arms on 16 octopi in exponential form. (Note: Octopi is the plural of octopus.)

19. A *googol* is a very large number defined as 10^{100}. Scientists think that this is greater than the number of atoms in the universe.
   a) Describe how you would find the number of zeros needed to write a googol in standard form.
   b) How many zeros are needed?

20. Different values of the megabyte are used in the computer industry. Go to **www.mcgrawhill.ca/links/math7** and follow the links to some sites with information on megabytes. Describe how different values of the megabyte are used in the computer industry.

Did You Know?

Mers Kutt invented the world's first personal computer in Ontario in 1973. It was called the MCM-70 Microcomputer. It had only 2 to 8 kilobytes of random access memory (RAM) and 14 kilobytes of read-only memory (ROM).

21. Find two whole numbers between 10 and 1000 that can be written as a power of 3 and as a power of 9.

Chapter Problem

22. A pond is polluted with fertilizer. Algae are growing on the water. The area of the algae on the pond is 1 m². The area of the algae doubles every week.
   a) What will the area of the algae be after 1 week, 2 weeks, and 3 weeks? Express each area as a power of 2.
   b) The pond has an area of 32 m². How long will it take for the pond to be completely covered in algae?

23. a) Express 16 as a power of 2.
   b) Express 16 as a power of 4.
   c) Express 16 as a power of 16.
   d) Find a whole number greater than 16 that can be expressed as a power of 2, a power of 4, and a power of 16. Write the three powers.

Extend

24. Large numbers are sometimes estimated to the nearest power of 10. For example, 8500 is about 10 000, or 10^4. Estimate each of the following to the nearest power of 10.

a) the number of people in the world
b) the height of the CN Tower, in centimetres
c) the length of the Trans-Canada Highway, in metres
d) your age, in minutes

Did You Know?

- The world's population was about 6.4 billion people at the end of 2003. What is it now?
- The CN Tower is 553 m high.
- The Trans-Canada Highway is the world's longest national highway. Its length is 7821 km.

Making Connections

Modelling Changes in Powers

Materials
- square blocks or tiles

1. a) Use tiles or blocks to model 1^2, 2^2, 3^2, 4^2, and 5^2.

Rearrange the tiles for each number into a straight line. Put the five lines of tiles next to each other. The diagram shows how to place the first two lines.

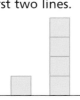

b) Describe the way that changing the base changes the size of the power.
c) Suppose you changed the exponent from 2 to 1. Would you see the same kind of change in the size of the power? Try it and describe what you see.

2. a) Use lines of tiles to model 2^1, 2^2, 2^3, 2^4, and 2^5. Put the lines of tiles next to each other.
b) Describe the way that changing the exponent changes the size of the power.
c) Suppose you changed the base from 2 to 1. Would you see the same kind of change in the size of the power? Try it and describe what you see.

7.4 Fermi Problems

Focus on...
- problem solving
- estimation

How many math textbooks would cover a football field?
How many tennis balls would fill your classroom?

Estimation problems like these are known as Fermi problems. They are named after a famous scientist, Enrico Fermi (1901–1954). He liked to ask his students to solve them.

Fermi problems may seem impossible at first. Different people may estimate different answers. As long as the answer makes sense, the process you follow is more important than your answer.

Discover the Math

How can you solve Fermi problems?

Example: Toonies

How many toonies are needed to cover the floor of a classroom?

Solution

Method 1: Use the Areas

Find out how many small objects will cover a large area.

1. Estimate the area of the toonie.
2. Find the area of the classroom floor.
3. Divide the area of the toonie into the area of the classroom floor.

I can do this in three steps.

Do It!

Strategies
Find needed information

1. Estimate the area of the toonie.

A toonie is about 2.8 cm across.

Strategies
Make an assumption

A toonie roughly covers a square with a side length of 2.8 cm.

Literacy Connections

Reading ≐
The symbol ≐ means "is approximately equal to."

Approximate area of one toonie = 2.8 cm × 2.8 cm
≐ 8 cm²

2. Find the area of the classroom floor.

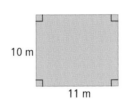

I measured the dimensions of my classroom floor. It is an 11 m by 10 m rectangle.

Convert the side lengths of the classroom floor to centimetres.

11 m = 11 × 100 cm   10 m = 10 × 100 cm
   = 1100 cm   = 1000 cm

The toonie is measured in centimetres. The classroom floor is measured in metres. To divide the areas, the units must be the same.

Area of floor = 1100 cm × 1000 cm
             = 1 100 000 cm²

3. Divide the area of the toonie into the area of the classroom floor.

Number of toonies needed = $\frac{1\ 100\ 000}{8}$

= 137 500   1100000 ÷ 8 = 137500.

About 137 500 toonies are needed to cover the floor of this classroom.

Look Back

Can you think of another way to solve the problem?
After you think about this, look at Method 2 on the next page.

Strategies
Solve a simpler problem

Method 2: Solve a Simpler Problem

I am going to estimate the number of toonies that would cover a 1 m by 1 m square.

The side length of the square is 1 m.
The diameter of a toonie is about 2.8 cm.

My measurements need to have the same units.

Change the side length of the square to centimetres.
1 m = 100 cm

Find the number of toonies that will fit along each side of the square.
100 ÷ 2.8 ≐ 36

The side length of the square is 100 cm. The diameter of the toonie is 2.8 cm. Now I can divide to find the number of toonies that will fit along one side of the square.

Find the total number of toonies inside the square.
36 × 36 = 1296
       ≐ 1300

I need to multiply the number of toonies along two sides of the square

The area of the classroom floor is
11 m × 10 m = 110 m²

This area is 110 times the area of the 1 m by 1 m square.

Find the number of toonies needed to cover the classroom floor.
110 × 1300 = 143 000

About 1300 toonies cover 1 m². The classroom floor covers about 110 m². I can multiply to find out how many toonies can fit on the classroom floor.

About 143 000 toonies are needed to cover the floor of this classroom.

Did You Know?

The value of the toonies on the floor of this classroom would be about $280 000!

The answers to Method 1 and Method 2 are both very large. They differ by only 5500. Both numbers round to 140 000, so both answers seem reasonable.

Key Ideas

- Fermi problems are estimation problems that may involve large numbers.
- To solve Fermi problems, you need to research missing information and make reasonable assumptions.
- A Fermi problem can have several solutions.

Communicate the Ideas

1. Fermi problems can have many "right answers." Explain why.

2. Suppose two students get two different answers to a Fermi problem. How might the students decide if one answer is more reasonable than the other?

Check Your Understanding

Apply

In each question, explain and justify the process that you used to get the result.

For help with question 3, refer to the Example.

3. How many loonies are needed to cover the floor of *your* classroom?

4. Estimate the number of table tennis balls that would fill a suitcase. Hint: Assume that the volume of the ball approximates the volume of a cube.

5. How many words are in this textbook?

6. Estimate the number of CD cases needed to cover the walls of your classroom.

7. How many math textbooks would it take to cover a football field?

8. How many tennis balls would fill your classroom?

9. How many bananas would fill a garbage bag?

10. How many of your footprints would cover the floor of your classroom? Solve this problem in three different ways.

   a) Assume that your foot approximates a rectangle. Calculate the area it covers. Complete the solution.

   b) Use a different method to find the area covered by your foot. Solve the problem again. Compare your solution with the one from part a). Is one solution better than the other? Explain.

   c) Use a method that does not involve calculating the area of your foot.

Extend

11. a) Write a Fermi problem of your own. Make sure that you can solve it.

    b) Have a classmate solve your problem.

    c) Compare the two solutions. If they are different, decide which method you prefer. Explain why. Which method do you think is more accurate?

CHAPTER 7 Review

Key Words

Unscramble the letters for each puzzle. Use the clues to help you solve the puzzles.

1. E P R O W
   a number in exponential form

2. F S R Q R E U C E A T E P
   a number whose square root is a natural number (2 words)

3. A S E B
   the factor you multiply

4. P O X N T E N E
   the number of factors you multiply

5. R O E S U T Q R O A
   a factor that multiplies by itself to give a number (2 words)

6. E O A P E X R M N O N T L F I
   a shorter method for writing numbers expressed as repeated multiplication (2 words)

7.1 Understand Exponents, pages 210–213

7. Find the area of each square.

   a)
   b)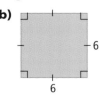

8. Find the volume of each cube.

   a)
   b)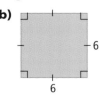

9. Use a calculator to evaluate.
   a) 16^2
   b) 1.3^2
   c) 8^3
   d) 11^3

10. Write each number in exponential form.
    a) 29×29
    b) $14 \times 14 \times 14$

11. Decide which number is greater in each pair.
    a) 3^3 or 5^2
    b) 14^2 or 6^3
    c) 3.2^2 or 2.2^3

12. a) Evaluate 11^2, 111^2, and 1111^2.
    b) Predict the values of $11\,111^2$ and $111\,111^2$.

13. A square pizza with a side length of 30 cm is cut into 9 equal square pieces.
    a) Draw a diagram to show this situation.
    b) What is the area of the top of each piece?

14. Suppose that 1 mL of paint covers an area of 10 cm². How much paint would you need to cover the outside of a wooden cube with an edge length of 5 cm?

7.2 Represent and Evaluate Square Roots, pages 214–217

15. Explain why 9 is a perfect square.

16. Evaluate.
    a) $\sqrt{16}$
    b) $\sqrt{169}$
    c) $\sqrt{6.25}$

17. The playing surface of a Snakes and Ladders game board has 100 small squares on it. Each small square has an area of 9 cm². What is the perimeter of the playing surface?

18. A square picture is glued to a square piece of construction paper. The construction paper is 4 times the area of the square picture mounted on it. The construction paper has a side length of 30 cm. What is the side length of the picture?

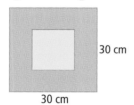

19. The first four diagrams in a pattern are shown. Each square has an area of 4 cm². Find the perimeter of the 30th diagram in the pattern.

7.3 Understand the Use of Exponents, pages 218–223

20. Write each number as a power. Do not evaluate.
  a) $8 \times 8 \times 8 \times 8 \times 8 \times 8 \times 8$
  b) $3.6 \times 3.6 \times 3.6 \times 3.6$

21. Write each power.
  a) 625 as a power of 5
  b) a billion as a power of 10
  c) 243 as a power of 3

22. Write the total number of legs on 27 tripods in exponential form.

23. Evaluate.
  a) 4^5
  b) 1^{12}
  c) 0.1^4

24. Write the following expressions in order from the least value to the greatest value.
$\sqrt{900}$ 30.5 3^3 29 $2 \times 2 \times 2 \times 2 \times 2$

25. Suppose that a bacterium takes 1 h to split into two bacteria. If there is 1 bacterium at the start, how many bacteria are there after 8 h?

26. Is there a number between 100 and 1000 that can be written as a power of 6 and as a power of 36? Explain.

7.4 Fermi Problems, pages 224–227

In each solution, explain and justify the process that you used to get the result.

27. How many telephone books would cover a basketball court?

28. Estimate the number of sheets of paper in a stack with a height of 1 km.

29. a) Estimate the number of coins you receive in change in a year.
  b) Estimate the volume of all those coins.

Chapter 7 Practice Test

Strand Questions	NSN 1–19	MEA	GSS	PA 5, 13, 18	DMP

Multiple Choice

For questions 1 to 5, choose the best answer.

1. The number 5^2 is expressed
   - **A** in standard form
   - **B** as a repeated multiplication
   - **C** in factored form
   - **D** in exponential form

2. The number 6^3 means
   - **A** 6×3
   - **B** $3 \times 3 \times 3 \times 3 \times 3 \times 3$
   - **C** $6 \times 6 \times 6$
   - **D** both B and C

3. A square has an area of 10 cm². Its side length is
   - **A** 100 cm
   - **B** between 3 cm and 4 cm
   - **C** over 1 cm
   - **D** between 2 cm and 5 cm

4. Which number is not a perfect square?
   - **A** 25
   - **B** 169
   - **C** 114
   - **D** 400

5. The first four numbers in a pattern are 1, 8, 27, and 64. The next number in the pattern is
   - **A** 125
   - **B** 128
   - **C** 216
   - **D** 192

Short Answer

6. Find the area of each square.

   a)   b)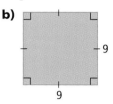

7. Find the volume of each cube.

   a) 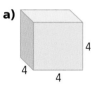   b)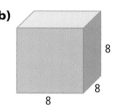

8. Evaluate.
   a) 5^2 b) 10^2
   c) 6^3 d) 7^3

9. Evaluate.
   a) $\sqrt{64}$ b) $\sqrt{400}$
   c) $\sqrt{1.44}$ d) $\sqrt{2.25}$

10. Decide if each number is a perfect square. Show how you know.
    a) 121 b) 47

11. Write as a repeated multiplication. Then, evaluate.
    a) 2^9 b) 3^6 c) 4^5

12. Write the following expressions in order from the least value to the greatest value. Show your reasoning.

    $3 \times 3 \times 3 \times 3 \quad 10^2 \quad \sqrt{6400} \quad 2^6 \quad 79.5$

13. The diagrams show the first four shapes in a pattern. Each cube has an edge length of 2 cm. What is the volume of the 5th shape?

14. A square photograph has an area of 361 cm². Will it fit inside a 30 cm by 18 cm rectangular frame? Explain.

15. a) Norah said, "2⁶ is greater than 2⁵, because 6 is greater than 5." Do you agree? Explain.
    b) Paul said, "1⁶ is greater than 1⁵, because 6 is greater than 5." Do you agree? Explain.

16. Express the number of pennies in $100 as a power of 100.

17. A 4 m by 3 m rectangular floor is covered by 300 square floor tiles. What is the side length of each tile?

18. a) Evaluate 9², 99², 999², and 9999².
    b) Predict the values of 99 999² and 999 999².

Extended Response

19. Describe how you would solve one of the following problems. In your description, list the missing information you would need and the assumptions you would make.
    a) How many years would it take to walk the distance from Earth to the Moon?
    b) How many cell phones would fill a backpack?
    c) How many litres of soft drinks do the students in your school drink in a year?

Chapter Problem Wrap-Up

In question 22, on page 222, you solved a problem involving algae on a pond. The following questions refer to three different bodies of water. Each one is polluted. In each case, the area of the algae doubles every week.

1. The area of the algae on a lake is 5 m². What will the area of the algae be after 4 weeks?
2. Algae covered an area of 2 m² in a square water-storage tank. Three weeks later the tank was completely covered. What was the side length of the tank?
3. It took 6 weeks for a pond to be completely covered by algae. What percent of the pond was covered after 4 weeks? Explain and justify your solution.

Geometry and Spatial Sense
- Identify, describe, compare, and classify geometric figures.
- Identify, draw, and construct three-dimensional geometric figures from nets.
- Use mathematical language effectively to describe geometric concepts, reasoning, and investigations.
- Recognize and sketch views of three-dimensional objects.

Measurement
- Develop and use the formula for finding the surface area and the volume of a rectangular prism.
- Understand the relationship between the dimensions and the volume of a rectangular prism.

Key Words

polyhedron

prism

pyramid

net

surface area

volume

CHAPTER 8

Three-Dimensional Geometry and Measurement

People often need to figure out how many items will fit in a given space. For example:

- When people move house, they need to pack their belongings carefully to avoid breakage and to reduce shipping costs.
- Store managers must determine how large a truck they need to deliver large items.
- You make choices about how to store your possessions. You may have collections of items that you arrange and store in special boxes.

Whether you are mailing presents to a friend or packing clothes to send to an earthquake region, an understanding of three-dimensional geometry will help you plan and use space well.

Chapter Problem

The packaging and transportation of consumer goods is a complex industry. It starts with the design of individual packages and ends with huge container loads of the product.

In this chapter you will consider some aspects of packaging and shipping orange juice.

Get Ready

Classify and Draw Polygons

A **polygon** is a closed shape whose sides are line segments. Polygons are classified by the number of sides they have.

In a **regular polygon**, the sides are all equal. Grid paper or triangle dot paper is helpful in drawing some regular polygons.

Name of Polygon	Number of Sides
Triangle	3
Quadrilateral	4
Pentagon	5
Hexagon	6
Heptagon	7
Octagon	8

equilateral triangle

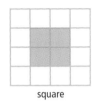
square

regular hexagon

1. Draw each polygon on triangle dot paper.
   a) an equilateral triangle with side length 3 units
   b) a regular hexagon with side length 2 units

2. Draw each polygon on grid paper.
   a) a square with side length 3 units
   b) a rectangle with length 4 units and width 3 units

Draw Three-Dimensional Objects

Artists use a variety of techniques to show three-dimensional objects on two-dimensional paper. The diagrams show how to draw a cube, with sides measuring 1 unit, on centimetre grid paper and on triangle dot paper.

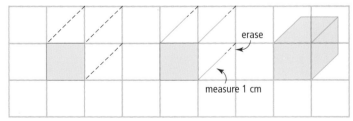

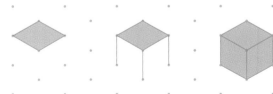

3. Draw a cube, with each side measuring 2 units,
   a) on centimetre grid paper
   b) on triangle dot paper

4. A box is drawn on triangle dot paper. Draw the box on centimetre grid paper.

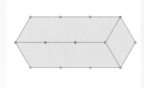

Vocabulary of Three-Dimensional Objects

Mathematicians describe three-dimensional objects in terms of their faces , edges , and vertices .

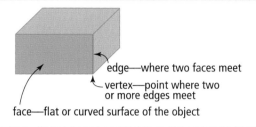

edge—where two faces meet
vertex—point where two or more edges meet
face—flat or curved surface of the object

5. Use your diagram from question 4. Label each part.
   a) Shade in one face.
   b) Outline one edge.
   c) Put a coloured mark on one vertex.

6. Use a real box or visualize the box that you drew in question 4. How many of each does a box have?
   a) faces   b) edges   c) vertices

Area of a Rectangle

Area is the number of square units needed to cover a surface.

The area of the rectangle shown is 6 cm².

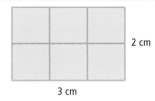

7. What is the formula for calculating the area of a rectangle? Find the area of each rectangle.

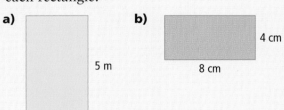

8. Which has the greater area? Show your calculations.
   • a square with dimensions 13 m by 13 m
   • a rectangle with dimensions 14 m by 12 m

9. The area of a rectangle is 64 cm². The length and width are both whole numbers of centimetres. What are the possible dimensions of the rectangle?

Get Ready • MHR   235

8.1 Explore Three-Dimensional Figures

Focus on...
- identifying and classifying geometric figures
- sketching three-dimensional figures

Some three-dimensional objects are made by combining simple shapes. Many three-dimensional objects have faces that are two-dimensional shapes. Look at the photograph. What three-dimensional shapes can you see? What two-dimensional shapes can you identify in each instrument?

Materials
- straws or stir sticks
- modelling clay

Optional
- BLM 8.1A Classifying Three-Dimensional Figures

Literacy Connections

Rectangular Prism
The everyday name for a rectangular prism is a box.

Discover the Math

How do you classify three-dimensional figures?

1. Here are the names of some three-dimensional figures. Look around your classroom. List examples for each type of three-dimensional figure. Check and combine your list with that of a partner.

cube   rectangular prism   triangular prism   square-based pyramid

sphere   cylinder   cone

2. The Egyptian pyramids were built in the shape of a square-based pyramid.

   a) How many faces does a square-based pyramid have, including its base?
   b) What is the shape of each face? Are any of them congruent?
   c) Construct a model of a square-based pyramid. Use straws or stir sticks for the edges and modelling clay to join them together.
   d) How many edges are there in the square-based pyramid? How many vertices?

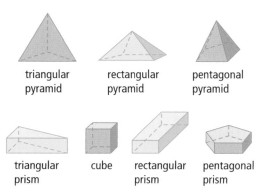

Making Connections

Geometry
Congruent means the same shape and size. You worked with congruent figures in Chapter 2.

3. a) Work in a group. Use the same materials as in step 2 to build a model of each **polyhedron** shown.
   b) List the shape and the number of each type of face for each polyhedron.

triangular pyramid   rectangular pyramid   pentagonal pyramid

triangular prism   cube   rectangular prism   pentagonal prism

polyhedron
- a three-dimensional figure with faces that are polygons
- plural is polyhedra
- *poly* means many, *hedra* means faces

4. **Reflect** Describe how pyramids and prisms are alike. Describe how they are different. What method might be used to name different types of pyramids and prisms?

Example 1: Classify Three-Dimensional Objects

Give the name of the three-dimensional figure that each object most resembles.

a)   b)   c)   d) 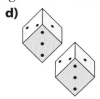   e)   f)

Solution

a) The basketball is a sphere.
b) The bookcase is a rectangular prism.
c) The can of pop is a cylinder.
d) The dice are cubes.
e) The doorstop is a triangular prism.
f) The garden stake is a square-based pyramid.

Example 2: Classify Polyhedra

Examine the figure shown. What type(s) of polygons are its faces? State the number of each type. Name the type of polyhedron.

Solution

There are two faces like this.
This is a hexagon.

There are six faces like this.
This is a rectangle.

The figure is a hexagonal prism.

Example 3: Properties of Polyhedra

A cereal box is a rectangular prism.
a) Name four edges of equal length.
b) Name three pairs of congruent faces.

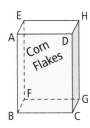

Solution

a) edges: AB = DC = HG = EF

There are two other ways to answer this question.
AD = BC = FG = EH
AE = DH = CG = BF

b) Faces ABCD and EFGH are congruent.
Faces ADHE and BCGF are congruent.
Faces ABFE and DCGH are congruent.

Key Ideas

- Three-dimensional figures can be classified according to their properties.

- A polyhedron is a three-dimensional figure. All faces of a polyhedron are polygons.

- **Prisms** and **pyramids** are named according to the shape of their bases. For example,

square-based prism

triangular pyramid

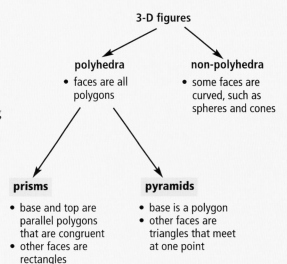

3-D figures
- polyhedra
  - faces are all polygons
- non-polyhedra
  - some faces are curved, such as spheres and cones

prisms
- base and top are parallel polygons that are congruent
- other faces are rectangles

pyramids
- base is a polygon
- other faces are triangles that meet at one point

238 MHR • Chapter 8

Communicate the Ideas

1. Which container is a polyhedron? Explain.

2. Compare the two figures shown. How are they alike? How do they differ? Classify each figure.

Check Your Understanding

Practise

For help with questions 3 to 5, refer to Example 1.

3. Give the name of the three-dimensional figure that each object most resembles.

   a)    b)

   c)    d)

4. Give the name of the three-dimensional figure that each object most resembles.
   a) a paperback book
   b) a baseball
   c) a new, unsharpened, six-sided pencil
   d) a nickel coin

5. Find and describe, or sketch, your own example of an object that has each shape.
   a) a cube
   b) a cylinder
   c) a rectangular prism
   d) a triangular prism
   e) a square-based pyramid
   f) a sphere
   g) a square-based prism

For help with questions 6 and 7, refer to Example 2.

6. Chocolates are packaged in the box shown.

   a) What type(s) of polygons are its faces? Give the number of each type.
   b) Name the type of polyhedron.

7. The jewel in a ring is shaped as shown.

a) Describe the polygon shapes that form its faces. Give the number of each type.
b) Name the polyhedron.

For help with questions 8 and 9, refer to Example 3.

8. A small tent has the shape shown.

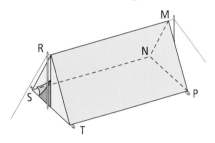

a) Name three edges of equal length.
b) Name one pair of congruent faces.
c) What type of polyhedron is this?

9. An A-frame cottage has the shape shown.

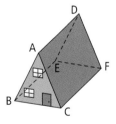

a) Name four edges of equal length.
b) Name two different pairs of congruent faces.
c) What type of polyhedron is this?

10. Use diagrams and words to show how a pyramid and a cone are alike and how they are different.

Apply

11. Give the name of the three-dimensional figure that each fish tank most resembles.

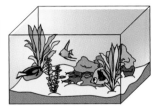

Describe two other objects that are made in the form of different figures.

12. The shape of this house is a pentagonal prism. Look closely and you will see that it can also be divided into two different figures. The attic and roof form a triangular prism. The rooms of the house form a rectangular prism.

Draw two other objects by combining two or more three-dimensional figures. Identify the parts.

13. What three-dimensional figure am I? Sketch and name one object that fits each description.

a) I have more than four faces.
b) I have only two flat faces.
c) I have two triangular faces.
d) All my faces are congruent.
e) I have one continuous surface, with no flat faces.

14. Make up some puzzles like the ones in question 13. Exchange them with another person to solve.

15.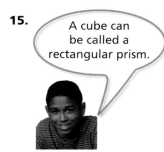

   "A cube can be called a rectangular prism."

   "No, it is a square-based prism."

   Explain why both students are correct.

16. **Try This!** Many office towers are rectangular prisms. Some are not. The photo shows the building at the corner of Front and Wellington Streets in Toronto. Sketch and identify the shape.

Extend

17. Use the Internet to search for famous buildings. Describe the polyhedra used in their design. Go to **www.mcgrawhill.ca/links/math7** and follow the links to some useful Web sites.

18. **a)** How many identical square faces are needed to form a regular polyhedron?
    b) Can any other regular polyhedron be formed with fewer faces using a different two-dimensional figure? Justify your thinking.

Did You Know?

The structure in question 16 is known as a "flatiron" building because the shape resembles an old-fashioned iron. There is a similar flatiron building in New York at Broadway and Fifth Avenue.

Making Connections

Regular Polyhedra

There are only five regular polyhedra. They are sometimes called the Platonic solids after the ancient Greek philosopher Plato. Plato, who lived in the fourth century, linked the shapes to basic elements in science. The motto inscribed over the door of Plato's school in Athens read "Let no one ignorant of geometry enter here."

Name of Regular Polyhedron	Shape of Each Face	Number of Faces	Science Connection
tetrahedron	equilateral triangle	4	fire
cube	square	6	earth
octahedron	equilateral triangle	8	air
icosahedron	equilateral triangle	20	water
dodecahedron	regular pentagon	12	universe

8.2 Sketch Front, Top, and Side Views

Focus on...
- sketching views of three-dimensional figures
- sketching three-dimensional objects from models and drawings

A bird's-eye view usually refers to what a bird sees as it flies overhead. Every object has a front view, a top view, and a side view.

Discover the Math

Materials
- various shaped objects
- centimetre grid paper

How do you draw views of objects?

1. Look around you. Make a list of five small objects. Include a variety of shapes, such as rectangular prisms, cones, and cylinders.

2. Choose one of the objects to examine. Draw the front view of the object as a two-dimensional drawing on grid paper.

3. Draw the top view of the object. What are you doing differently here than in step 1?

4. Draw the side view of the object. Is there more than one way to do this?

5. Repeat steps 2 to 4 for the other four objects on your list.

6. **Reflect** Compare your drawings with those of other students. How many views do you need to be able to identify the object? Explain.

Example 1: Draw Front, Top, and Side Views

Models of two houses are built using linking cubes. Draw the front view, top view, and side view of each house.

a)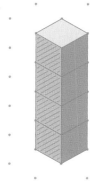

b)

Solution

a)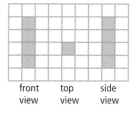

front view   top view   side view

Visualize a stack of four cubes

b)

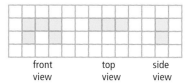

front view   top view   side view

Strategies
Make a model

Example 2: Draw Views of a More Complex Shape

Here is a third house built with linking cubes. Draw its front view, top view, and side view.

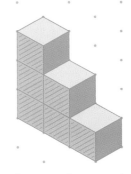

Solution

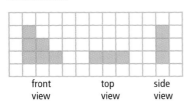

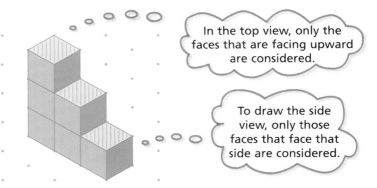

In the top view, only the faces that are facing upward are considered.

To draw the side view, only those faces that face that side are considered.

Key Ideas

- Drawings of the front view, top view, and side view of a three-dimensional figure show how the figure appears from each of these viewpoints.

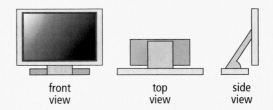

- Front views, top views, and side views are two-dimensional drawings.

Communicate the Ideas

1. A building is made of five cubes. Its front, top, and side views are shown. Describe what the actual building looks like.

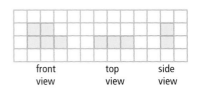

2. The front view, top view, and side view of an object are all identical circles. What is the object?

Check Your Understanding

Practise

For help with questions 3 and 4, refer to Example 1.

3. Draw the front view, top view, and side view for each object. Consider the shaded face as the front.

a) 　　b)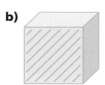

4. Draw the front view, top view, and side view for each object.

a) 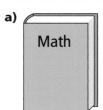　　b)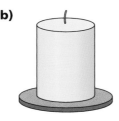

For help with questions 5 to 7, refer to Example 2.

5. Draw the front view, top view, and side view for each object. Consider the shaded face as the front.

a) 　　b)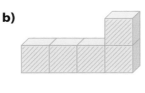

6. Draw the front view, top view, and side view for each object.

a) 　　b)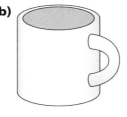

7. a) Draw the front view, top view, and side view of your eraser.
   b) Exchange drawings with a partner. Draw or describe the shape of your partner's eraser.

Apply

8. The diagrams show the front view, top view, and side view of an object. Sketch the three-dimensional object.

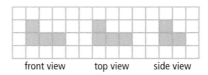
front view　　top view　　side view

9. The diagrams show the front view, top view, and side view of an object. Sketch the three-dimensional object.

front view　top view　side view

10. What kind of figure am I?
   a) My top view is a circle. My front view and my side view are congruent rectangles.
   b) All three views of me are circles.
   c) My front view is a rectangle. My side view is a thinner rectangle. My top view is an equilateral triangle.

> **Did You Know?**
>
> **3-D Movies and Computer Games**
> The first 3-D film was shown in New York on June 10, 1915. To see the three-dimensional effects, the audience wore red and green glasses. Since the 1980s, new technology has allowed much more realistic 3-D effects to be seen using special polarized lenses.
>
> Video games use a technique called adaptive 3-D geometry. Shapes are represented by millions of tiny polygons, giving very realistic illustrations.

11. Work with a partner.

   a) Use six identical cubes to build a three-dimensional figure. Do not let your partner see the model.
   b) Draw the front, top, and side views.
   c) Show your partner only two of the three views. Your partner is to build the figure with another set of six cubes. Does the model match yours?
   d) Give your partner all three views of the model. Does this let your partner duplicate your model? Try it and see.
   e) Is it possible to have a hidden block or empty space not obvious in the drawings? Explain.

12. The side views drawn so far in this section have always been right-side views. Are left-side views the same as right-side views? Give examples to support your answer.

13. Imagine you are a fly on the ceiling of a room at home. Draw a top view of the room, including the top view of the furniture. Label the objects.

Extend

14. A mat plan is another way of representing a three-dimensional building made of cubes. On a mat plan, the top view is shown with a number telling how many cubes are in each position.

   a) This is the mat plan for one of the houses in the Examples at the beginning of this section. Which one is it?

   | 3 | 2 | 1 |

   b) Make a mat plan for this building.

   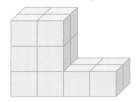

   c) Interpret this mat plan. Draw the front, top, and right-side views of the object.

	2	
3	2	1
3	2	1
	2	

Literacy Connections

Floor Plans

A top view of a room is called a floor plan. Designers use floor plans to help decide where the furniture should go in a room.

To try this, draw the top view of all pieces of furniture in your room. Label each and cut it out.

Draw an outline of the room on grid paper. Include the location of doors and windows. Place the top views of the furniture on your floor plan. Move the top views around until you like the way the furniture is arranged.

Did You Know?

Other Dimensions
Scientists think there may be more than three dimensions. Some consider time to be the fourth dimension. Others have suggested that Earth might be a sort of shadow of a four-dimensional entity, just as your shadow is a 2-D version of your 3-D self.

8.3 Draw and Construct Three-Dimensional Figures Using Nets

Focus on...
- building three-dimensional figures from nets
- sketching three-dimensional figures from nets

Sweat shirts can be made from a single pattern piece with a hole cut out. Only two seams are needed. Each seam closes up one side and sleeve. Why might manufacturers use this method?

Discover the Math

Materials
- small rectangular cardboard boxes

What is a net?

1. Look at a cardboard box.

   a) How many faces are there?
   b) What shapes are the faces?
   c) Are there any congruent faces? If so, which ones?

2. Carefully cut along some of the edges of the box. Cut just enough edges to allow you to open the box and lay it flat on your desk. The resulting two-dimensional shape is called a **net**. Draw your net in your notebook.

net
- a single flat pattern piece
- can be folded to form a three-dimensional object

3. **Reflect** Compare your net with others in the class. What is the same about all of the nets? How do they differ?

Example 1: Draw Nets

A small box of raisins measures 4 cm by 3 cm by 2 cm. Draw two possible nets for the box. Label the faces and measurements on your net.

Solution

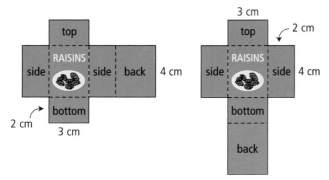

Strategies
What other nets can be used for this box?

Example 2: Sketch Three-Dimensional Objects From Nets

Examine the net. Sketch the three-dimensional object that can be made from it. What type of figure is formed?

Solution

Method 1: Trace and Build the Object

The three-dimensional figure formed is a hexagonal pyramid.

Strategies
Make a model

Method 2: Visualize the Object
The net has one regular hexagon and six congruent triangles. When the triangles are folded up, the three-dimensional figure formed is a hexagonal pyramid.

Literacy Connections

Prisms and pyramids are two classes of three-dimensional figures. A particular prism or pyramid is named according to the shape of its base.

Key Ideas

- A net is a two-dimensional drawing that can be folded up to form a three-dimensional figure. It is a single pattern piece that shows all the faces of the figure.

Communicate the Ideas

1. Will the nets shown make two identical three-dimensional figures? Explain.

2. Here are three possible nets.

   Which net will not form a cube? Justify your choice.

Check Your Understanding

Practise

For help with questions 3 and 4, refer to Example 1.

3. A small juice box measures 11 cm by 5 cm by 4 cm. Draw a net for the box. Show the measurements of each face.

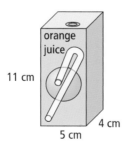

4. Playing cards are sold in a box measuring 6 cm by 9 cm by 2 cm. Draw a net for the box. Show the measurements of each face.

For help with questions 5 to 9, refer to Example 2.

5. Make a larger copy of the net on centimetre grid paper. Cut out the net and fold along the dotted lines. Tape the edges to form a three-dimensional figure. What type of figure is it?

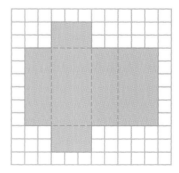

6. Make a larger copy of the net on centimetre grid paper. Cut out the net and fold along the dotted lines. Tape the edges to form a three-dimensional figure. What type of figure is it?

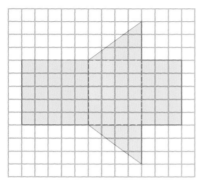

7. a) Sketch the three-dimensional figure that can be made from this net.

b) Name the type of polyhedron.
c) Describe an item at the store that is packaged in this way.

8. a) Sketch the three-dimensional figure that can be made from this net.

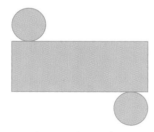

b) What can be packaged using this type of net?

9. Sketch the three-dimensional figure that can be made from this net. What type of figure is formed?

Apply

10. Two nets are shown. What three-dimensional figure will each net form? How are the two figures alike? How are they different?

A

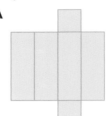

B

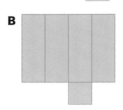

11. Decide which edges of a square-based pyramid you would need to cut to form its net.

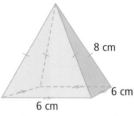

a) Draw a net for the square-based pyramid on centimetre grid paper.
b) Construct a three-dimensional model of the square-based pyramid using your net.

Did You Know?

Most people are familiar with the ancient pyramids in Egypt. Many other cultures built in this shape too. Pyramid structures, or partial remains of them, are found in India, Peru, Mexico, Greece, China, Japan, Tibet, Cambodia, Brazil, Bolivia, and Spain. Perhaps you know of other locations too.

Chapter Problem

12. A single-serving orange juice box measures 10 cm by 7 cm by 5 cm. Draw a net for the box. Print the words ORANGE JUICE on each face except the bottom, so the words will read the correct way when the box is assembled. Add artwork if you like. Mark the position of the hole for the straw.

13. Design a small paper-recycling bin. It is to have a square base and is open at the top. The square top is larger than the base.

a) What shape is each of the side faces?
b) Draw a net for the bin.
c) Cut out the net and tape it together to form the bin.

14. Use a net to construct a model of each pyramid. Sketch three-dimensional drawings of them in your notebook.

a) a triangular pyramid that is made up of four equilateral triangles
b) a triangular pyramid where only the base is an equilateral triangle and the other three faces are congruent isosceles triangles

 15. Design a net for a new rectangular package for crackers. Which net will you use? Sketch it. Explain why you chose this design.

Extend

16. There are 11 different nets for a cube. Try to draw them all. Make sure they are actually different. Some may be the same when you turn or flip them over. How can you check that they are all different?

Making Connections

Gift Box Set

The nets used to form the packaging of some items, such as toys and dishes, are quite complex and may include cut-out holes to hold breakable parts. Most gift boxes are designed to have a top lid that fits over the bottom container. Sometimes the lid is separate, sometimes not.

- Design a special gift box. Think of an item you want to package. Then, decide on the shape of box you will use.
- Draw the net(s) for your gift box. Cut out the net(s) and make the gift box. Test and modify your design until you are pleased with it.

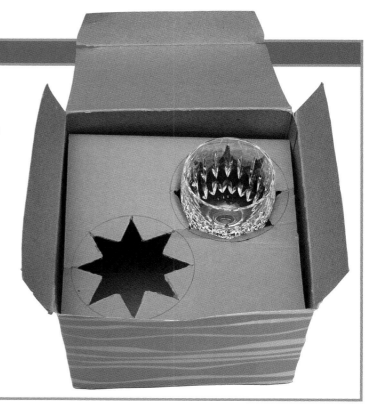

8.4 Surface Area of a Rectangular Prism

Focus on...
- developing the formula for the surface area of a rectangular prism

Essex County, in southwestern Ontario, has the greatest concentration of greenhouse agriculture in all of North America. In the Leamington area, some greenhouses are as large as 400 m by 200 m. A greenhouse allows solar energy to pass through its glass and become heat energy.

Discover the Math

Materials
- centimetre cubes or other identical cubes
- centimetre grid paper

Optional
- BLM 8.4A Surface Area

surface area
- the number of square units needed to cover an object

How can you find the surface area of a rectangular prism?

1. **a)** Make a prism using 12 cubes.
   b) Sketch the prism on grid paper. Label the dimensions.
   c) Draw a net for the prism. Label the dimensions.
   d) Count the number of square faces on the outside of the prism. This represents the **surface area** of the prism. Check the count using the net.

2. Repeat question 1 to obtain data for as many different prisms as you can make using 12 cubes.

3. How many faces are there in a net of any rectangular prism? Which pairs of faces are congruent? How many pairs are there?

4. **Reflect** The surface area of an object is the sum of the areas of its faces. Describe how to find the surface area of a rectangular prism with length l, width w, and height h.

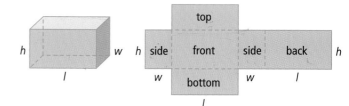

Example 1: Find Surface Area

The figure is made with four centimetre cubes. What is its surface area?

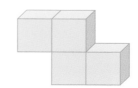

> **Strategies**
> What other method might be used?

Solution

There are many ways to solve this problem. Here are two methods.

Method 1: Build the Figure

I can count the cube faces on the model.

> **Strategies**
> Make a model

There are 18 exposed cube faces. Each face has an area of 1 cm². The surface area of the figure is 18×1 cm², or 18 cm².

Method 2: Count the Number of Cube Faces on Each View
front view and back view: 4 + 4 = 8
top view and bottom view: 3 + 3 = 6
side views: 2 + 2 = 4
Total number of cube faces: 8 + 6 + 4 = 18
Each face has an area of 1 cm².
The total surface area of the figure is 18 cm².

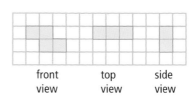

front view   top view   side view

Example 2: Calculate Surface Area

Sarah makes her own gift boxes. She glues wrapping paper onto the outside to make them look nice. One box has length 25 cm, width 20 cm, and height 10 cm. How much paper does she need to cover all six faces of the box?

Solution

Method 1: Find the Area of the Different Faces

Front face:

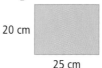

$A = l \times w$
$A = 10 \times 25$
$A = 250$
The back face has the same area as the front face, 250 cm².

Top face:

$A = 20 \times 25$
$A = 500$
The base has the same area as the top face, 500 cm².

Left-side face:

$A = 10 \times 20$
$A = 200$
The right-side face has the same area as the left-side face, 200 cm².

Surface area = front + back + top + base + left side + right side
S.A. = 250 + 250 + 500 + 500 + 200 + 200
= 1900

Sarah will need at least 1900 cm² of wrapping paper to cover the box.

Method 2: Use a Formula
$l = 25, w = 20, h = 10$
S.A. $= 2(l \times w) + 2(l \times h) + 2(w \times h)$
$= 2(25 \times 20) + 2(25 \times 10) + 2(20 \times 10)$
$= 1000 + 500 + 400$
$= 1900$

Sarah will need at least 1900 cm² of wrapping paper to cover the box.

> **Strategies**
> Choose a formula

Key Ideas

- The surface area of a prism is the sum of the areas of all of its faces.
- The formula for the surface area of a rectangular prism is
  S.A. $= 2(l \times w) + 2(l \times h) + 2(w \times h)$
- Surface area is measured in square units.

Communicate the Ideas

1. A rectangular prism and its net are shown.

   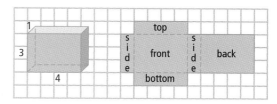

   Jasmine looks at the net. She calculates the surface area of the box by finding the area of each of the six rectangles and then adding these areas. Her work shows

   S.A. = left side + top + front + bottom + right side + back
   = $1 \times 3 + 4 \times 1 + 4 \times 3 + 4 \times 1 + 1 \times 3 + 4 \times 3$

   How can Jasmine simplify her work? Explain.

2. Sianni says she looks at the sketch of the box and uses the formula for surface area. Her work shows
   S.A. = $2(3 \times 4) + 2(4 \times 1) + 2(3 \times 1)$
   Is Sianni right? Explain.

Check Your Understanding

Practise

For help with questions 3 and 4, refer to Example 1.

3. The figure is made with four centimetre cubes. What is its surface area?

   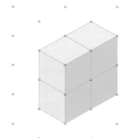

4. The figure is made with five centimetre cubes. What is its surface area?

   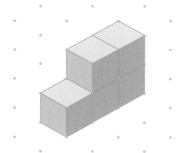

For help with question 5 to 7, refer to Example 2.

5. Find the surface area of the rectangular prism made from each net.

   a)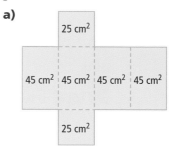

   b)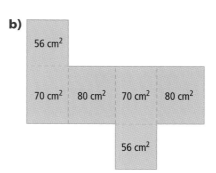

8.4 Surface Area of a Rectangular Prism • MHR   255

6. Calculate the surface area of each rectangular prism.

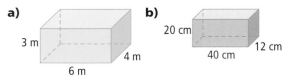

7. Measure the length and width of the front cover of this textbook, to the nearest centimetre. Find the area of cardboard needed to make the front and back covers.

Apply

8. Match each cube-shaped object with its approximate surface area. Three possible areas are given for each. Explain how you reached your conclusions.
   a) an ice cube: 6 cm², 54 cm², 300 cm²
   b) a small room: 24 m², 100 m², 12 m²
   c) one number key on a touch-tone phone: 1 cm², 6 cm², 20 cm²

9. Willem wants to protect his favourite paperback book by covering it with sticky transparent paper. He will cover the front, spine, and back of the book. What area of sticky paper will he need?

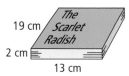

10. Jean plans to build a wooden chest. The length is 65 cm, the width is 45 cm, and the height is 30 cm. The lid has a height of 6 cm and is attached with hinges so it fits perfectly with no overlap. How much wood does Jean need to build the rectangular chest?

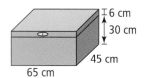

11. Kallil helps in his mother's office. He is asked to wipe away the dust from a filing cabinet. The cabinet measures 0.35 m by 0.40 m by 0.92 m. Its back is against the wall. What is the exposed surface area that Kallil has to wipe?

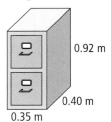

12. Freya wants to paint her bedroom door and the closet door, but she is not sure if she has enough paint. Both doors measure 2.0 m tall, 0.7 m wide, and 0.05 m thick. Freya has one can of paint that says it covers 3.5 m². Does she have enough to paint all sides, except the bottoms, of both doors?

13. For what type of rectangular prism is the area of each face the same? Sketch the prism. Write a simplified formula for its surface area.

Chapter Problem

14. Single-serving orange juice boxes measure 10 cm by 7 cm by 5 cm.
    a) After it is filled, each box is covered with a thin coat of wax. This coating seals the box and helps to preserve the juice. Find the area of each juice box to be coated.
    b) Packs of three boxes are shrink-wrapped for sale. Find the area of plastic needed to wrap three boxes.

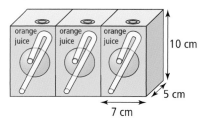

15. Use eight cubes. Can you construct different rectangular prisms that have the same surface area? If so, sketch the figures and draw nets for them. If not, explain why not.

16. George has a cube-shaped box that he wants to gift-wrap. Each edge of the cube measures 30 cm. A local stationery store sells sheets of wrapping paper in three sizes.

A: 60 cm by 60 cm
B: 80 cm by 80 cm
C: 90 cm by 90 cm

Which size should George buy? Explain your choice.

17. Pat built a doghouse. She cut out a square hole in the front for the entrance. Pat wants to paint the outside walls blue. She will not paint the roof or the triangular gable ends. What surface area does she plan to paint?

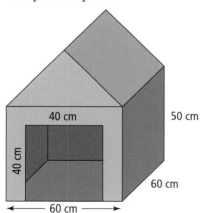

Extend

18. The top and bottom of Kelly's jewellery box both measure 12 cm by 12 cm. The other four faces are congruent to each other. The surface area of the box is 768 cm². What is the height of the jewellery box?

19. A box with dimensions of 20 cm by 10 cm by 10 cm has a 2-cm-wide decorative border along all its edges. What is the surface area not covered by the border?

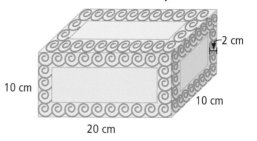

20. Cora bought a CD player, 2 CDs, and a CD holder for her son's birthday.

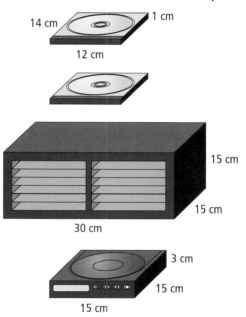

a) Suggest possible sizes for each gift box, if the items are placed into
   i) one box
   ii) two boxes
   iii) three boxes
b) How many boxes would you use? Explain why.

8.5 Volume of a Rectangular Prism

Focus on...
- developing a formula for the volume of a box

The truck is being loaded with supplies for a store. The supplies have been packaged into crates. Each crate measures 1 m by 1 m by 1 m. There are 24 crates to be shipped. The truck box has length 4 m, width 2 m, and height 3 m. How can you find out whether all of the crates will fit in one load?

Discover the Math

Materials
- centimetre cubes

How can you find volume?

1. Consider the truck box to be a rectangular prism with dimensions 4 m by 2 m by 3 m.
   a) Build a model of the truck box using centimetre cubes. Let one centimetre represent one metre.
   b) How many cubes did you use to form the bottom layer of the rectangular prism? What is the area of the floor of the truck box?
   c) How many layers of cubes are there?
   d) How many cubes did you use altogether to build the rectangular prism? This is its **volume**.

2. What if the truck box measured 5 m by 2 m by 3 m? Complete parts a) to d) of step 1 for this truck box.

volume
- the amount of space occupied by an object
- measured in cubic units

258 MHR • Chapter 8

3. What if the truck box measured 6 m by 2 m by 3 m? How many cubes would you need to construct a model of this truck box?

4. Compare the three truck boxes. How is the number of cubes used related to the area of the base (the floor of the truck box) and the height?

5. **Reflect** Describe how to find the volume of a rectangular prism in terms of the area of its base and its height.

Example 1: Calculate Volume

The area of the base of a box of toothpicks is 12 cm². The height of the box is 2 cm. Find the volume of the box.

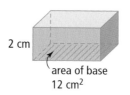

Literacy Connections

Cubic Units
The volume 24 cm³ is read "twenty-four cubic centimetres." The raised 3 refers to cubic units. Cubic units are used for the volume of a three-dimensional space.

Solution

Volume = area of base × height
$V = 12 \times 2$
$V = 24$
The volume of the box of toothpicks is 24 cm³.

Example 2: Solve a Problem Involving Volume

Students in the drama club plan to store costumes in two lockers. They need to know how much space they have. What is the volume of the two lockers shown?

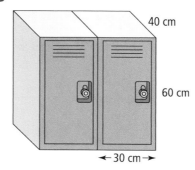

Solution

Find the volume of one locker.
$l = 40$, $w = 30$, $h = 60$
Volume = area of base × height
$V = 40 \times 30 \times 60$
$V = 72\ 000$

The base is a rectangle, 40 cm by 30 cm.

The volume of one locker is 72 000 cm³.

The volume of two lockers is 2 × 72 000 cm³, or 144 000 cm³.

Strategies
Choose a formula

Key Ideas

- Volume is the amount of space occupied by an object.
- The formula for the volume of a rectangular prism is
  Volume = area of base × height
- Volume is measured in cubic units.

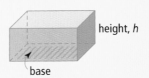

Communicate the Ideas

1. Sugar cubes are packaged in a box. On the bottom layer, there are 5 rows of 10 cubes. There are 500 cubes altogether in the box. How many layers are there? Explain.

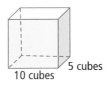

2. Another formula for the volume of a rectangular prism is
$V = l \times w \times h$

Explain why this formula is equivalent to the formula
Volume = area of base × height

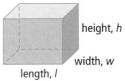

Check Your Understanding

Practise

For help with questions 3 and 4, refer to Example 1.

3. The area of the base and the height of each box are given. Calculate the volume of each box.

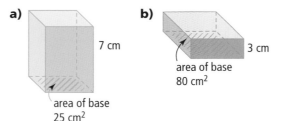

4. A sandbox has a base area of 2.5 m². What volume of sand is needed to fill the sandbox to each depth?

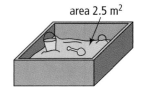

a) 0.4 m   b) 0.5 m
c) 0.24 m

For help with questions 5 to 7, refer to Example 2.

5. Calculate the volume of each rectangular prism.

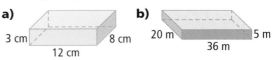

6. Calculate the volume of each rectangular prism.

   a) a box for a tube of toothpaste measures 19 cm by 5 cm by 3 cm
   b) a box of tissues measures 22 cm by 12 cm by 7 cm
   c) a classroom measures 10 m by 8 m by 3 m

Apply

7. What does 1 m^3 look like? Choose the most likely object. Justify your choice.

   A a microwave oven
   B a washing machine
   C a toaster

8. What does 10 cm^3 look like? Choose the most likely object. Justify your choice.

   A a dictionary
   B a lunch box
   C an eraser

9. Ashley is at summer camp and is feeling homesick. She asks her mother to send her old teddy bear. The bear fits snugly into a box measuring 24 cm by 10 cm by 18 cm. What is the volume of the box?

10. Ben decides to take a cooler of drinks with him to the beach. He fills one third of the space in the cooler with ice. The cooler measures 65 cm by 30 cm by 40 cm. What volume is left for his drinks?

11. Sanjay bought a fish tank that has length 40 cm, width 25 cm, and height 30 cm.

    a) What is the volume of the tank, in cubic centimetres?
    b) One litre of water has a volume of 1000 cm^3. How many litres of water are needed to half fill the tank?

Chapter Problem

12. Single-serving orange juice boxes measure 10 cm by 7 cm by 5 cm. Packs of three boxes are shrink-wrapped for sale at the grocery store. Find the volume of juice in a three-pack.

13. Usaf works at a bakery. His job is to carefully pack items for shipping to stores. Five small cakes take up 600 cm^3. How many of these cakes can be put in a crate that measures 120 cm by 80 cm by 50 cm?

14. Jenna is moving to a new apartment. She wants to have all her books shipped so she has decided to buy some storage boxes. There are three different sizes available.

    A: 60 cm by 50 cm by 45 cm
    B: 70 cm by 45 cm by 40 cm
    C: 65 cm by 40 cm by 40 cm

    a) Draw a sketch of each box and label its dimensions.
    b) Find out as much as you can about the boxes. Consider the volume and the surface area of each size.
    c) What are the advantages and the disadvantages of each size?
    d) Which size should Jenna choose to move her books? Give reasons for your answer.

Extend

15. Vijnya bought a picnic basket that is 32 cm long and 20 cm wide. The storekeeper told her that the basket has a volume of 7360 cm^3. Make a sketch of the basket. What is the picnic basket's depth?

16. A certain cube has the same number of cubic units for its volume as the number of square units it has for its surface area. What are the possible dimensions of this cube?

CHAPTER 8 Review

Key Words

For questions 1 to 6, copy the statement and fill in the blanks. Use some of these words: net, polygons, prism, pyramid, rectangles, space, square, surface area, two, three

1. A polyhedron is a ▇-dimensional figure with faces that are ▇.
2. A ▇ is a two-dimensional pattern that can be folded up to form a three-dimensional figure.
3. A ▇ is a polyhedron that has one base and the same number of triangular faces as there are sides on the base.
4. A ▇ is a polyhedron that has two parallel and congruent polygon bases, with rectangular faces connecting them.
5. Volume is the amount of ▇ occupied by an object.
6. The ▇ of an object is the sum of the areas of its faces.

8.1 Explore Three-Dimensional Figures, pages 236–241

7. Sketch and name a polyhedron that fits each description.
   a) one square face and four triangular faces
   b) 12 edges of equal length
   c) six vertices

8. Can you find more than one solution to any of the parts in question 7? Show other solutions where possible.

9. The toy wagon shown consists of different three-dimensional figures.

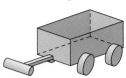

   a) Name the different parts of the wagon and the three-dimensional figure each part most resembles.
   b) Draw two other objects that are combinations of two or more three-dimensional figures.

8.2 Sketch Front, Top, and Side Views, pages 242–246

10. Draw the front view, top view, and side view for each figure.
    a)
    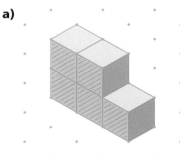

    b)
    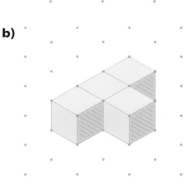

262 MHR • Chapter 8

11. Sketch the three-dimensional figure that has the following front view, top view, and side view.

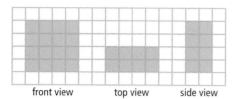

front view   top view   side view

8.3 Draw and Construct Three-Dimensional Figures Using Nets, pages 247–251

12. Name the polyhedron that can be made by folding each net.

a)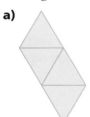

b)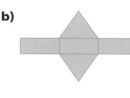

13. Draw a possible net for each polyhedron.
 a) a square-based pyramid
 b) a square-based prism

14. Which net will fold to make the standard die shown? Explain why the other two will not.

8.4 Surface Area of a Rectangular Prism, pages 252–257

15. Find the surface area of the video box shown.

16. Kristin works for a CD packaging plant. Each CD case measures 14.2 cm by 12.5 cm by 1.0 cm, and is tightly covered with plastic wrap.

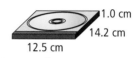

 a) How much plastic wrap is needed to cover one CD case?
 b) How much plastic wrap should Kristin order to wrap 2000 CD cases?

8.5 Volume of a Rectangular Prism, pages 258–261

17. Danny ordered a pizza for lunch. It was delivered in a box measuring 30 cm by 30 cm by 5 cm. What is the volume of the pizza box?

18. Ken bought a vase for a wedding gift. He wants to put the vase in a box, along with some foam chips, before he has it gift-wrapped. The box measures 12 cm by 12 cm by 25 cm. If the vase takes up half of the box, what volume of foam chips are needed?

19. Maki is packing for a long trip. She wants to take all the clothes that are stuffed in her dresser drawer. The dimensions of the drawer are 65 cm by 55 cm by 30 cm. Does she have enough room for the clothes if her suitcase measures 80 cm by 45 cm by 25 cm?

CHAPTER 8 Practice Test

Strand	NSN	MEA	GSS	PA	DMP
Questions	3, 4, 8, 11, 12	3, 4, 7, 8, 11, 12	1, 2, 4, 5, 7, 9–12		

Multiple Choice

For questions 1 to 4, select the correct answer.

1. A wedge of cheese most resembles a

   A square-based pyramid
   B rectangular prism
   C triangular prism
   D cone

2. The side view of a cylinder, such as the hockey puck shown, is a

   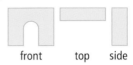

   A circle
   B square
   C triangle
   D rectangle

3. The surface area of the rectangular prism shown is

   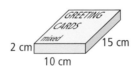

   A 100 cm^2
   B 200 cm^2
   C 300 cm^2
   D 400 cm^2

4. The volume of a classroom that measures 8 m by 10 m by 4 m is

   A 84 m^3
   B 196 m^3
   C 320 m^3
   D 384 m^3

Short Answer

5. Draw a sketch of the three-dimensional figure that has the following front view, top view, and side view.

   front   top   side

6. a) Sketch a polyhedron that has six faces. Name the polyhedron.
   b) Draw a net for this polyhedron.
   c) Sketch a different polyhedron that also has six faces.

7. What is the total surface area of the two fuzzy dice? They are identical in size. Each edge measures 9 cm.

8. Aaron plans to make a recycling bin out of cardboard. He wants it to have length 18 cm, width 12 cm, and height 20 cm. The bin does not need a lid.

   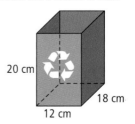

   a) What area of cardboard does he need to make the recycling bin?
   b) How much space will the bin take up?

9. a) Draw a prism with two square bases and four rectangular faces.
   b) Is it possible to have a prism that consists of two rectangular faces and four square faces? Use sketches or models to help you explain your answer.

Extended Response

10. a) Sketch two different three-dimensional figures you can build using two of these triangular prisms.

   b) Draw the front view, top view, and side view of each figure.
   c) Draw a net for each figure.

11. Rhianne plans to build a sandbox. She has 1.6 m³ of sand. If she makes the box square, with each side 2.0 m long, how deep should she make the sandbox? Sketch a diagram of the sandbox and explain your result.

12. Suppose you want to wrap two objects, each with a volume of 12 cm³. One object is a key chain in a box that is almost a cube. The other is a pen in a long thin box.
   a) Which box will need more wrapping paper? Explain your reasoning.
   b) What strategy or strategies work well to help you answer this problem?

Chapter Problem Wrap-Up

Single-serving orange juice boxes measure 10 cm by 7 cm by 5 cm. Packs of three boxes are shrink-wrapped for sale at grocery stores. The packs are shipped from the wholesaler in cartons containing 48 three-packs.

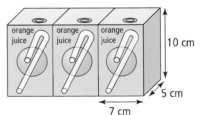

1. Design a carton that can be used by the wholesaler. Use pictures, words, and numbers to describe your design to a possible buyer. Include all important measurements.
2. Is your design the only possible one? If others are possible, show them.
3. Which carton design do you think is best for shipping the juice boxes? Why?

Making Connections

Convert Among Units

You can use the basic length conversions to find the related conversions for area and for volume.

Length: 1 m = 100 cm

Area: A square with side length 1 m has an area of 1 m × 1 m, or 1 m².

In centimetres, the area of the square is 100 cm × 100 cm, or 10 000 cm².

So, 1 m² = 10 000 cm².

Volume: A cube with side length 1 m has volume of 1 m × 1 m × 1 m, or 1 m³.

In centimetres, the volume of the cube is 100 cm × 100 cm × 100 cm, or 1 000 000 cm³.

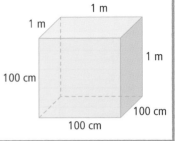

Making Connections

What's math got to do with architecture?

Architects combine technical skills with artistic sense and creativity. They draw several plans to provide different views of a new building's design. Sometimes they build a scale model.

1. Research architects' drawings on the Internet. Go to **www.mcgrawhill.ca/links/math7** and follow the links to some useful Web sites. Does each architect's view show the same amount of detail?
2. Which building trades use architects' drawings? Describe how they use them.
3. What three-dimensional shapes can you identify in the design of the Northern Ontario cottage shown here?

Design a Stage

The Prisms, a four-person band from Ireland, are coming to your school. You have the job of designing the stage. The school stage is square and has an area of 36 m².

The Prisms' stage manager sends the following e-mail message.

> Do for the drummer:
> • build a raised rectangular platform
> • pack with something to muffle vibrations
> • cover with black cloth
> • decorate edges
> Thanks
> P.S. drummer likes silver

1. Design the platform.

2. Decide what materials you will need to build it. Calculate the amount of each material.

3. Make a model of the drummer's platform.

4. Use a diagram to show your complete plan and where you will place the platform on the school stage. Use words, pictures, and numbers to justify your response.

Chapters 5–8 Review

Chapter 5 Fractions, Decimals, and Percents

1. Klaus scored 35 out of 50 on his geography test, and 30 out of 40 on his science test.
   a) What was Klaus's score on each test as a percent?
   b) On which test did Klaus do better?

2. Three friends are comparing their progress on a science project.

   a) Write the progress of each student as a fraction.
   b) Who is the closest to being finished? Explain.
   c) Can you order estimates in the same way as exact values? Discuss.

3. On a very hot day, Jackie's dog, Bowzer, runs 9 laps around his backyard, instead of his usual 12. What percent of his daily workout does Bowzer *not* complete?

4. Alina and Kristoff had to answer 20 multiple-choice questions for a driver-training test. Their scores are given as percents.
   Alina   90%
   Kristoff 75%

   To pass, you must get 16 out of 20 correct answers. Who passed the test? Describe your method.

Chapter 6 Patterning

5. Copy and complete each pattern by replacing each ■ with the appropriate value.
   a) 6, ■, 14, ■ (addition sequence)
   b) 45, ■, 23, ■ (subtraction sequence)
   c) 7, ■, 28, ■ (multiplication sequence)
   d) ■, 64, ■, 16 (division sequence)

6. Pennies can be arranged to make triangle shapes, as shown.

   a) How many pennies do you need to make the next triangle shape in this pattern? Explain your reasoning.
   b) How many pennies would you need for the tenth triangle shape in the pattern? for the fortieth?

7. a) Plot the points given in the table. Describe the pattern.
   b) What is the next point in the pattern?

x	y
0	3
1	4
2	5
3	6
4	7

8. a) Copy and complete this table.

Number of Jugs of Water	Volume in Litres
1	4
2	8
3	12
4	
5	

b) Give a pattern rule for the volume of water.
c) Write a variable expression for the volume, in litres, in n jugs of water.
e) Use your expression or your pattern rule to find the volume of water in 25 jugs.

Chapter 7 Exponents

9. Write each expression as a power.
    a) $3 \times 3 \times 3 \times 3$ b) $8 \times 8 \times 8$

10. Find the area of a square that has each perimeter.
    a) 60 cm   b) 26 m

11. Evaluate.
    a) $\sqrt{25}$ b) $\sqrt{49}$ c) $\sqrt{900}$

12. Write these expressions in order from the one with the greatest value to the one with the least value.
    3^3 33 $\sqrt{169}$ 1^{13} 13^2

13. Zoltan is tiling his kitchen floor. How many 25 cm by 25 cm square tiles does he need
    a) to cover a 1 m by 1 m square?
    b) to cover an area of 9 m²?

14. Estimate the total number of words spoken in your school in one day.

Chapter 8 Three-Dimensional Geometry and Measurement

15. What closed three-dimensional figure am I? Sketch and name one object that fits each description.
    a) All my faces are rectangles.
    b) Two of my faces are flat, the other is curved.
    c) I have four faces.
    d) I have six congruent faces.

16. The diagrams show the front view, top view, and side view of a three-dimensional object. Sketch the object.

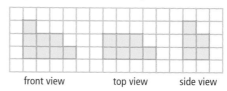

front view   top view   side view

17. a) Calculate the surface area of the tissue box shown.

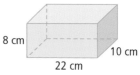

b) What is the volume of the box?

18. Adam's favourite kind of cookie is sold in boxes of two different sizes.

A: measures 12 cm by 12 cm by 15 cm, costs $3

B: measures 10 cm by 16 cm by 18 cm, costs $5

a) Which box is the better buy? What assumption did you have to make?
b) Draw a net for each box.
c) Calculate the surface area of each box. Which box is the more efficient design? Explain.

Data Management and Probability
- Collect, organize, describe, and interpret data on tally charts, stem-and-leaf plots, and frequency tables.
- Use symbols, titles, and labels when displaying data on bar graphs, pictographs, and circle graphs with and without technology.
- Read and interpret data on bar graphs, pictographs, and circle graphs.
- Search databases for information and interpret the numerical data.
- Explore with technology to find the best presentation of data.
- Understand the difference between a spreadsheet and a database.

Number Sense and Numeration
- Solve problems that involve converting between fractions, decimals, and percents.

Key Words

primary data
frequency table
secondary data
stem-and-leaf plot
circle graph
database
spreadsheet
pie chart

CHAPTER 9

Data Management: Collection and Display

What types of books do you like to read? Do you have a favourite author? What types of magazines do you like? What features of these magazines do you like?

Collecting and organizing this kind of information can help publishers, advertisers, and writers sell their work. In this chapter, you will learn a variety of ways to organize data and assess which type of display is best for presenting the data.

Chapter Problem

You are the publisher of a new magazine for students in grades 7 to 9. You are planning the first issue. Your editor has surveyed several grade 7 students. The results of the survey are shown.

Topic	Tally
Global issues	III
Entertainment	IIII IIII IIII II
Health and fitness	IIII
People and careers	IIII III
Sports	IIII IIII
Technology	IIII IIII II

How could you organize and display the data?

How might the data help you plan your first issue?

Get Ready

Tally Charts

A tally chart is a table used to record experimental or survey data. In the tally chart shown, 8 people picked hockey as their favourite sport.

Favourite Sport	Tally
Hockey	IIII III
Football	IIII
Baseball	III
Basketball	IIII I
Golf	II

Use the Favourite Sport tally chart to answer questions 1 and 2.

1. a) How many people picked football?
  b) How many people picked basketball?
  c) Which sport is the least popular? How many people picked it?

2. How many people were surveyed about their favourite sport?

3. Survey 20 students about their favourite sport. Make up a tally chart to show the data.

Bar Graphs and Pictographs

This bar graph displays numeric data from the Favourite Sport tally chart. The height of a bar is equal to the total for that category.

Label the vertical axis and the horizontal axis.

The bars are evenly spaced and of equal width.

Include a title for the graph.

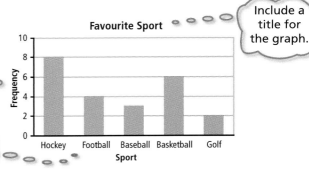

This pictograph displays the same data by using a symbol that is related to the data. The symbols can be interpreted to give the total for each category.

represents 2 people

4. Use the Favourite Sport bar graph and pictograph to answer these questions.
   a) Which sport is the most popular?
   b) Which sport is the least popular?
   c) How many more people chose hockey than football?
   d) Explain how each graph shows the answers to parts a), b), and c).

5. A group of young teens identified their favourite flavour of ice cream. Draw a bar graph and a pictograph to show the data.

Flavour	Number
Chocolate	5
Butterscotch	8
Cookies and cream	10
Mint chocolate chip	6
Other	4

Simple Circle Diagrams and Angles

A fraction can be represented by a circle diagram.

For example, $\frac{1}{8}$ can be shown by shading one-eighth of a circle.

You can use a protractor to measure the size of an angle inside a circle. The angle shown measures 45°.

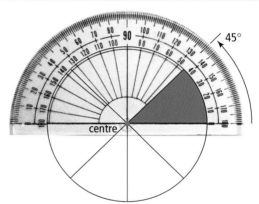

6. Use a protractor to measure each angle.
   a)
   b)
   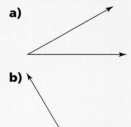

7. Use a ruler and a protractor to draw an angle with each measure.
   a) 40°   b) 90°   c) 150°

8. Draw a circle diagram to represent each set of fractions.
   a) $\frac{1}{4}, \frac{3}{4}$ b) $\frac{1}{3}, \frac{2}{3}$

9. For each circle diagram in question 8, measure the size of the angle that represents the smallest fraction.

9.1 Collect and Organize Data

Focus on...
- tally charts and frequency tables
- bar graphs and pictographs

School dances can be lots of fun, especially if the right kind of music is played. Data from a survey are often used to make informed decisions like this.

primary data
- data you collect yourself, such as from a survey

The results from Anand's survey provided **primary data**. What question might Natalie ask to collect primary data this year?

Discover the Math

How can you organize data that you collect?

Example 1: Use a Frequency Table to Draw a Bar Graph

frequency table
- a table that shows the count, or frequency, for each survey choice or experimental outcome

a) Organize the data using a **frequency table**.
b) Draw a bar graph to show the data.
c) How can this information help Natalie choose music for the school dances?

Type of Music	Tally
Dance	||||| ||||| ||
Rock	|||||	
Hip-hop	||||| ||||
Country		
Alternative	||||| ||

Solution

a)

Type of Music	Tally	Frequency
Dance	||||| ||||| ||	12
Rock	|||||		6
Hip-hop	||||| ||||	9
Country			1
Alternative	||||| ||	7

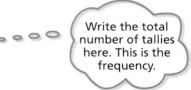

Write the total number of tallies here. This is the frequency.

274 MHR • Chapter 9

b)

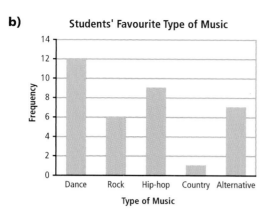

c) Dance music is the most popular. Country is not very popular. Natalie should choose mostly dance songs. She might include a mix of hip-hop, alternative, and rock songs.

Example 2: Use a Frequency Table to Draw a Pictograph

In a wildlife study, researchers recorded the population of rabbits in a certain field over time. They collected primary data.

A newspaper reporter found the data on the Internet. She used the **secondary data** in a story about the rabbits.

a) Draw a pictograph to show the data.
b) Describe at least one advantage and one disadvantage of using a pictograph for the data.

Hop into May
Researchers say there are more rabbits in May than...

Rabbit Population

Month	Frequency
January	10
February	20
March	35
April	50
May	87
June	63

secondary data
• data obtained from someone else, such as in published research

Literacy Connections

Primary and Secondary Data
How can you tell the difference between the two?

Primary data is information you collected by surveying people or counting things.

Secondary data is information someone else collected. You may read or hear about it.

Solution

a) Use a rabbit symbol. Let one 🐰 symbol represent 10 rabbits. Divide by 10 to find the number of 🐰 symbols to use for each value.

Rabbit Population

January 🐰
February 🐰🐰
March 🐰🐰🐰🐰
April 🐰🐰🐰🐰🐰
May 🐰🐰🐰🐰🐰🐰🐰🐰🐰
June 🐰🐰🐰🐰🐰🐰

🐰 represents 10 rabbits

b)

+	−
• rabbit symbol makes graph more interesting • reader sees information is about rabbits	• difficult to accurately read some of the populations

Key Ideas

- Primary data consist of information you collect by surveying or counting.
- Secondary data consist of information obtained from other sources.
- Tally charts and frequency tables are useful for recording data.
- A bar graph compares the frequencies of different parts of a data set.
- A pictograph also shows frequencies. Pictographs are visually appealing, but may not represent the data precisely.

Communicate the Ideas

1. What type of data is each person collecting? Explain your choices.

2. Describe the information shown in the table. Is this a frequency table? Explain why or why not.

Favourite Car	Tally
Convertible	|||| ||
Sport utility vehicle	||||	
Sports car	|||| ||||
Minivan	||
Truck	|||| ||

3. Use a plus/minus chart to list one advantage and one disadvantage of using each of the following to display data.
   a) bar graph b) pictograph

	+	−
Bar graph		
Pictograph		

Check Your Understanding

Practise

For help with questions 4 to 9, refer to Example 1.

4. Make a frequency table for the data.

Favourite Animal	Tally
Cat	||||	
Dog	|||| ||
Horse	||||
Rabbit	|||| |||

5. Make a frequency table for the data.

Favourite Insect	Tally
Butterfly	|||| ||||
Spider	||||
Fly		
Ant	||||

6. "Who will win the Stanley Cup this year?" When a number of hockey fans were asked this question, they responded as shown.

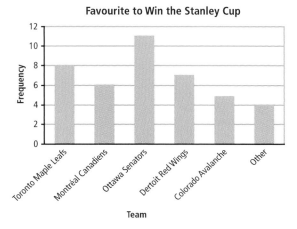

a) Based on the data, who is the favourite to win the Cup?
b) Make a frequency table for the data.
c) How many people were surveyed?

7. a) How many methods of getting to school are shown in the bar graph?
b) Make a frequency table for the data.

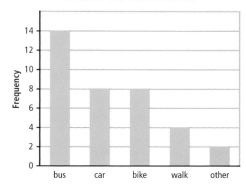

8. Draw a bar graph for the data.

Ice Cream	Tally	Frequency
Chocolate	|||| |||	8
Strawberry	||||		6
Cookies and cream	|||| ||||	10
Bubble gum	||||	4
Vanilla	||	2

9. A juice company conducted a taste test among young teenagers to find their favourite flavours.

Juice	Tally	Frequency
Apple	||||	
Orange	||||		
Strawberry-kiwi	|||	
Citrus surprise	|||| |||	
Crazy berry	|||| ||||	

a) Copy and complete the frequency table.
b) Display the data using a bar graph.
c) What is the most popular flavour?
d) What is the least popular flavour?
e) How many teens took the taste test?

For help with questions 10 to 12, refer to Example 2.

10. The pictograph shows the most popular costume types at Talia's party.

Costume	Number of People
Pirate	👓 👓
Monster	👓 👓 👓
Super-hero	👓 👓
Movie/pop star	👓 👓 👓 👓 👓 ◠
Other	👓 👓 ◠

👓 represents 2 people

a) How many friends dressed up as pirates?
b) How many friends dressed up as movie or pop stars?
c) Make a frequency table for the data.
d) How many people attended the party?

11. Draw a pictograph for the data.

Ice Cream	Tally	Frequency
Chocolate	|||| |||	8
Strawberry	||||		6
Cookies and cream	|||| ||||	10
Bubble gum	||||	4
Vanilla	||	2

12. Points for the teams in a recreational soccer league are shown.

Team	Points
Dizzy Dogs	50
Big Red Machine	45
Toe-Tappers	30
Lawn Mowers	58
Inter-Nets	22

a) Draw a pictograph to show the data.
b) Explain how you decided how many points each symbol should represent.
c) The top three teams advance to the playoffs. Identify these teams.

13. Classify each of the following as primary data or secondary data.
a) Randy conducts a survey about sports.
b) Kuzana researches the bald eagle population of Ontario using resources in the school library.
c) Jordan measures the height of each student in his classroom.
d) Sarah looks up hockey statistics on the Internet.

Apply

14. a) Survey ten friends about their favourite kind of music. Use a tally chart to record their responses.
b) Draw a graph of the results.
c) What does the graph tell you about the music your friends prefer?

> **Making Connections**
>
> You worked with probability in Chapter 4.

15. When you roll a number cube, what number is most likely to turn up? Suppose you roll a number cube 50 times and display the results using a bar graph.

a) Predict what the graph would look like. Describe and sketch your prediction.
b) Carry out the experiment. Record your results in a frequency table.
c) Draw a bar graph or pictograph to show your results.
d) Compare your results with your prediction. Explain any differences that you see.

16. Look at question 15. What do you think will happen if you roll the number cube 150 more times?

   a) Collect secondary data from three friends. Draw a bar graph to show data for all 200 trials.

   b) Compare this graph with the one in question 15c). Describe what you notice.

   c) Do any numbers have a higher probability of turning up? Use the results of this experiment to explain your answer.

17. a) Pose a question on a topic of interest to you that you are likely to find secondary data on. Some examples are professional sports, health, and celebrity information.

   b) Search the Internet, a library, or other sources to find data. Record the source of your data.

   c) Display the data using a bar graph or a pictograph.

   d) Explain why you chose the type of graph that you did.

Chapter Problem

18. What magazine is read by the most grade 7 students at your school?

   a) Design a survey.

   b) Try to get at least 20 responses. Use a frequency table to organize your results.

   c) Create a graph to display the data.

   d) What can you say about the magazines read by students at your school based on the information you collected?

19. Write an e-mail or a letter to a friend that missed class, describing the difference between primary data and secondary data. Support your explanation with examples.

20. a) Design a simple survey on a topic of interest to you.

   b) Conduct the survey with at least 10 friends, classmates, or family members. Use a table to record your results.

   c) Display your data using a bar graph.

   d) Display your data using a pictograph.

   e) Which graph displays the data more effectively? Explain why.

Extend

21. Clive's boss has been nagging him to keep a tidier desk. One day Clive spilled coffee on an important graph he had been working on. He did not have a backup copy of the graph on his computer.

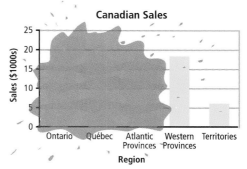

That night at home, Clive found the original frequency table that he had used to draw the graph. Unfortunately, Clive's dog ate part of it.

Region	Sales ($1000s)
Ontario	20
Québec	15
Atlantic provinces	
Western provinces	
Territories	
TOTAL	71

Use the information that Clive found to reconstruct the frequency table and graph.

9.2 Stem-and-Leaf Plots

Focus on...
- stem-and-leaf plots

Canadian heroes like Manon Rheaume and Hayley Wickenheiser are partly responsible for the increasing interest in Canadian women's hockey. As the game becomes more popular, the amount of data increases. How can you organize the data you collect about *your* sports heroes?

Discover the Math

How can you reorganize data into numeric groups?

The table shows the final regular season standings in the National Hockey League (NHL) for a recent year.

Team	Points	Team	Points
Anaheim	95	Montréal	77
Atlanta	74	Nashville	74
Boston	87	New Jersey	108
Buffalo	72	NY Islanders	83
Calgary	75	NY Rangers	78
Carolina	61	Ottawa	113
Chicago	79	Philadelphia	107
Colorado	105	Phoenix	78
Columbus	69	Pittsburgh	65
Dallas	111	San Jose	73
Detroit	110	St. Louis	99
Edmonton	92	Tampa Bay	93
Florida	70	Toronto	98
Los Angeles	78	Vancouver	104
Minnesota	95	Washington	92

1. Explain how the data values in the table are organized.

2. a) Which team had the greatest point total?
   b) Which team had the least point total?
   c) Which team had the fifth greatest point total?
   d) Describe how you found your answers.

3. a) How many teams had points in the 60s? List the scores.
   b) How many teams had points in the 70s? List the scores.
   c) How many teams had points in the 80s? List the scores.
   d) How many teams had points in the 90s? List the scores.

4. **Reflect** In what ways might you organize data to help you find numerical information?

Example 1: Read and Interpret a Stem-and-Leaf Plot

A **stem-and-leaf plot** arranges data into groups of increasing order. The following stem-and-leaf plot shows the NHL data from the Discover.

Stem (tens)	Leaf (ones)
6	1 5 9
7	0 2 3 4 4 5 7 8 8 8 9
8	3 7
9	2 2 3 5 5 8 9
10	4 5 7 8
11	0 1 3

For this data set, the stems represent the tens digit. The leaves represent the ones digit.

stem-and-leaf plot
- a way of organizing numerical data by representing part of each number as a stem and the other part of the number as a leaf

a) How many teams scored in the 60s? What were their point totals?
b) What stem contains the most data?
c) What was the most common point total?

Solution
a)
Stem (tens)	Leaf (ones)
6	1 5 9

↑ Stem   ↑ Leaves

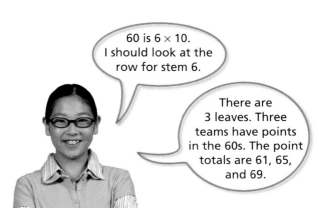

60 is 6 × 10. I should look at the row for stem 6.

There are 3 leaves. Three teams have points in the 60s. The point totals are 61, 65, and 69.

b) Stem 7 has the most data. It has 11 leaves.

c) There are three 8s in stem 7.

7 | 0 2 3 4 4 5 7 (8 8 8) 9

These represent 78s. So, 78 is the most common total.

Example 2: Create a Stem-and-Leaf Plot

Here are scores for a figure skating competition.

6.2	7.1	5.7	9.3	7.7	6.4	4.7	5.2	7.0	8.5
7.5	5.3	8.2	9.6	7.1	7.3	5.9	6.7	8.8	6.3
5.5	6.8	7.9	7.1	4.6	6.0	8.1	8.4	7.6	6.2

a) Create a stem-and-leaf plot to show the data.
b) How many skaters scored in the 5s?
c) How many skaters scored in the 8s or higher?
d) In what numeric group did more skaters score than any other?
e) What does the shape of the stem-and-leaf plot tell you about the overall scores?

Solution

a)

I will use the ones digit for the stem. The numbers after the decimal will be the leaves. First, I'll group all the 4s together.

Stem (ones)	Leaf (tenths)
4	6 7
5	5 3 7 9 2
6	2 8 4 0 7 3 2
7	5 1 9 1 7 1 3 0 6
8	2 1 4 8 5
9	3 6

My leaves are not in order. I'll rewrite to show them in order.

Stem (ones)	Leaf (tenths)
4	6 7
5	2 3 5 7 9
6	0 2 2 3 4 7 8
7	0 1 1 1 3 5 6 7 9
8	1 2 4 5 8
9	3 6

b) Five skaters scored in the 5s.

Scoring in the 5s means a score from 5.0 to 5.9. I should look at stem 5.

c) Seven skaters scored in the 8s or higher.

Look at stem 8 and stem 9. Count the total number of leaves in these two stems.

d) Stem 7 has the most leaves. Nine skaters scored in the 7s.

282 MHR • Chapter 9

e) The shape of the plot shows a large number of scores clustered near the middle. There are fewer scores at the high and low ends. This suggests that most skaters scored in the 5 to 8 range.

Stem (ones)	Leaf (tenths)
4	6 7
5	2 3 5 7 9
6	0 2 2 3 4 7 8
7	0 1 1 1 3 5 6 7 9
8	1 2 4 5 8
9	3 6

Key Ideas

- A stem-and-leaf plot is used to organize and order large sets of numeric data.
- To create a stem-and-leaf plot, organize the data into groups (stems). Then, order the data within each stem and write the leaves.

Communicate the Ideas

1. The stem-and-leaf plot shows people's ages.
  a) Describe how the stem and leaf of a value are related.
  b) What ages are shown?

Stem (tens)	Leaf (ones)
1	1 2 5
2	3 6
3	0 4

2. What's wrong? The prices for 6 pairs of running shoes, rounded to the nearest dollar, are $69, $74, $79, $79, $85, and $89. Describe the error. Explain how to fix it.

Stem (tens)	Leaf (ones)
6	9
7	4 9
8	5 9

Check Your Understanding

Practise

For help with questions 3 to 5, refer to Example 1.

3. a) How many stems are in the plot?
  b) Which stems have two leaves?
  c) Which stem has the most data?
  d) List the scores shown.

Stem (tens)	Leaf (ones)
1	4 7
2	0 3 3 8
3	1 5
4	2

4. The stem-and-leaf plot shows the average monthly rainfall, in millimetres, for a particular city.
  a) What amounts of rainfall are represented in stem 7?
  b) How many rainfall measurements are recorded in the stem-and-leaf plot?

Stem (tens)	Leaf (ones)
2	8
3	2
4	8
5	4
6	0
7	2 5
8	5 6 8
9	3

5. The stem-and-leaf plot shows the ages of people at a family picnic.

Stem (tens)	Leaf (ones)
0	7 9
1	1 3 3 7
2	
3	6 7 8
4	1 2 2
5	8 9
6	2 3

a) How many children are under 10? How old are they?
b) How many teenagers are there? How old are they?
c) Four grandparents are present. How old do you think they are? Explain.

For help with questions 6 to 9, refer to Example 2.

6. Complete the stem-and-leaf plot by organizing the leaves in increasing order.

Stem (tens)	Leaf (ones)
1	1 6 2
2	7 0 3
3	8 5 1
4	3 2 7 3
5	6 9

7. Organize the following data using a stem-and-leaf plot.

10 21 32 47 12 22 34 47
14 25 36 18

a) What will the stem values represent?
b) What will the leaf values represent?
c) Create the stem-and-leaf plot.
d) Write one question about your stem-and-leaf plot. Answer your question.

8. a) Organize the following scores using a stem-and-leaf plot.

22 36 18 24 41 55 15 27 22 38
44 32 36 22 13 45 50 20 37 40

b) Write one question about your stem-and-leaf plot. Answer your question.

9. Victor rounded his grocery bills to the nearest dollar. During the past two months, he spent $67, $81, $73, $64, $66, $73, $82, and $59.

a) Organize the data using a stem-and-leaf plot.
b) What is Victor's most common grocery bill amount?
c) How much has Victor spent on groceries over the past two months?

Apply

10. A group of students wrote a test. Their scores, out of 50, are shown.

Stem (tens)	Leaf (ones)
1	8 9
2	2 4 5 8 8 9
3	0 1 2 2 3 6 8 8 8 9
4	2 3 5 5 8
5	0

a) How many students scored in the 40s? What were their scores?
b) In which stem did more students score than any other?
c) Did any student write a perfect test? Explain.
d) How many students scored below 50%? What were their scores?

11. The numbers of wins for the teams in a minor baseball league are 45, 61, 57, 90, 88, 80, 95, 49, 53, 80, 85, 92, 103, 85, 77, 73, 85, 68, 74, and 82.

a) Organize the data using a stem-and-leaf plot.
b) How many teams won more than 90 games?
c) What was the most common number of wins?

12. a) Collect age data from 15 to 20 family members, friends, and acquaintances.
   b) Create a stem-and-leaf plot to show the data.

13. The masses, in grams, of samples of a particular chemical are 1.2, 1.3, 0.8, 1.0, 0.8, 1.4, 0.7, and 1.0.
   a) Create a stem-and-leaf plot for the data.
   b) Explain what the stems and leaves represent.
   c) What is the difference between the greatest mass and the least mass?

14. The times for 50-m sprinters are shown in a stem-and-leaf plot.

Stem	Leaf
6	8 3 2
7	0 1 1 3 5

   a) Organize the plot.
   b) What do the stems and leaves represent? Justify your answer.
   c) Explain five things that the data set tells you.

Extend

15. Weekly payroll data values for a small company are given. The stem shows the hundreds digit and the tens digit.

Stem	Leaf
28	0 5
29	0 0 0 4 8
30	5 5
31	8
32	
33	
34	8

   a) How many employees work for the company? Justify your response.
   b) What is the most common weekly wage?
   c) What is the highest weekly wage?
   d) What is the company's total weekly payroll expense? Explain how you found this.
   e) The boss wishes to hire a very experienced and talented worker. What is the highest weekly wage the boss can offer the worker without exceeding the payroll budget of $3640? If the new worker accepts, will that person be the highest paid employee? Explain.

Literacy Connections

Prepare to Write
Use a fishbone organizer to help you plan paragraph answers. The examples shown describe bar graphs and pictographs. Use the point-form information to write two paragraphs.

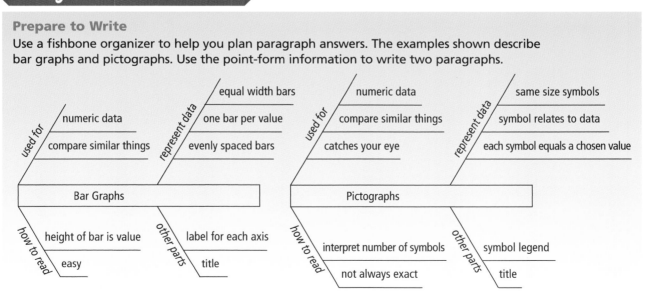

9.3 Circle Graphs

Focus on...
- circle graphs

Three friends agree to build a quarter-pipe skateboard ramp for the owner of a small sports store. The friends decide to divide the job into three individual tasks. The store owner wants to pay the group $240. How can you show how much each person should be paid?

Discover the Math

Materials
- grid paper
- manipulatives (e.g., coloured counters)
- compasses
- protractor
- pencil crayons

How can you show how part of a data set compares to the whole data set?

1. The friends decide to divide the money according to the number of hours they each worked.

Worker	Hours Worked
Fumio	2
Suriya	4
Justin	6

   a) Draw a circle to represent the whole project.
   b) Using estimation, divide the circle into three pieces to show each person's contribution to the project.
   c) Explain how you decided on the size of each person's section.

2. a) How much do you think each person should get paid?
   b) Explain how you decided on these amounts.

3. How can a **circle graph** help you show how parts of a whole are related?

circle graph
- a graph that uses sections of a circle to show how a data set is divided into parts

286 MHR • Chapter 9

Example 1: Draw a Circle Graph

A group of 100 grade 7 students were surveyed to find their favourite hobbies.

Draw a circle graph to display the data.

Hobby	Number of People
Read books/comics	25
Watch television	50
Surf the Internet	10
Play video games	15

Solution

Method 1: Work With the Fractions You Know

Hobby	Number of People	Fraction	Size of Section
Read books/comics	25	$\frac{25}{100}$	This is the same as one quarter. I can show one quarter of a circle.
Watch television	50	$\frac{50}{100}$	This is the same as one half. I can show half the circle.

Making Connections

For help with converting fractions to decimals, refer to Chapter 5.

Method 2: Calculate Section Angles You Do Not Know

To do this:
- express each category as a fraction
- write as a decimal
- since there are 360° in a circle, multiply each decimal value by 360 to find the section angle

Hobby	Number of People	Fraction	Decimal	Section Angle
Surf the Internet	10	$\frac{10}{100}$	$10 \div 100 = 0.1$	$0.1 \times 360° = 36°$
Play video games	15	$\frac{15}{100}$	$15 \div 100 = 0.15$	$0.15 \times 360° = 54°$

Use the decimal shown on your calculator to complete the calculation for the section angle.

Draw a circle. Use a protractor to measure each section.

Favourite Hobbies of Grade 7 Students

Shade or colour the sections if you want to.
Then, label the sections and write a title.

Example 2: Read and Interpret a Circle Graph

The circle graph shows what Mary does with her monthly earnings from baby-sitting.
a) Which is Mary's greatest monthly expense? How much is it?
b) How much does Mary spend on movies and CDs each month?
c) How much of Mary's monthly budget goes to "Other" expenses?

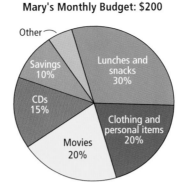

Solution

a) The largest section is "Lunches and Snacks." Find 30% of $200.
30% of $200 = 0.3 × $200
= $60
Mary spends $60 per month on lunches and snacks.

b) She spends 20% on movies and 15% on CDs.
Movies: CDs:
20% of $200 = 0.2 × $200   15% of $200 = 0.15 × $200
= $40   = $30
Movies and CDs:
$40 + $30 = $70
Mary spends a total of $70 per month on movies and CDs.

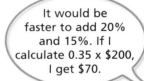

It would be faster to add 20% and 15%. If I calculate 0.35 × $200, I get $70.

c) First, find the percent for this category. Find the sum of all expenses except "Other."
30% + 20% + 20% + 15% + 10% = 95%
The circle must total 100%. Subtract.
100% − 95% = 5%

Mary spends 5% of her monthly earnings on "Other" expenses.
5% of $200 = 0.05 × 200
= 10
Mary spends $10 per month on "Other" expenses.

Making Connections

For help with finding a percent of a number, refer to Chapter 5.

Strategies

What other method might you use to find this answer?

Key Ideas

- A circle graph shows how each part of a data set compares to the whole.
- To create a circle graph, use the following steps:
  - To create a circle graph using simple fractions:
    - Express each category as a fraction of the whole.
    - Show each fraction of the circle.
  - To create a circle graph with more difficult fractions:
    - Express each fraction as a decimal.
    - Use the decimal to find the size of each section angle.
    - Use a protractor to measure and draw each section angle.
  - Add section labels and a title to the circle graph. You may wish to shade the sections.

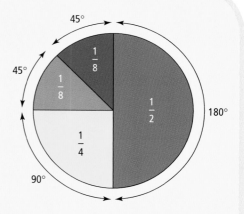

Communicate the Ideas

1. Compare a circle graph with a bar graph. Use an organizer to help you. Describe one advantage and one disadvantage of each type of graph.

2. Which set of data can best be presented in a circle graph? Explain.

   a)
Favourite Pet	
Cat	35%
Dog	45%
Bird	15%
Rabbit	10%
Other	5%

   b)
	Score out of 10
Quiz 1	8
Quiz 2	7
Quiz 3	9
Quiz 4	8
Quiz 5	10

3. What is wrong with each circle graph? Describe how it can be corrected.

   a)

   b)

Check Your Understanding

Practise

For help with questions 4 to 7, refer to Example 1.

4. Draw a circle graph for each set of section angles.
   a) 120°, 120°, 120°
   b) 90°, 90°, 90°, 90°
   c) 60°, 60°, 60°, 60, 60°, 60°

5. Draw a circle graph for each set of section angles.
   a) 60°, 60°, 60°, 45°, 45°, 90°
   b) 30°, 30°, 30°, 135°, 90°, 45°

6. In a grade 7 class, there are 12 boys and 20 girls. How could you find the section angles to represent the data?
   a) Copy and complete the table of calculations.

Grade 7 Students	Number	Fraction	Decimal	Section Angle
Boys	12			
Girls	20			
TOTAL				

   b) Draw a circle graph to show the data.

7. Three friends worked together on a science project. Melissa worked 5 h, Zach worked 4 h, and Cecilia worked 3 h. How could you find the section angles to represent the data?
   a) Copy and complete the table of calculations.

Person	Hours	Fraction	Decimal	Section Angle
Melissa	5			
Zach	4			
Cecilia	3			
TOTAL				

   b) Draw a circle graph to show the data.

For help with questions 8 and 9, refer to Example 2.

8. a) Which student group is the largest?
   b) What percent of the student population is in grade 8?

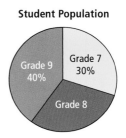
Student Population
Grade 9 40%
Grade 7 30%
Grade 8

9. a) What type of food does Claude eat the most?
   b) If Claude consumes 1000 g of food per day, how much of it is meat and alternatives?
   c) How much of the 1000 g of food per day is dairy products?

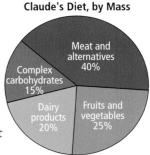
Claude's Diet, by Mass
Meat and alternatives 40%
Complex carbohydrates 15%
Dairy products 20%
Fruits and vegetables 25%

Apply

10. A group of hockey fans were surveyed to identify their favourite Canadian team in the NHL.

Team	Number of People
Calgary Flames	15
Edmonton Oilers	15
Montréal Canadiens	8
Ottawa Senators	5
Toronto Maple Leafs	7
Vancouver Canucks	10

   a) Estimate the size of the three largest sections of a circle graph for the data.
   b) Draw a circle graph. How close were your estimates?
   c) This survey was conducted in only one province. Which province do you think it was? Explain why you think so.

11. a) Which type of mammal has the greatest population? What fraction of the total mammal population is this?

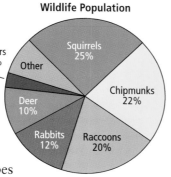

Wildlife Population

b) Which three types of mammal account for about $\frac{2}{3}$ of the total mammal population?

Chapter Problem

12. How many of the pages in your favourite magazine are advertisements?
  a) Count all the pages in your favourite magazine with ads.
  b) Count all the pages without ads.
  c) What other method might you use to find the number of pages without ads?
  d) Create a circle graph that shows the number of pages with ads and without ads.
  e) How could this information be useful to a magazine publisher?

13. Draw a circle graph that shows the probability of rolling each number (1 to 6) on a standard number cube.

 14. a) Search newspapers, magazines, textbooks, or the Internet. Find an example of a circle graph that is used to show data.
  b) Briefly describe what the graph shows.
  c) Comment on the graph's effectiveness in
     • catching the reader's attention
     • making a point
  d) Pose and answer two questions about the data that can be answered by looking at the graph.

Extend

15. a) Search various media, such as the Internet, magazines, or newspapers, for a bar graph that can also be represented as a circle graph. Create a circle graph to represent the same data.
  b) Describe any advantages either graph has over the other.

16. a) Find a circle graph and transform it into a bar graph.
  b) Comment on the advantages of each graph.

Making Connections

Materials
• BLM 9.3A Percent Circle

Percent Circle

A percent circle is a circle that is divided into 100 equal sections. Each section represents 1%. For example, to show 8%, colour in 8 sections of the percent circle.

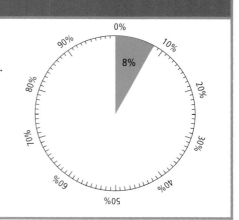

How can you use a percent circle to draw a circle graph?

Using a percent circle, create a circle graph to display the data in Example 1. Compare your graph with the graph that was made. Are they the same? Explain.

9.4 Use Databases to Find Data

Focus on...
- databases

What programs start at 6 P.M.? On what channels? You could use this **database** to find out. You could check all the program choices for a certain time slot. You could also find out when and on what channel your favourite show is.

Thursday, July 12 — TV Schedule

6:00 **8 3 V** NEWS
6:30 **K** LETTUCE TALK
Radishes—the next big thing or just another vegetable?
9 SUPERDUDE
Superdude gets audited and it looks like he's going to be fighting crime from the inside for a while.
7:00 **Y** ACQUAINTANCES
Russ dyes his hair blue. Mona discovers a quantum singularity in her purse.
7:30 **5** CRAFT TIME
Build your own tool shed from discarded cereal boxes.
18

8:00 **2** STAR LURK
Datum discovers the emotion of sadness after the Captain confines him to the brig for getting Rigelian muffin crumbs in the ship's navigation controls.
8:30 **7** THUNDER PIGEONS
Sally cuts up Commander Bob's credit cards.
9:00 **A** MOVIE–Camp Afoul
Teens run afoul of a mad wounded chicken in an abandoned summer camp.
9:30 **A** THEY WEREN'T ALL GOOD TIMES
Wally falls asleep in his parked car and rolls into a van bound for Mexico.

database
- an organized collection of information
- often stored electronically

Materials
- TECH 9.4A Accessing the CANSIM Database

Optional
- TECH 9.4B Printing Graphs From CANSIM

Alternative
- BLM 9.4A Databases Without Technology

Discover the Math

How can you use a database to find data?

Statistics Canada has created a giant database that includes information about Canada's people and resources. The database is called CANSIM: the Canadian Socio-economic Information Management System. E-STAT is the Internet link between you and the CANSIM database.

1. Go to **www.mcgrawhill.ca/links/math7** and follow the links to access E-STAT. Click **English**.

   On the next Web page, click **Accept and enter**.

2. Find data about adolescent students from the Table of Contents Web page.

   Under People, select **Education**.

   On the next Web page, click **Data**.

   Then, under CANSIM, select **Students**.

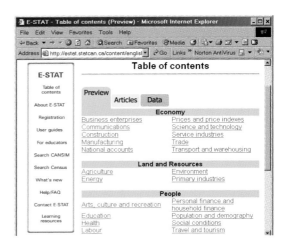

3. Look at the list of tables. Briefly describe some of the information that looks interesting to you.

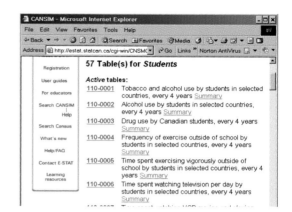

4. Find data about the television-watching habits of Canadian youths.

   Select table **110-0006**.

   Next, select **Canada**, **Males** or **Females**, and **13 years**. Then, for **Time spent**, click **Select all**. Finally, click **Retrieve as a Table**.

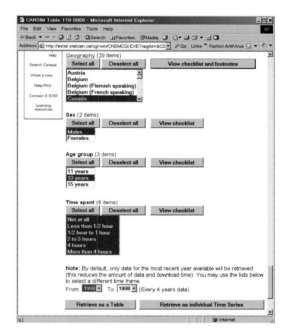

5. Select a type of graph to display the data.

   Choose **Bar graph (vertical)**, **Bar graph (horizontal)**, or **Pie chart**. Then, click **Go**.

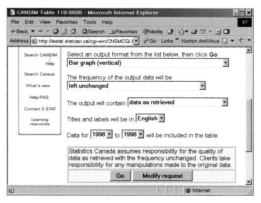

> **Technology Tip**
> - To compare graphs, you could copy and paste them into a word-processing document.

9.4 Use Databases to Find Data • MHR   293

6. **a)** Look at the data. What does the graph tell you?
   b) Use the Internet browser's **Back** button to return to the previous page. Select a different type of graph to present the data, and click **Go**. Does this new graph give any more information? Which type of graph do you prefer? Why?

7. Go back and look at the data for the opposite gender. Are there any noticeable differences between the television viewing habits of boys and girls? Explain.

8. **a)** Does age make a difference? Go back and look at data for 11-year-olds and 15-year-olds. Compare the three age categories.
   b) Make notes on anything interesting that you see.
   c) How do the habits of Canadian youths compare to those from other countries?

9. **Reflect** Explain how to access data in the CANSIM database.

I can use this skill on my science project. I'll type "earthquakes, Richter scale" into a Web search engine. I'll look for interesting data. How can you use what you learned here?

Key Ideas

- A database is an organized collection of information.
- Electronic databases allow you to select specific information or sort the information in different ways.

Communicate the Ideas

1. Describe three things you discovered about the television-watching habits of Canadian youths.

2. Identify three topics that you can use E-STAT to find data for.

3. What databases do you use regularly? Classify them as print or electronic databases.

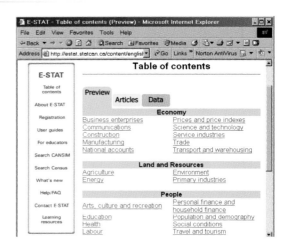

Check Your Understanding

Practise

4. Telephone directories are print databases. Check out the white pages for your community.
   a) How are the listings organized?
   b) What information does each listing provide?

5. Look at the yellow pages for your community.
   a) How are the listings organized?
   b) What information does each listing provide?

Go to www.mcgrawhill.ca/links/math7 and follow the links to access E-STAT.
Use E-STAT for questions 6 to 8.

6. Use E-STAT to find data that describe the weekly exercise habits of Canadian youths. Follow these steps:
   - From the **People** section of the Table of Contents Web page, select **Arts, culture, and recreation**.
   - On the next page, select **Data**.
   - Then, under **CANSIM**, select **Recreation**.
   - Finally, select table **110-0004**.
   a) On the Table 110-0004 Web page, what choices need to be selected to retrieve data on 13-year-old Canadian males?
   b) Select your choices from part a). Click **Select all** for the **Frequency of exercise** choice. Next, click **Retrieve as a Table**.
   c) Display the data in a bar graph or a pie chart. What does the graph tell you?

7. a) Go back and select the same data as in question 6 for females.
   b) Display the data in the same graph type as for question 6c). What does the graph tell you?
   c) Are there any noticeable differences in the weekly exercise habits of boys and girls? Explain.

8. Search CANSIM by subject.
   - Go to the Table of Contents page. Click **Search CANSIM** on the left side of the page.
   - Select **Subject**. Click **Continue**.
   - Select a topic that interests you.

   Write a couple of sentences to describe your findings.

Apply

What music is popular today? Go to www.mcgrawhill.ca/links/math7 and follow the links to a music database.
Use this database for questions 9 to 11.

9. Nielsen SoundScan gathers data on the most popular current songs. The chart shows the top 100 singles in Canada. It also lists the number of weeks each song has been on the chart, where the song ranked two weeks ago, and its ranking last week.
   a) What are the top three songs today? How accurate was your prediction?
   b) Pick any two songs from the chart. For each song, identify the name of the song, the artist, the number of weeks on the chart, and the current chart position.

10. Go back to the top of the Web page for question 9. From the **Top Albums: City by City** pull-down menu, select **Toronto**. A chart of the top 20 albums is displayed.

   a) Identify the song and artist that hold the number one spot.

   b) What chart position does this song hold on the national chart?

   c) From the **Top Albums: City by City** pull-down menu, select **Ottawa**. Answer parts a) and b) for this city.

11. Go back to the top of the Web page for question 9. From the **Other Charts** pull-down menu, select **Top 100 CD's of all time**.

   a) For each of the top three Canadian singles of all time, identify the title of the song, the artist, and the year it was produced.

   b) Scan the top 20 singles of all time. Copy and complete the frequency table.

	Tally	Frequency
Songs I recognize		
Artists I recognize		
Artists I have never heard of		

   c) Do the results in your table surprise you? Explain why or why not.

   d) Few new artists appear on this list. Why do you think this is? Do you think this may change over time? Explain.

12. Look at television listings for this week. Make up a plan to compare how two Canadian channels slot different types of programs from 6 P.M. to 10 P.M. What conclusions can you make? Hint: Canadian channels include CBC, CTV, CITY, and Global.

Go to *www.mcgrawhill.ca/links/math7* and follow the links to an NHL database. Use this database for questions 13 and 14.

13. The three fields at the top of the Web page allow you to access various statistics. Select

   the Most Current Season

   Entire League

   Skater Summary

Abbreviations Legend
GP = Games Played
G = Goals
A = Assists
PTS = Points
+/– = Plus/Minus Rating
PIMS = Penalties in Minutes
MINS = Minutes Played

   a) By selecting one of the headings, you can change the way the data set is organized. Identify the top three ranked players in goals, the top three in assists, and the top three in points.

   b) What do the data values in the **MINS/GP** column tell you? Find the top three ranked players in this category. Explain what this tells you about their play.

14. Go back to the top of the Web page for question 13. Change some of the field selections to explore data for other regular seasons, playoffs, or specific teams.

   a) Write two questions or problems that can be answered by looking at the data you found.

   b) Answer your questions.

   c) Trade with a partner and answer each other's questions. Check to make sure that you each found the same answers.

Which were the most watched TV shows last week? Go to **www.mcgrawhill.ca/links/math7** and follow the links to a television ratings site. Use this Web site for question 15.

15. a) Click **Nielsen Top 10 Ratings** in the top right corner of the Web page. For each of the top three **Broadcast TV Programs**, list the name of the program, the rating (percent of households that watched the program), and the number of viewers.
  b) Go back one Web page. List the same data as in part a) but for the top three **Cable TV Programs**.
  c) How do you think Nielsen obtains its data?

Chapter Problem

16. Your school or public library probably uses a computerized database to store its book information. Ask for assistance from the librarian if you need it.
  a) Pick the title of a book that you like or are interested in. Enter the title into the database and find the name of the author.
  b) Enter the name of the author to find other books that he or she has written. How many titles are in this list?
  c) Write down the names of two other books written by this author that you are interested in, or have not read.
  d) Use the database to find out if these books are in the library.

17. Interview two adults about how databases are used in their workplace.

18. Use an organizer to make point-form notes about what you have learned about databases. Use your notes to plan a one-minute talk.
  • Describe what a database is.
  • Explain how a database is organized.
  • Give examples of information you found in a database.

What kind of music is popular today? Go to **www.mcgrawhill.ca/links/math7** and follow the links to a music database. Use this Web site for question 19.

 19. Use the links to explore other parts of the music database.
  a) Describe two things that you found interesting.
  b) Write two questions or problems that can be answered using this database.
  c) Find the answers to your questions.
  d) Trade with a partner and answer each other's questions. Check your answers.

Extend

20. a) Search the Internet to find a database of information that interests you. Some possible topics are sports, entertainment, movies, and astronomy.
  b) Write three questions that can be answered by retrieving information from the database.
  c) Use the data to answer your questions.
  d) In a few sentences, briefly describe the database. Use examples to support your description.
  e) Display the data for one of your questions.

9.5 Use a Spreadsheet to Display Data

Focus on...
- spreadsheets

spreadsheet
- a software tool for organizing and displaying numeric data
- software packages include AppleWorks, ClarisWorks, Microsoft® Excel, and Quattro® Pro

An effective presenter uses dynamic visuals to capture and hold an audience's attention. With the help of technology, you can quickly produce a number of different graphs that will add that professional touch to your presentation.

A **spreadsheet** allows you to enter data into a computer in organized rows and columns and then display the data in a variety of ways. How could you use a spreadsheet to create the graph in the photo?

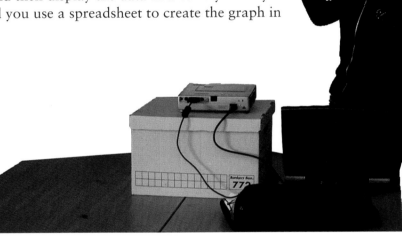

Materials
- TECH 9.5A Using AppleWorks 6.2

Optional
- TECH 9.5B Using AppleWorks 5.0
- TECH 9.5C Using Quattro® Pro 10
- TECH 9.5D Using Microsoft® Excel 2002

Alternative
- BLM 9.5A Spreadsheets Without Technology

Discover the Math

How can you use a spreadsheet to display data?

The table shows the results of a math test.

Result	Number of Students
Below Level 1	1
Level 1	3
Level 2	5
Level 3	11
Level 4	12

1. Open a new spreadsheet document in AppleWorks 6.2.

   a) Enter the categories and values, as shown. Click a cell. Type in the text or value. Press **Enter**.

   b) What does "< Level 1" mean?

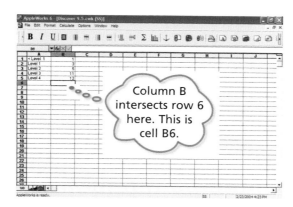

Column B intersects row 6 here. This is cell B6.

Technology Tip
- You can use the arrow keys or your mouse to move from one cell to another.

2. Create a **pie chart** to show this set of data.

   Select the data set. Click cell **A1** and drag to cell **B5**.

   From the **Options Menu**, select **Make Chart**.

   From the **Chart Options, Gallery** list, select **Pie**.

   Click **OK**.

   Right click inside the pie chart and select **Chart Options…** from the pop-up menu. Click on the **Labels** tab. In the **Title** box, type Math Grades.

   Click **OK**.

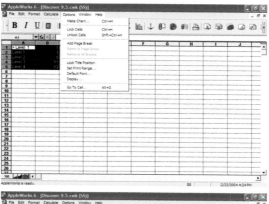

pie chart
- the same as a circle graph

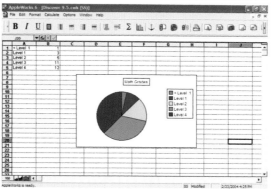

3. Explode the pieces of the pie chart.

   Right click inside the pie chart and select **Chart Options…** from the pop-up menu. Click on the **Series** tab.

   In the **Edit Series** box, select **All**.

   In the **Display as** area, click **Explode slice**. Click **OK**.

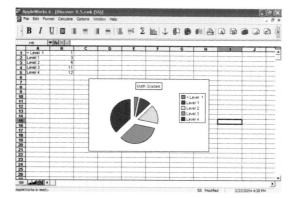

4. **a)** Look at the pie chart. Estimate what percent of the students scored
   - at Level 3
   - below Level 4

   b) Right click the pie chart and select **Chart Options…** from the pop-up menu. Click on the **Series** tab. In the **Display as** area, unclick **Explode slice**. Click **Label Data** and **% in legend**. Click **OK**.

   c) Compare the percent values with your estimate in part a).

   d) Explain how a circle graph is useful for showing the comparison of part of a data set to its whole.

> **Technology Tip**
> - If, after you right click, you don't see the option you are looking for, move the cursor position. Try to click a clear spot.

5. **a)** Click and drag the pie chart underneath the data.

   b) Right click on the pie chart. Experiment with the different formatting options available for the pie chart.

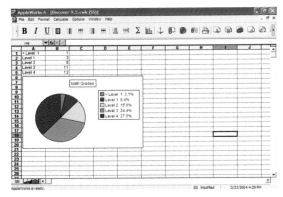

6. **a)** Experiment with different chart types, including vertical and horizontal bar graphs, line graphs, and any other types that look interesting.

   Right click the pie chart and select **Chart Options…** from the pop-up menu.

   From the **Gallery** tab, select a different type of chart. Click **OK**.

   b) Describe what each type of graph looks like. Use words and simple sketches.

7. **a)** Which type of graph do you think is the easiest to read? Explain why.

   b) Which type of graph did you find the most interesting? Explain why.

   c) Which type or types of graphs were not suitable for displaying the data? Explain why not.

8. **Reflect** List the various ways you can display data using a spreadsheet.

"I want to graph some information for my geography project. I like the bar graph format that shows the bars in 3D. Maybe I'll use that.

What was your favourite graph type? How can you use it?"

Key Ideas

- A spreadsheet is a software tool used to organize and display numeric data.
- A spreadsheet can be used to develop various types of graphs.
- The best choice of graph type depends on the nature of the data and the types of comparisons you wish to focus on.
  - Use a pie chart to show how each part of a data set compares to the whole or to show percents.

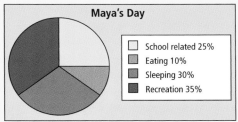

  - Use a bar graph to show how different parts of a data set are related or to compare similar things.

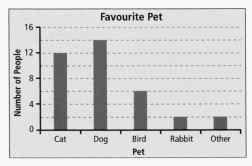

Communicate the Ideas

1. What type of graph would you use to display each set of data? Explain why.

a)
School Population	
Grade 7	35%
Grade 8	30%
Grade 9	35%

b)
Hours Spent Reading	
Monday	2
Tuesday	3
Wednesday	3

2. a) Describe three different ways you can format a spreadsheet graph.
   b) Identify advantages and disadvantages of using these different formats.

3. Who is right? Explain.

A spreadsheet and a database are the same thing.

They are not the same. I can search a database to find information. I use a spreadsheet to record and display numeric data.

Check Your Understanding

Practise

4. A soccer team has 12 wins, 8 losses, and 4 ties. The data set is displayed as a bar graph and a pie chart. Which graph do you prefer? Why?

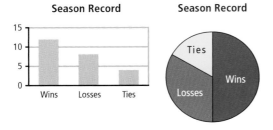

5. Marks, in percent, for the five students competing for an academic scholarship are shown.

Student	Percent
Sheila	92
Olaf	90
Ahmed	95
Lena	97
Scott	94

The data set is displayed as a bar graph and a pie chart. Which graph clearly identifies the top students? Explain your answer.

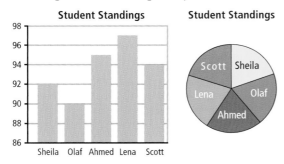

6. What is another name for a pie chart? Explain how you can remember both terms.

7. Marisa spent $50 while at the local amusement park.

Item	Price ($)
Admission	5
Ride tickets	30
Food	15

The data set is displayed as a bar graph and a pie chart. Which graph clearly identifies how Marisa spent her $50? Explain your answer.

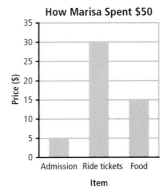

Apply

8. The table shows waterslide park ticket sales by age.

Age Group	Tickets
Under 4	50
4 to 60	225
Over 60	50

   a) Enter the data into a spreadsheet program.
   b) Create a graph.
   c) Explain your choice of graph in part b).

9. Ellen collects comic books. She has grouped her collection into several categories, as shown.

Type	Number of Comics
Super-hero	80
Fantasy	35
Comedy	40
Graphic novels	16
Horror	26
Other	12

   a) Enter the data into a spreadsheet program.
   b) Create a graph.
   c) Explain your choice of graph in part b).

10. a) Use the Internet or a library to find the population of each of Canada's provinces.
    b) Use a spreadsheet to create a graph to illustrate the data.
    c) Explain why you chose the type of graph you did.

11. Use an organizer of your choice to compare spreadsheets and databases. Show similarities and differences.

Try This!
12. a) Pick a topic that interests you. Design a survey to collect numeric data about this topic. Ask three or four questions.
    b) Survey several students. Collect and organize the data.
    c) Enter the data into a spreadsheet.
    d) Create two or three effective graphs for your data.
    e) Explain why you chose the types of graphs that you did.

Extend

13. a) Explore other types of graphs that you can produce with a spreadsheet.
    b) Pick one type of graph that interests you. What is it called?
    c) For what kind of data would this type of graph be suitable?
    d) Describe some advantages of using this type of graph.
    e) Find a suitable set of data and use a spreadsheet to create this type of graph. Explain why your graph is effective.

Making Connections

What do databases have to do with sports reporting?
The commentators on televised professional sports events seem to know a lot about the teams and athletes involved. How can they remember all those facts and figures?

Sports statisticians work behind the scenes. They retrieve interesting data and relay it quickly to the commentators. For example, suppose an NHL player is having a tremendous game, scoring a lot of points. The sports statistician will quickly search a database to find the most points ever scored by one player in an NHL game.

Did You Know?
Darryl Sittler of the Toronto Maple Leafs scored 10 points in one game on February 7, 1976! This included 6 goals and 4 assists. No NHL player has ever beaten this record.

Chapter 9 Review

Key Words

For questions 1 to 5, copy the statement and fill in the blanks. Use some of these words.

primary data frequency table
secondary data pie chart
stem-and-leaf plot circle graph
spreadsheet database

1. Sarah used a ▢ on the Internet to look up hockey statistics. This type of data is called ▢.

2. Data can be organized into stems and leaves using a ▢.

3. Randy conducts a survey about sports. He records tally marks in a ▢. This type of data is called ▢.

4. A ▢ is a software tool that lets you display data as different types of graphs.

5. A ▢ is another name for a circle graph.

9.1 Collect and Organize Data, pages 274–279

6. The cook at a school cafeteria conducts a survey to find out which weekly special is the most popular.

Meal	Tally	Frequency
Hamburger	|||||	
Chili		
Submarine sandwich	||||| |||	
Pizza	||||| ||||| ||||	
Fish and chips	||||| |||||	

a) Complete the frequency table.
b) Draw a bar graph to display the data.
c) Which weekly special would you suggest the cook replace? Why?

7. The pictograph shows the total flight time of four pilots.

Total Flight Time

Yamaguchi ✈ ✈ ✈ ✈ ✈ ✈
O'Connor ✈ ✈ ✈ ✈ ✈ ✈ ✈ ✈
Lindgren ✈ ✈ ✈ ✈
Ziffareto ✈ ✈ ✈

✈ represents 10 h

a) How many hours has Yamaguchi flown?
b) Which pilot has the most flight experience? How many hours is this?
c) Approximately how many more hours has O'Connor flown than Ziffareto? How can you tell?

9.2 Stem-and-Leaf Plots, pages 280–285

8. Test scores, as percents, are shown.

71 62 83 76 49 60 73 55 89 62
91 58 63 70 81 50 66 62 73 80

a) Create a stem-and-leaf plot to display the data.
b) What is the most common score?
c) How many students scored at Level 3 (70% to 79%)?

9. The stem-and-leaf plot shows the number of matches won by tennis club members.

Stem (tens)	Leaf (ones)
2	8 9
3	0 4 4 7
4	1 3

a) Two players won the same number of matches. How many matches did each win?
b) How many matches did the best player win? Explain.
c) How many more matches did the top player win than the person ranked eighth?

9.3 Circle Graphs, pages 286–291

10. Three friends worked together on a geography project. Kelly did 33% of the work. Vaughn did 25% of the work. Martika did the rest.

  a) Draw a circle graph to show each person's effort.

  b) What percent of the project did Martika do? Explain how you found this.

11. Drago is preparing for grade 9 exams. Draw a circle graph to show a breakdown of Drago's study time.

Course	Study Time
English	30%
Mathematics	35%
Geography	25%
Physical education	10%

9.4 Use Databases to Find Data, pages 292–297

Go to www.mcgrawhill.ca/links/math7 and follow the links to access E-STAT.

12. Use E-STAT to find data about the average amount spent on books per household by province.

- From the **People** section of the Table of Contents Web page, select **Personal finance and household finance**.
- On the next page, select **Data**.
- Then, under **CANSIM**, select **Consumer spending**.
- Finally, select table **203-0011**.

  a) On the Table 203-0011 Web page, what choices need to be selected to retrieve the data asked for?

  b) Display the data in a bar graph. What does the graph tell you?

9.5 Use a Spreadsheet to Display Data, pages 298–303

Test results for a class are shown. Use the data for question 13 or 14.

	A	B	C	D	E
1	< Level 1	1			
2	Level 1	4			
3	Level 2	5			
4	Level 3	9			
5	Level 4	6			
6					
7					

13. a) The heading "< Level 1" represents "below Level 1." How many students achieved below Level 1? Identify the cell where you found this.

  b) How many students wrote the test? Explain how you know.

14. a) Use a spreadsheet to display the data.

  b) Which type of graph is best suited to display the data? Explain why you think so.

15. a) Use examples to explain the difference between a database and a spreadsheet. How are they similar?

  b) Describe a situation in which you would use a database.

  c) Describe a situation in which you would use a spreadsheet.

CHAPTER 9 Practice Test

| Strand Questions | NSN 4, 5, 9, 10 | MEA | GSS | PA | DMP 1–10 |

Multiple Choice

For questions 1 to 6, select the correct answer.

Hamburger Sales

 represents $200 in sales

1. According to the pictograph, which restaurant sold $400 in hamburgers?

   A McRobert's B Sally's
   C Burger Buddy D Fast Fry

The stem-and-leaf plot shows the number of goals scored by the forwards of a hockey team. Use the data to answer questions 2 to 4.

Stem (tens)	Leaf (ones)
0	7 9
1	1 4 4 9
2	0 2 6
3	3 8
4	8

2. How many forwards scored more than 30 goals?

   A 2 B 3
   C 4 D 5

3. How many goals did the two lowest scoring forwards get in total?

   A 7 B 9
   C 14 D 16

4. Two forwards scored the same number of goals. How many?

   A 9 B 14
   C 38 D 48

The circle graph shows how Raymond spent a $2500 scholarship. Use the information to answer questions 5 and 6.

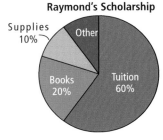

5. According to the circle graph, what did Raymond spend the most on?

   A supplies B tuition
   C books D other

6. According to the circle graph, how much did Raymond spend on "Other" items?

   A $250 B $500
   C $800 D $1200

Short Answer

7. The daily high temperatures, in degrees Celsius, for a two-week period in the summer are given.

   27 31 28 33 30 25 24
   19 27 27 34 31 29 22

   a) Create a stem-and-leaf plot to show the data.
   b) What was the most common daily high temperature? How many days had this as the daily high temperature?

8. a) Use an organizer to show the difference between a database and a spreadsheet.
   b) Why did you choose that type of organizer?

Extended Response

9. A group of 24 students were surveyed about their favourite colour.

Colour	Tally	Frequency
Blue	‖‖ ‖‖‖	
Green	‖‖‖‖	
Red	‖‖ ‖	
Purple	‖‖‖	
Other	‖‖‖	

a) Copy and complete the frequency table.
b) Display the data using a bar graph.
c) Display the data using a circle graph.
d) Which graph displays the data best? Explain.

10. A basketball coach uses the circle graph to plan her team's practices. Practices run for 90 min.

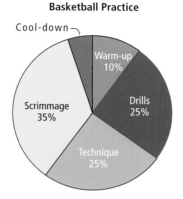

a) Determine the amount of practice time spent on each activity. Round your answers to the nearest minute.
b) Write up a practice schedule that begins at 4:00 P.M. Include the start time for each activity.

Chapter Problem Wrap-Up

In question 18 on page 279, question 12 on page 291, and question 16 on page 297, you explored collecting and organizing data about books and magazines.

You are the publisher of a new magazine for students in grades 7 to 9. You are planning the first issue. Your editor has surveyed several grade 7 and grade 8 students. The results of the survey are shown.

1. Organize and display the data.

2. What organization method did you use, and why?

3. Did you use technology? Explain why or why not.

4. What information can you get from your data display?

Grade 7 Students

Topic	Tally
Global issues	‖‖‖
Entertainment	‖‖‖‖ ‖‖‖‖ ‖‖‖‖ ‖‖
Health and fitness	‖‖‖‖
People and careers	‖‖‖‖ ‖‖‖
Sports	‖‖‖‖ ‖‖‖‖
Technology	‖‖‖‖ ‖‖‖‖ ‖‖

Grade 8 Students

Topic	Tally
Global issues	‖‖‖‖
Entertainment	‖‖‖‖ ‖‖‖‖ ‖‖‖
Health and fitness	‖‖‖‖
People and careers	‖‖‖‖ ‖‖‖‖
Sports	‖‖‖‖ ‖‖‖‖
Technology	‖‖‖‖ ‖‖‖‖ ‖‖‖

Data Management and Probability

- Collect, organize, describe, and interpret displays of data and present the information using mathematical terms.
- Evaluate data and make conclusions from the analysis of data.
- Identify and describe trends in graphs.
- Describe information presented on stem-and-leaf plots and frequency tables.
- Understand that each measure of central tendency gives different information about the data.
- Describe data using calculations of mean, median, and mode.
- Analyse bias in data-collection methods.
- Make inferences and convincing arguments that are based on data analysis.
- Evaluate arguments that are based on data analysis.

Key Words

measure of central tendency
mean
median
mode
bias

CHAPTER 10

Data Management: Analysis and Evaluation

Have you ever wondered how book and magazine publishers choose their covers? How do they decide on the right photo?

Magazine and book publishers collect, analyse, and evaluate data about their readers. This information can be used to make decisions about cover photos that will attract readers.

By the end of this chapter, you will be able to collect, analyse, and evaluate your own data. This skill will be useful in many subjects, including math, science, and geography. It will also help you make decisions around the school.

Chapter Problem

You are the publisher of a magazine for students in grades 7 to 9. You are planning a special sports issue for the spring. You decide to survey readers about possible cover photos.

Would you use this survey question? Explain.

> Perdita Felicien is an Olympic level athlete. I think we should put her on the cover. Don't you?
>
> YES   NO

Get Ready

Frequency Tables

A frequency table is used to organize survey or experimental data. For example, six of the students surveyed chose "Go to the movies" as a fun Saturday activity.

Fun Saturday Activity	Tally	Frequency
Go to the movies	IIII I	6
Play a sport	IIII II	
Go shopping	IIII	
Play video games	IIII	
Other	III	

Write the total number of tallies here. This provides the frequency.

1. a) Copy and complete the frequency table for fun Saturday activities.
   b) What was the most popular activity?
   c) What was the least popular activity?

2. How many students were surveyed about their fun Saturday activities?

Line Graphs

A line graph is used to show changes in data over time. The line graph shows the changes in the population of a small town from 1996 to 2003.

Population of Dream Point

Year	Population (1000s)
1996	12
1997	14
1998	15
1999	17
2000	20
2001	22
2002	25
2003	27

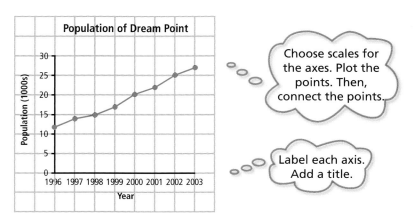

Choose scales for the axes. Plot the points. Then, connect the points.

Label each axis. Add a title.

3. Look at the line graph for the population of Dream Point.
   a) What does the graph tell you when it goes up to the right?
   b) For how many years has the population been increasing?
   c) Do you think the trend will continue? Explain.

4. The table shows the mass of a kitten over several months.

Age (months)	Mass (g)
1	100
2	150
3	220
4	280
5	320

   a) Draw a line graph for the data.
   b) How is the mass of the kitten changing?
   c) Do you think this trend will continue?
   d) How do you think the graph will eventually change? Explain why.

Mean of a Set of Data

A group of students compared their heights, in centimetres.

150   152   160   152   154   156   147

To calculate the mean, find the sum of all the values and then divide by the number of values.

$$\text{mean} = \frac{150 + 152 + 160 + 152 + 154 + 156 + 147}{7}$$
$$= \frac{1071}{7}$$
$$= 153$$

Add these first. Then, divide by 7.

Average is another word for mean.

The mean is 153 cm.

5. Calculate the mean of each set of data.
   a) 10   15   12   10   13
   b) 40   20   35   30   40   40   35
   c) 75   68   57   68   78   82   62
   d) 2.3   2.1   2.3   1.9   1.9   2.3   2.6

6. Several students want to compare their shoe sizes.

   6   8   5   7   6   7   9   7   8

   Find the mean of the shoe sizes.

10.1 Analyse Data and Make Inferences

Focus on…
- recognizing trends in data
- describing trends in data

When do you think you will stop growing? Do boys and girls stop growing at the same age? How can graphs be used to explore questions like these?

Discover the Math

How can you recognize and describe trends in data?

The line graphs show the heights of two friends measured over time.

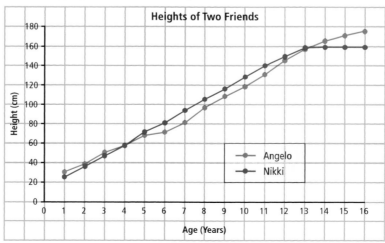

1. a) Describe how Angelo's height has changed.
   b) Describe how Nikki's height has changed.

2. a) How are these trends similar?
   b) How are they different?

3. a) Who was taller at age 6? Approximately how much taller was that person?
   b) Who was taller at age 15? Approximately how much taller was that person?

4. a) At what age were the two friends the same height?
   b) Explain how you found your answer.

5. a) Estimate how tall each friend was at birth.
   b) Explain how you found your answers.

6. a) Copy the graphs and then extend to age 25.
   b) Explain your predictions.

7. **Reflect** How does a line graph help you find trends in a data set?

Literacy Connections

Identifying Trends in Graphs

A trend is the general direction that a line graph is going. The following graphs show the sales of computer games.

- Sales are increasing.

- Sales are decreasing.

- Sales are staying the same.

- Sales increased, then levelled off.

Example 1: Analyse a Trend

Analyse the sales trends for each movie.
a) Which movie is the most popular?
b) Which movie do you think will stop playing soon? Why?
c) Which movie slowly gained in popularity and then levelled off?

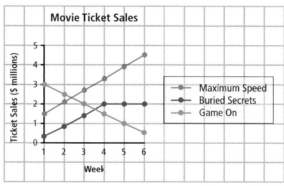

Solution

a) *Maximum Speed* is the most popular movie. Its ticket sales increased steadily.
b) *Game On* will probably stop playing soon. The ticket sales have been dropping steadily.
c) *Buried Secrets* increased in sales and then levelled off.

Example 2: Analyse a Data Set

Alysia has a baby-sitting job. She records how much she earns every week.

$25 $25 $23 $25 $26 $24 $25 $27 $26

a) Use a frequency table to organize the data.
b) Describe the data.
c) Estimate how much Alysia can earn in a month.

Solution

a)

Earnings ($)	Tally	Frequency
23	I	1
24	I	1
25	IIII	4
26	II	2
27	I	1

The smallest value is 23. The largest value is 27. The most common value is 25.

b) Alysia earns between $23 and $27 per week. She earns $25 most often.

c) 4 × $25 = $100
Alysia can earn about $100 in a month.

Most of the time Alysia earns $25 in a week. A month has about 4 weeks. So, multiply by 4.

Key Ideas

- Trends in line graphs can be analysed to solve problems and make predictions.
- A frequency table can be used to organize a set of data. This makes it easier to describe the data.

Communicate the Ideas

1. Which set of data can be represented by a line graph? Do you think drawing a line graph for the other data set would be meaningful? Explain.

A

Week	Plant Height
Week 1	1 cm
Week 2	2.2 cm
Week 3	3.4 cm
Week 4	4.2 cm
Week 5	4.5 cm

B

Animal	Population
Fox	2
Deer	15
Beaver	4
Raccoon	8
Porcupine	3

2. Sketch a line graph that shows each type of trend.
   a) does not change
   b) decreasing
   c) increasing
   d) decreased, then levelled off

3. The daily maximum temperatures for one month are recorded. How can this frequency table help you describe the data set?

Temperature (°C)	Tally	Frequency
23	IIII	4
24	HH I	6
25	HH II	7
26	HH HH	10
27	III	3

Check Your Understanding

Practise

For help with questions 4 to 7, refer to Example 1.

4. Describe each trend.
   a)   b)

5. Describe each trend.
   a)   b)

6. Match each description of a student's grade performance with the graph that shows each trend.

   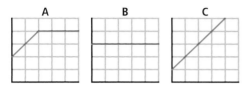

   a) Karen has shown steady improvement.
   b) Jacques's performance improved for a while, and then levelled off.
   c) Lita's marks have not changed.

7. Sergio works at a pet store. Part of his job is to train new kittens to use a litter box. Sergio records their weekly progress.

   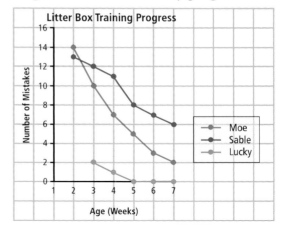

   a) How old was each kitten when it started litter box training?
   b) Which kitten do you think was already litter trained before Sergio started keeping records?
   c) Which of Moe and Sable seems to be the faster learner? Explain.

10.1 Analyse Data and Make Inferences • MHR 315

For help with questions 8 to 11, refer to Example 2.

8. Use the frequency table to describe the data set.

Male Student Heights (cm)	Tally	Frequency
147	IIII	4
148	III	3
149	HHT II	7
150	HHT	5
151	II	2

9. Use the frequency table to describe the data set.

Female Student Heights (cm)	Tally	Frequency
153	HHT	5
154	III	3
155	HHT I	6
156	IIII	4
157	II	2

10. A group of teens worked for a farmer one day, picking cherries. The number of baskets each teen picked is shown in the table.

Number of Baskets	Tally	Frequency
9	I	1
10	I	1
11	IIII	4
12	II	2
13		0
14	I	1

a) What is the least number of baskets picked by a teen?
b) What is the greatest number of baskets picked by a teen?
c) What is the most common number of baskets picked by the group of teens?

11. Dale's weekly baby-sitting earnings are shown.

$20 $20 $18 $22 $20 $24 $20

a) Create a frequency table for the data set.
b) What does the frequency table tell you about Dale's weekly baby-sitting earnings?

Apply

12. Charlie took a taxi to work and back home again. He did this for five days. Here are the fares for the 10 taxi rides.

$4.50 $5.00 $4.00 $4.50 $5.50
$3.50 $5.00 $4.50 $4.50 $4.50

a) Use a frequency table to describe the data set.
b) What is the most common taxi fare?
c) Estimate how much Charlie would spend on taxi fares per month. Explain how you solved this.

13. The graph shows a hit song's weekly chart position.

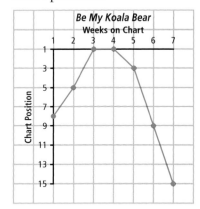

a) Describe the trend.
b) When did the song hit #1 on the chart?
c) How long did it stay at #1?
d) Look at the vertical scale of the graph. Why do you think the numbers are placed in reverse order?

14. The graph shows T-shirt sales at Diane's T-Shop.

a) When do you think the T-shirt craze started?
b) When did T-shirt sales reach a peak? Describe what happened after this.
c) Explain why this kind of information is useful to the store owner.

15. The graph shows the population of two towns.

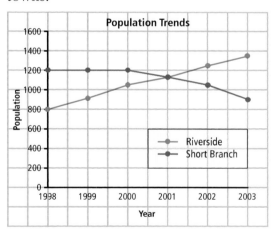

a) Describe the trend in each town's population.
b) Which town had a greater population in 1999?
c) Which town had a greater population in 2002?
d) When did the two towns have an equal population?
e) Predict each town's population in 2005. Explain your predictions.

16. The graph shows the number of television viewers on a typical weekday.

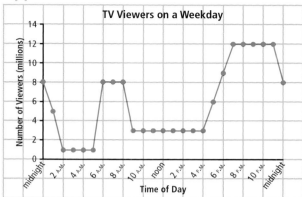

a) Which hours have the least number of viewers? Give a possible reason for this.
b) Describe the viewing trend from 4 P.M. to 11 P.M.
c) Advertisers usually pay more for commercial spots between 7 P.M. and 11 P.M. Why do you think this is?

Extend

17. The graph shows the ranking of two National Hockey League teams over the first half of the season.

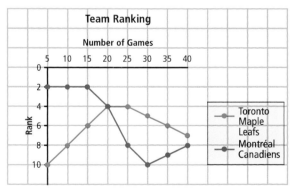

Write a half-page news story that describes the performance of each team. Include in your story
• possible explanations for each trend
• predictions for the rest of the season (you must finish 8th or higher to make the playoffs)

10.2 Measures of Central Tendency

Focus on...
- understanding mean, median, and mode
- identifying which measure of central tendency best describes a data set

How tall were most of the towers last year, Ms. Liza?

What was the middle height? What was the mean height?

measure of central tendency
- a value that a data set tends to be centred around
- the mean, median, or mode

A group of Ms. Liza's science students are entering a challenge to build the tallest tower using simple materials.

The kind of information Ms. Liza's students are asking for is related to the overall performance of last year's group. Each is asking for a **measure of central tendency**.

Discover the Math

Materials
- centimetre cubes or interlocking cubes

What do the measures of central tendency tell you about a set of data?

Last year's tower-building results for students in Ms. Liza's class are shown.

1. Build the towers. Arrange them in order from tallest to shortest.

2. Robert wants to know the height that occurred most often. This is called the **mode**. What is the mode for the towers?

mode
- the value that appears most often in a set of data

3. Monica asked about the middle height, which is the **median**. Carefully remove the tallest and shortest towers. Repeat this until there is only one tower left. What is the median height?

median
- the middle value when a set of data is arranged in order

Competitor	Tower Height (cm)
Nora	19
Miguel	14
Asra	7
Ryan	16
Erika	14
Seth	15
Kaitlyn	24
Chen	21
Alexis	14

4. Devon asked for the **mean** height. You can find this by levelling out the towers until each is the same height.

   a) Move blocks from the taller towers to the shorter towers. Make sure you keep all nine towers. Measure the height of the towers after they have been levelled out.

   b) Calculate the mean height of the towers.

   c) Compare the answer you got from levelling out the towers to the calculated mean.

5. **Reflect** Compare the values of the mean, median, and mode for this set of data. What do these values tell you about the data?

mean
- the sum of the values divided by the number of values in a set

Example 1: Use a Stem-and-Leaf Plot to Find the Mode and Median

Here are students' scores, in percent, for a report in Mr. McCuddle's English class.

74 76 65 66 72 81 57 85 71
87 71 82 91 53 71 66 80 48

a) Create a stem-and-leaf plot for the data.
b) Find the mode.
c) Find the median.

Solution

a)
Stem (tens)	Leaf (ones)
4	8
5	7 3
6	5 6 6
7	4 6 2 1 1 1
8	1 5 7 2 0
9	1

Use the tens digit for the stem. Then, the ones digit will be the leaves.

Stem (tens)	Leaf (ones)
4	8
5	3 7
6	5 6 6
7	1 1 1 2 4 6
8	0 1 2 5 7
9	1

Now arrange the leaves in increasing order.

b) The mode is 71. It appears three times.

c)
Stem (tens)	Leaf (ones)
4	8
5	3 7
6	5 6 6
7	1 1 1 2 4 6
8	0 1 2 5 7
9	1

Cross off pairs of least and greatest values until you reach the middle value(s).

There are two middle values, 71 and 72. The median is 71.5, halfway between them.

Example 2: Understand Median, Mode, and Mean

Sabra is teaching her puppy to catch a ball. For each training session, she records the number of tries it takes her puppy to catch the ball.

15   6   15   5   3   4   1

a) Find the median. Explain how you know.
b) Find the mode. Explain what this tells you.
c) Find the mean. Explain what the value tells you about the puppy's ball-catching skill.
d) Which measure of central tendency best describes the data? Explain.

Solution

a) *Method 1: Use a Model*

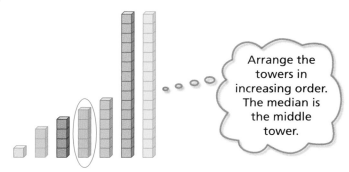

Arrange the towers in increasing order. The median is the middle tower.

The median is 5. This is the middle value.

Method 2: Arrange the Numbers

Arrange the data in increasing order.

1   3   4   5   6   15   15

The median is the middle value. Cross off pairs of least and greatest values until you reach the middle.

1̸   3̸   4̸   5   6̸   1̸5̸   1̸5̸

The median is 5. This is the middle value.

b) The mode is 15. In two cases, it took the puppy 15 tries.

Literacy Connections

Central Tendency
The mean, median, and mode are measures of central tendency. Each gives some sense of how a set of data is clustered around a centre.

- add all values in set
- divide by number of them
 → mean

- look for values that repeat
 → mode

measures of central tendency

→ median
- order the values
- locate the middle one

c) Method 1: Make a Model

The mean is 7. This means that it took the puppy 7 tries, on average, to catch the ball.

Method 2: Calculate the Mean

To calculate the mean, add the data values and then divide by the number of values.

$$\text{mean} = \frac{15 + 6 + 15 + 5 + 3 + 4 + 1}{7}$$
$$= \frac{49}{7}$$
$$= 7$$

The mean is 7. This means that it took the puppy 7 tries, on average, to catch the ball.

d)

The mean is 7 tries. The puppy took fewer tries in five of the training sessions. I would not choose this value.

The mode is 15 tries. This is the greatest number of tries for the puppy. The puppy was probably just starting to learn to catch a ball. I would not choose this value.

The median is 5 tries. The puppy took fewer tries about half of the time. It needed more tries about half of the time. I would choose this value.

The median is the best measure of central tendency in this case. It describes a typical training session for Sabra's puppy.

Key Ideas

- The three main measures of central tendency are the median, mode, and mean.

The median describes the middle value in a data set.

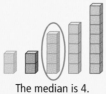

The median is 4.

The mode describes the most common value in a data set.

The mode is 2.

The mean describes the average value of a data set.

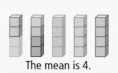

The mean is 4.

Communicate the Ideas

1. Use an organizer to show what the median, mode, and mean show about a set of numeric data.

2. Cheryl says: "The mode for the towers in Key Ideas is 2. I think this value best describes the towers."
   Michel says: "The median for the towers is 4. I think this value best describes the towers."
   Fareeha says: "The mean for the towers is 4. I think this value best describes the towers."
   Who is right? Explain.

Literacy Connections

Organizers
Tables, flow charts, and mind maps are useful organizers. There is a sample of a flow chart on page 238. A sample mind map is on page 320.

Check Your Understanding

Practise

For help with questions 3 to 6, refer to Example 2.

3. Model each set of data using blocks or a diagram. Then, find the median, mode, and mean.
   a) 4, 2, 9, 6, 4
   b) 11, 5, 8, 11, 10
   c) 4, 5, 8, 5, 6, 9, 5

Use the table to answer questions 4 to 6.

A basketball coach records the number of successful free throws for her teams. Each player gets 10 shots. The results are shown for nine boys and eight girls.

Boys' Team Baskets	6	4	8	7	6	9	5	6	2
Girls' Team Baskets	7	6	7	8	9	3	5	7	

4. a) Model the set of data for the boys' team using blocks or diagrams.
   b) Find the median, mode, and mean. Explain what each value tells you about the free-throw skills of the boys.

5. a) Model the set of data for the girls' team using blocks or diagrams.
   b) Find the mode, median, and mean. Explain what each value tells you about the free-throw skills of the girls.

6. a) Which team do you think has a better record?
   b) Compare the median, mean, and mode for each team.

For help with questions 7 to 9, refer to Example 1.

7. Find the median and the mode for each data set.

   a) | Stem (tens) | Leaf (ones) |
      |---|---|
      | 3 | 2 5 |
      | 4 | 1 1 7 8 |
      | 5 | 3 6 9 |

   b) | Stem (tens) | Leaf (ones) |
      |---|---|
      | 6 | 0 |
      | 7 | 2 5 6 |
      | 8 | 0 3 3 |

8. Find the median and the mode for each data set.

   a) | Stem (tens) | Leaf (ones) |
      |---|---|
      | 4 | 4 7 |
      | 5 | 1 1 2 3 |
      | 6 | 2 3 5 9 |

   b) | Stem (tens) | Leaf (ones) |
      |---|---|
      | 5 | 5 |
      | 6 | 2 3 8 |
      | 7 | 0 2 2 4 |

9. Create a stem-and-leaf plot for each data set.
   a) 30, 45, 37, 42, 35, 47
   b) 41, 53, 49, 67, 52, 41, 63

Apply

10. Pina plays defence for her school's hockey team. Her time on the ice for each game is recorded, in minutes.

    21, 22, 19, 24, 23, 19, 20, 24

    a) Find the median and mean.
    b) In this case, there are two modes. What are they?
    c) Which measure of central tendency best describes a typical game for Pina? Explain.

11. The final marks for a physical education class are given.

    | 84 | 68 | 71 | 55 | 66 | 63 | 82 |
    | 92 | 70 | 75 | 64 | 58 | 73 | 88 |
    | 65 | 73 | 76 | 73 | 62 | 83 | |

    a) Create a stem-and-leaf plot for the data.
    b) Find the median, mode, and mean.
    c) Which measure of central tendency best describes the marks? Explain.

Chapter Problem

12. The table shows the sales of your teen magazine in one sports store.

Month	Number
January	6
February	6
March	12
April	20
May	25
June	35

    a) Find the median, mode, and mean. Explain what each value tells you about the monthly magazine sales.
    b) Which measure of central tendency best describes the monthly magazine sales? Explain.

13. A clothing store needs to place a monthly order for winter jackets. The sizes that sold last month are shown in the table.

Size	Tally	Frequency
30	I	1
32	I	1
34	III	3
36	IIII II	7
38	IIII	4
40	II	2
42	I	1
44	I	1

    a) Find the three measures of central tendency. Explain how you found them.
    b) Which of these is the most important to the store manager? Explain why.

14. Try This! Choose one of the following characteristics:
    - height
    - hand width
    - arm span
    - shoe size

    a) Measure this characteristic for 10 to 20 friends and family members.
    b) Display the data.
    c) Find the measures of central tendency.
    d) State any conclusions you can make from your findings.

Extend

15. Zack has five major tests in geography, all marked out of 50. His scores for the first four tests are 33, 42, 38, and 44.

    a) What is Zack's current mean test score?
    b) What score must Zack get on the last test to raise his mean score to 80%?
    c) Explain how you solved part b).

Use Technology

Focus on...
- calculating mean, median, and mode

Use technology to find the measures of central tendency. This is another way of doing Example 1 on page 319.

Materials
- AppleWorks 6.2 or other spreadsheet software
- computers

Optional
- TECH 10.2A Measures of Central Tendency (AppleWorks 6.2)
- TECH 10.2B Measures of Central Tendency (AppleWorks 5.0)
- TECH 10.2C Measures of Central Tendency (Quattro® Pro 10)
- TECH 10.2D Measures of Central Tendency (Microsoft® Excel 2002)

Find Measures of Central Tendency With a Spreadsheet

Sabra is teaching her puppy to catch a ball. For each training session, she records the number of tries it takes her puppy to catch the ball.

15, 6, 15, 5, 3, 4, 1

1. Open a new spreadsheet document in AppleWorks 6.2.
   Enter the text and values, as shown.
   - Click a cell.
   - Type in the text or value.
   - Press **Enter**.

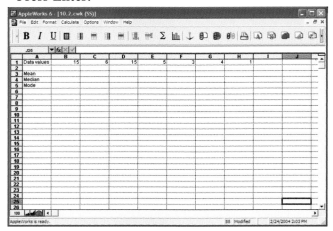

2. Use the **AVERAGE** function to calculate the mean of the data values.
   - Click cell B3.
   - Type **=AVERAGE(B1..H1)**.
   - Press **Enter**.

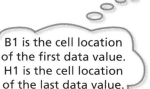

B1 is the cell location of the first data value. H1 is the cell location of the last data value.

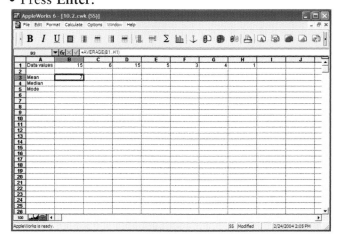

324 MHR • Chapter 10

3. Arrange the data values in increasing order.
   - Click cell B1.
   - Hold the Shift key down and click cell H1.
   - From the **Calculate Menu**, choose **Sort...**.
   - In the dialogue box, select **Horizontal**, then click **OK**.

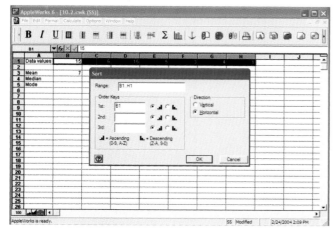

4. Your data should now be sorted with B1 containing the smallest value and H1 the largest. Find the median and the mode of the data values.
   - The median is the middle value. There are seven data values. The median is the fourth value. Click cell B4. Type the value located in cell E1.
   - The mode is the most common value. Scan the sorted list. Notice that 15 occurs twice. Click cell B5. Type **15**.

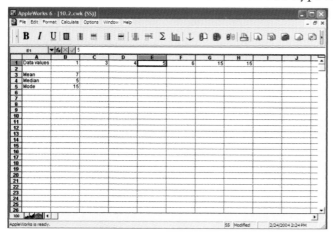

I can use technology to find the measures of central tendency for the number of parents attending our annual concert. This will help me decide on the number of chairs to put out. How might you use it?

5. **Reflect** How can you use technology to calculate the mean, median, and mode of a data set?

10.3 Bias

Focus on...
- identifying bias in questions and responses
- rewording questions to remove bias

When conducting a survey, how can you make sure that the questions do not influence the responses?

Consider the following questions:
"Do you like my new hairstyle?"
"My brother's new band is really great! What do you think?"
"What's your favourite vegetable: broccoli, turnip, or Brussels sprouts?"

The questions above contain **bias**. How each question is worded or the way that it is asked can affect the responses.

bias
- encourages a certain response by the wording of a question

Discover the Math

How can you ensure that survey questions and responses are free of bias?

Example 1: Bias in Survey Responses

Ms. Mason is the school cafeteria cook. She decides to gather feedback on a new special.

Explain why Ms. Mason's survey on yesterday's special may get biased responses. What could she do to reduce the bias?

Solution

Students may not answer honestly. They do not want to hurt Ms. Mason's feelings. To reduce the bias, Ms. Mason could place survey cards on the cafeteria tables. Students could then put their responses into a box.

Example 2: Bias in Survey Questions

Read each survey question. Decide whether it contains bias. If there is bias, explain how it could be removed.

a)
What is your favourite sport?
A   Hockey
B   Golf
C   Baseball
D   Other _____

b)
Should schools have a more reasonable schedule by changing to a four-day week?
   YES   NO

c)
Do you think professional athletes are paid too much, about right, or too little?

Solution

a) More people might select the three sports listed. This is easier to do than write in a different choice. This question contains bias. To remove the bias, ask the question without the choices: "What is your favourite sport?"

b) The first part of the question tries to make you think that the current school schedule is unreasonable. Then, more people might answer yes. This question contains bias. To remove the bias, reword the question: "Should schools change to a four-day-week schedule?"

c) The choices in this question seem to be fair. This question does not seem to contain bias.

Literacy Connections

Bias and Tone of Voice

Your tone of voice when asking a question can also show bias.

Ask the question from part c) of Example 2 using various tones of voice.

Emphasize different words to show bias.

Key Ideas

- Survey questions should be free from bias.
- How a question is asked can influence responses.
- The wording of a question can encourage certain responses more than others.

Communicate the Ideas

1. Who is right? Explain why.

2. Ms. Mason is the school cafeteria cook. She decides to gather feedback on a new special. She places survey cards on the cafeteria tables. Will the results be free from bias? Explain why or why not.

Check Your Understanding

Practise

For help with questions 3 to 6, refer to Example 1.

A middle school is having an art contest for a new school logo. Two grade 7 students, Wes and Faye, enter the contest. To get an idea of who might win, the following questions are asked of fellow students. How do you think the students will answer questions 3 to 6? Explain why.

3. Wes asks, "Do you like my design the best?"

4. A friend of Faye's asks, "Do you like my friend Faye's design?"

5. Faye asks, "Do you *really* like Wes's design?"

6. A friend of Wes's asks, "Which design do you like best?"

For help with questions 7 to 10, refer to Example 2.

7.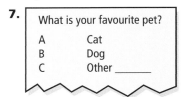

   a) What choices do you think are expected?
   b) How does the wording of this question influence your choice?

8.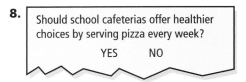

   a) What answer do you think is expected?
   b) How does the wording of this question encourage you to pick that answer?

9. What response is expected? Explain why.

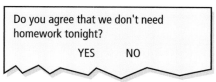

10. What response is expected? Explain why.

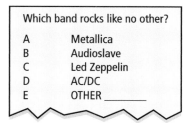

Apply

11. Consider the survey question shown.

Which band rocks like no other?
A   Metallica
B   Audioslave
C   Led Zeppelin
D   AC/DC
E   OTHER _____

   a) Is this question free from bias? Why or why not?
   b) Is it clear what the question is asking?
   c) If your answers to parts a) and b) are both No, reword the question to remove the bias and make it easy to understand.

12. a) Describe a survey method that will produce biased responses.
    b) Explain why you think the bias exists.
    c) Describe how the bias can be removed.

13. Read each survey question. Decide whether it contains bias. If there is bias, rewrite the question to remove the bias.

   a) Due to the horrible start of a professional baseball team's season, do you think the manager should be fired?
      YES   NO

   b) Do you agree or disagree that fighting in professional hockey should be eliminated?

   c) Who is the most famous Prime Minister of all time?
      A   Pierre Trudeau
      B   John A. MacDonald
      C   Jean Chrétien
      D   Other _____

14. Consider the following three survey questions.

   "Since they give students a sense of pride, should school uniforms be made mandatory?"

   "Since they take away a student's freedom of expression, should school uniforms be made optional?"

   "Should school uniforms be made optional or mandatory?"

   a) Which questions contain bias? Explain.
   b) Which question do you think was written by a teacher? Explain why.
   c) Which question do you think was written by a student? Explain why.
   d) Which is the best survey question to use? Explain why.

15. A television magazine conducts a survey to find out the most popular television show.

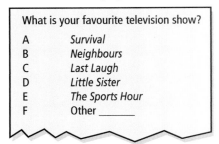

a) Describe the bias in this survey question.
b) How might a television broadcasting station benefit from the bias in this survey question?
c) How could the question be changed to remove the bias?

Chapter Problem

16. Which question does not contain bias? Explain.

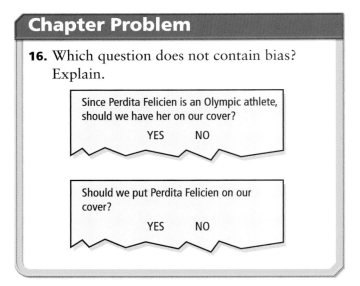

17. a) Create a short comic strip or cartoon that illustrates a biased survey question.
b) Trade with a classmate. Identify the bias in each other's survey question.

18. a) Write a survey question that contains bias.
b) What response(s) do you expect?
c) Conduct the survey with at least 10 friends, classmates, or family members. Use a table to record your results.
d) Compare your survey results with your answer to part b).
e) Reword your survey question to remove the bias.

Extend

19. Search the Internet, newspapers, or magazines to find an article that contains a biased survey method. Identify the article, its location, and the type of bias present. Suggest how the bias could have been removed.

Making Connections

Media Bias

Many media stories include bias. Look at these sample news headlines. Which ones suggest bias? Explain.

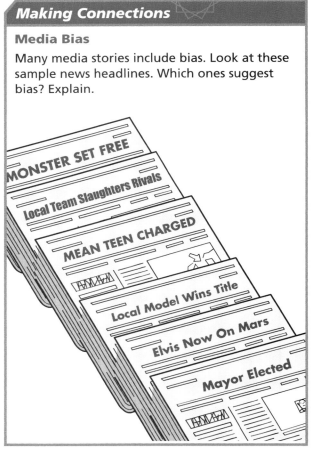

10.4 Evaluate Arguments Based on Data

Focus on...
- evaluating statements based on data
- recognizing misleading graphs

Your low-fat alternative.
Now with just 5 grams of fat and only 300 calories!*

*Addition of cheese or condiments alters nutritional value.

Some fast food restaurants now offer healthier menu choices. How do you know that the meal is actually healthier?

Advertisers sometimes make bold statements based on data. Often their goal is to convince you to buy something. Can you believe everything in an advertisement?

Discover the Math

How can you evaluate statements that are based on data?

Literacy Connections

Bias and Advertisements
Advertisements can show bias by providing misleading information. The information is not incorrect. It is displayed in a manner that gives an incorrect impression.

1. How is this advertisement trying to convince you to buy the sandwich shown?

2. a) Read the fine print at the bottom of the ad. What information does this tell you?
   b) Compare this information with the picture of the sandwich. What does this tell you about the advertiser's claim?

3. Suppose you add cheese and mayonnaise.
   a) Use the table to find the total fat content and number of calories of your sandwich.
   b) Compare the total fat content and number of calories with those claimed in the ad. By how much have these values changed?

Item	Fat (g)	Calories
Basic Sub	5	300
Cheese	8	100
Mayonnaise	32	300
TOTAL		

4. **Reflect** In what ways do you think the advertisement misleads people? Why do you think the advertiser only used certain facts?

Example: Misleading Graphs

Glide On skateboards used the graph shown as part of an advertisement.

How does the graph exaggerate the savings if you buy a Glide On skateboard?

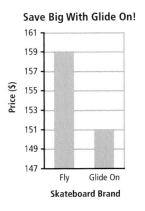

Solution

One bar looks more than three times as tall as the other. You might think that the savings are over 50%.

By showing just part of the vertical scale, the advertiser has exaggerated the difference in prices.

Here is the entire graph. It shows how small the price difference really is.

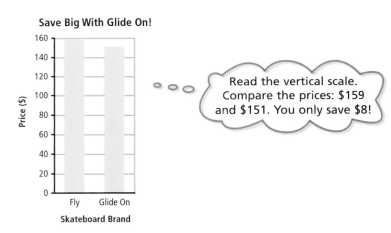

Read the vertical scale. Compare the prices: $159 and $151. You only save $8!

Key Ideas

- The media often make statements that are based on data. Sometimes important points are intentionally left out.
- A misleading graph can be used to exaggerate a point.

Communicate the Ideas

1. Is this graph misleading? Explain.

2. Why do you think an advertiser would use a misleading statement or graph?

Cool Flavours

Dairy Tasty

$4.5 million $4.3 million

Cool Flavours! It's the ice cream kids scream for!

Check Your Understanding

Practise

For help with questions 3 to 6, refer to the Example.

3. The two graphs show the increase in the price of hockey tickets over the past few years.

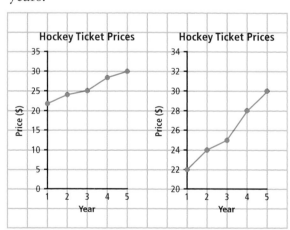

a) How are the graphs different?
b) What impression does each graph give about the price increases?

4. The two graphs show the number of chocolate bars sold by two students.

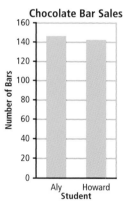
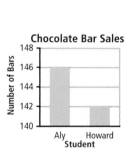

a) How are the graphs different?
b) What impression does each graph give about how many chocolate bars were sold?

5. The two graphs show the number of bags of leaves raked by two friends.

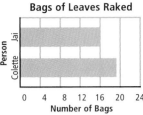

a) How are the graphs different?
b) What impression does each graph give about how many bags of leaves were raked?

6. The graph shows the progress of friends playing a video game.

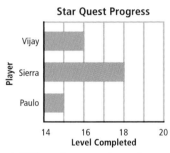

a) What impression does the graph give?
b) Explain why this graph is misleading.

Apply

7. Draw a new graph for the data in question 6 that is not misleading.

8. This graph was used in a cat food ad.

Meow Chow! Because your cat *says* so!

a) Which company do you think created this ad? Explain.
b) Why is this graph misleading? What effect has been achieved?

9. Two television show producers are competing for sponsorship from a major advertiser. The ratings are in the table below. The ratings are a percent of the entire viewing audience.

Show	Year 1	Year 2	Year 3	Year 4
Happy Times	20	17	18	16
Buddies	12	13	14	15

The advertiser is shown two different graphs, one prepared by each television producer.

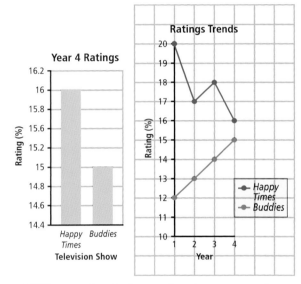

a) What rating information is presented in each graph?
b) Which graph do you think was created by the producer of *Happy Times*? Explain.
c) Which graph do you think was created by the producer of *Buddies*? Explain.

10. How might this graph mislead people?

Where Should We Sell T-Shirts?

11. Is this graph misleading? Explain.

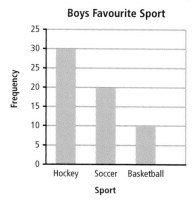

12. Results of a music video showdown are displayed for the television viewing audience.

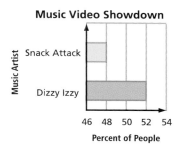

a) What impression does the graph give?
b) Is this graph misleading? Explain.
c) Draw a new graph for the data that is not misleading.
d) Compare the two graphs. Describe what you notice.

13. Is this graph misleading? Explain.

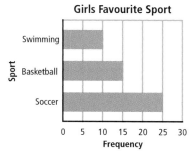

14. A group of teens were surveyed about the amount of time they spend using a computer.

Time	Tally	Frequency
Daily	|||| ||||		
1 to 6 times a week	|||| |||| ||||	
Less often	|||| |||	

a) Copy and complete the frequency table.
b) Draw a graph that does not distort the data.

 15. Find some data that you are interested in.

a) Draw a graph that does *not* distort the data.
b) Explain why your graph is not misleading.

Extend

16. Look at the sports page of a newspaper, or use the Internet to find some current sports data. Pick a team.

a) Create a misleading graph based on the data you found. Make it look like the team is doing really well. Write one or two sentences based on your graph to explain why the team is doing so well.
b) Use the same data, or other data, and create a new misleading graph. Make it look like the team is doing poorly. Write one or two sentences based on your graph to explain why the team is struggling.
c) Show these graphs to a parent, guardian, or older sibling. Can you convince the person that you are right in both cases?

CHAPTER 10 Review

Key Words

For questions 1 to 5, copy the statement and fill in the blanks. Use some of these words: mode, mean, measures of central tendency, bias, median, unbiased

1. 6 is the _____ of 7, 2, 6, 8, 2.

2. 5 is the _____ of 7, 2, 6, 8, 2.

3. Questions that encourage you to answer a certain way contain _____.

4. Mean, median, and mode are all _____.

5. 2 is the _____ of 7, 2, 6, 8, 2.

10.1 Analyse Data, pages 312–317

Use the graph for questions 6 to 8. The graph shows the population for the three high schools in a city.

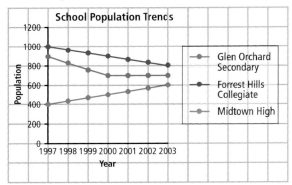

6. a) Describe the trend for each high school population.
   b) Which school had the greatest population for the time period shown?

7. Predict which school will have the greatest population in two years. Justify your prediction.

8. Estimate the population of each high school for each year.
   a) 1999
   b) 2005

Use the data for questions 9 and 10. Oswald grew tomatoes in his garden. He recorded the number of tomatoes picked from each plant.
8, 11, 9, 7, 8, 10, 6, 8, 9, 8

9. a) Create a frequency table for the data set.
   b) Use the frequency table to describe the data set.

10. a) How many tomatoes were picked in total?
    b) Oswald would like a total of 100 tomatoes next year. Estimate how many more tomato plants Oswald should plant. Explain your solution.

10.2 Measures of Central Tendency, pages 318–323

11. Phyllis books appointments at a dentist's office. She recorded the number of appointments that were made one week in a table.

Monday	Tuesday	Wednesday	Thursday	Friday
16	15	13	3	3

a) Find the mean, median, and mode. Explain what each value tells you about the appointments.
b) Which measure of central tendency best describes the appointments at the office? Explain your choice.

12. The final marks for a science project are shown.

   78 74 80 66 65 75 47
   68 60 85 77 92 61 84
   81 55 87 53 62 67 92

   a) Create a stem-and-leaf plot to organize the data.
   b) Find the median and the mode.
   c) Which value from part b) best describes the overall performance of the entire class? Explain.

10.3 Bias, pages 326–330

13.

   a) Describe the bias in the interviewer's question.
   b) Who do you think may have written the question? Explain.
   c) Reword the question to remove the bias.

14. One of the judges in a talent show is also a parent of a child in the show.
   a) Explain the possible bias in this situation.
   b) Describe how the bias could be removed.

15.

   a) Describe the bias in this survey question.
   b) How could the question be changed to remove the bias?

10.4 Evaluate Arguments Based on Data, pages 331–335

16. The results of a bake-sale competition between a grade 7 and a grade 8 class are shown.

   a) What impression does the graph give?
   b) Is this graph misleading? Explain.

17. People were surveyed about their favourite submarine sandwich shop.

Shop	Tally	Frequency
Sub Lubber's	‖‖‖ ‖‖‖ I	
Hungry Cat	‖‖‖ ‖‖‖ II	
Hero Plus	‖‖‖ III	
Other	‖‖‖ II	

   a) Copy and complete the frequency table.
   b) Draw a graph that exaggerates the popularity of Hungry Cat.
   c) Draw a similar graph that does not distort the data.
   d) Which graph might the owner of Hungry Cat use? Why?

CHAPTER 10 Practice Test

| Strand Questions | NSN | MEA | GSS | PA | DMP 1–9 |

Multiple Choice

For questions 1 to 4, select the correct answer.

1. Which of the following is the best choice for displaying trends in data?
   A circle graph
   B bar graph
   C pictograph
   D line graph

Use the data for questions 2 to 4.
Greg recorded the number of hours he spent watching television each day last week.

   3, 2, 1, 0.5, 1, 6, 4

2. What is the mean of the number of hours Greg spent watching television?
   A 1
   B 1.5
   C 2
   D 2.5

3. What is the mode of the number of hours Greg spent watching television?
   A 1
   B 1.5
   C 2
   D 2.5

4. What is the median of the number of hours Greg spent watching television?
   A 1
   B 1.5
   C 2
   D 2.5

Short Answer

5. The graph shows game attendance for two basketball teams.

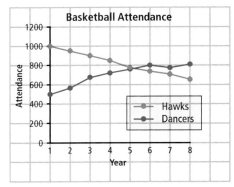

   a) Describe the trend for each team's game attendance.
   b) Predict the attendance for each team's next game.

6. Parents at a children's soccer game are asked to fill out a survey. The results will be used to pick players for an all-star game.
   a) Describe the possible bias in this survey.
   b) How could the bias be removed?

7. The company that makes Bubbles pop used the graph shown in an advertisement.

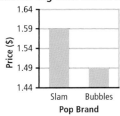

   a) Is the graph misleading? Explain.
   b) How do you think the company will benefit from this ad?

Extended Response

8. A group of teens worked for a farmer one day, picking cherries. The number of baskets each teen picked is recorded in the table.

Number of Baskets	Tally	Frequency
9	I	1
10	I	1
11	IIII	4
12	II	2
13		0
14	I	1

a) How many teens were picking cherries?
b) How many baskets did these teens pick, in total?
c) Estimate how many more workers the farmer should hire to get a total of 150 baskets per day. Explain your thinking.

9. Rena collected the mathematics test scores from 11 of her friends and added her own. This is the data set she collected. The test was out of 85.

66 80 68 72 75 64
81 73 73 74 75 75

a) Use a frequency table to describe the data set.
b) Rena received the mean score. What mark did she get?
c) Find the other two measures of central tendency.

10. Explain how a graph can be used to mislead the viewer. Create your own data and misleading graph to support your answer.

Chapter Problem Wrap-Up

You are the publisher of a magazine for students in grades 7 to 9. You need information to help you plan the magazine covers for future issues.

1. Design a reader survey.

2. Collect, organize, and display data from possible readers.

3. Analyse the data you have collected. What decisions might you make? Explain why.

Making Connections

What does data analysis have to do with saving animals at risk?

As our world changes, some species of animals and plants become extinct. Two species that are at risk in Ontario are the wolverine and the polar bear.

The wolverine is a tough member of the weasel family. It is about the size of a dog and lives in the forests of Northern Ontario. Logging and hunting are two possible factors in the decline of the wolverine population.

The polar bear spends much of its time hunting seals on ice caps. When the ice caps melt, the bear is forced to live on land, where it often goes hungry. Global warming, hunting, and pollution may also be factors in the decline of the polar bear.

Wildlife researchers can use population data trends to track species at risk. Then, by making predictions, the researchers can make recommendations to protect these valuable creatures.

 Go to www.mcgrawhill.ca/links/math7 and follow the links to find out more about animals and plants at risk in Ontario.

Making Connections

What do your buying habits have to do with trends in the consumer market?

The products you buy influence what products are available. There are about 2.5 million kids between the ages of 9 and 14 in Canada. They spend almost $2 billion themselves and influence over $20 billion in household purchases.

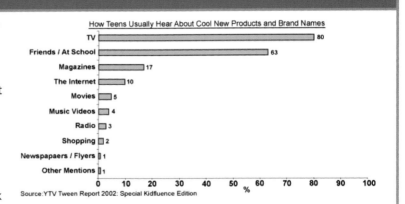

The Canadian youth television network YTV does research on Canadian kids. They collect information on topics such as leisure and sports activities, spending habits, and media habits.

Why would this information be useful to YTV?

What other companies might find this information helpful?

Go to www.mcgrawhill.ca/links/math7 and follow the links to find out more about "Kidfluence" and Kid Trends.

Plan a Television Schedule

You are the programmer for a television station. Your boss asks you to plan a new fall schedule for Wednesday evenings. Evening programming starts at 4 P.M. and ends at 10 P.M.

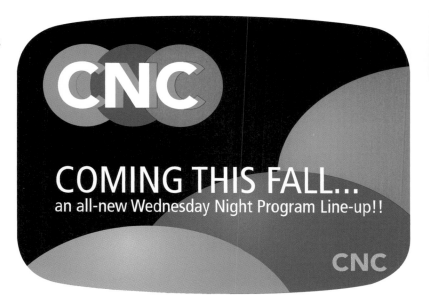

1. Design a survey to find out
   - what types of shows you might air
   - specific shows your audience would like to see
   - the best time slot for a news program
   - the best time slots to air other popular shows

2. Conduct your survey.

3. Organize and present your data. Be sure to use some measures of central tendency.

4. Based on your data, plan one evening's programming and provide a viewing guide.

5. After reading your viewing guide, your boss sent the following email.

   FROM: KimikaR
   TO: PaulG
   SUBJECT: shows and viewers
   - Why did you pick the shows you did?
   - How did you come up with the time slots?
   - Why do you think viewers will tune in to our station?
   - What type of viewer will watch us?

   Answer your boss's questions. Support your answers with appropriate data.

Number Sense and Numeration
- Compare and order integers.
- Represent integers using counters, a number line, and symbols.
- Add and subtract integers, with and without the use of manipulatives.
- Use a calculator to solve problems that involve large numbers.

Patterning
- Identify and use a pattern.

Data Management and Probability
- Make predictions from graphed data.
- Read and create bar graphs and scatterplots.

Key Words
integer
positive integer
negative integer
opposite integers
zero principle

CHAPTER 11

Integers

Have you ever felt warm one moment and then colder because of a gust of wind? The change in temperature was caused by wind chill. For example, the thermometer may report a temperature of 5°C on a bright winter day. But if there is a wind, it can *feel* as cold as –2°C.

Understanding integer concepts will help you to read and understand wind chill values. Go to **www.mcgrawhill.ca/links/math7** and follow the links to find out more about wind chill.

Chapter Problem

The thermometer reads –10°C. The wind is from the northwest at 25 km/h. The radio weather forecaster says that this gives a wind chill of –19. This means it feels as cold as it would on a calm day with a temperature of –19°C.

What patterns or trends can you see in the wind chill chart? Start by looking down the –10°C temperature column and across the 25 km/h wind speed row.

Wind Chill Chart

Wind Speed (km/h)	Air Temperatures (°C)							
	5	0	–5	–10	–15	–20	–25	–30
5	4	–2	–7	–13	–19	–24	–30	–36
10	3	–3	–9	–15	–21	–27	–33	–39
15	2	–4	–11	–17	–23	–29	–35	–41
20	1	–5	–12	–18	–24	–31	–37	–43
25	1	–6	–12	–19	–25	–32	–38	–45
30	0	–7	–13	–20	–26	–33	–39	–46
35	0	–7	–14	–20	–27	–33	–40	–47
40	–1	–7	–14	–21	–27	–34	–41	–48
45	–1	–8	–15	–21	–28	–35	–42	–48
50	–1	–8	–15	–22	–29	–35	–42	–49
55	–2	–9	–15	–22	–29	–36	–43	–50
60	–2	–9	–16	–23	–30	–37	–43	–50

Get Ready

Identify Integers

A debt of $5 can be represented by the integer –5.
A temperature of 3°C above freezing can be represented by the integer +3.
A depth of 4 m below sea level can be represented by the integer –4.

These integers are shown on the number line.

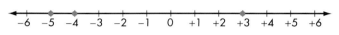

1. Represent each situation using an integer. Show the integers on a number line.
   a) a loss of $6
   b) a temperature of 5°C below freezing
   c) sea level
   d) a gain of 2 kg

2. Write a simple situation that each integer might represent. Use a different type of situation for each.
   a) –10   b) +3   c) –17

3. What integer is represented by each point on the number line?

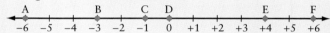

4. Describe how the positions of the integers in each pair are related to 0.
   a) +1 and –1
   b) –5 and +5

Compare and Order Numbers

Earth has one moon. The table shows the number of moons other planets have.

You can use one of the symbols >, <, or = to compare the number of moons that Jupiter and Uranus have. Jupiter has more moons than Uranus. 16 > 11.

The numbers of moons written in increasing order are 0, 2, 11, 16, 18.

Planet	Number of Moons
Jupiter	16
Mars	2
Saturn	18
Uranus	11
Venus	0

5. Compare the numbers in each pair using > or <.
   a) 12 ■ 15   b) 32 ■ 23
   c) 20 ■ 0   d) 33 ■ 42
   e) 29 ■ 30   f) 4 ■ 40

6. Write the numbers in each set in increasing order.
   a) 8, 10, 5, 19, 2
   b) 20, 11, 35, 6, 0, 12
   c) 16, 27, 30, 1, 23, 15, 12

Find the Mean of a Set of Data

Ten students were asked how many hours of television they watch per week.
Their responses were 15, 12, 0, 4, 18, 7, 12, 6, 7, 3
Calculate the mean number of hours.

$$\text{Mean} = \frac{\text{sum of data}}{\text{number of pieces of data}}$$
$$= \frac{15 + 12 + 0 + 4 + 18 + 7 + 12 + 6 + 7 + 3}{10}$$
$$= \frac{84}{10}$$
$$= 8.4$$

The mean number of hours of television watched per week is 8.4.

7. Calculate the mean of each set.
   a) 2, 0, 3, 7, 8
   b) 10, 30, 50, 40, 30, 40, 10
   c) 27, 13, 22, 46, 18, 55, 0, 19
   d) 5.2, 6.3, 8.2, 10.8, 11.5

8. Calculate the mean of each set of data.
   a) Marks, in percent: 78, 56, 92, 43, 66
   b) Ages, in years: 3, 5, 3, 7, 6, 10, 3, 2, 6, 5
   c) Cost of a hamburger, in dollars: 1.49, 1.35, 2.79, 0.99, 2.45, 1.25, 1.85, 2.29

Find the Median of a Set of Data

The price of a movie ticket in seven cities is recorded.
$7.50, $10.00, $9.50, $8.00, $8.50, $9.00, $8.50
What is the median price?

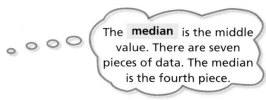

The **median** is the middle value. There are seven pieces of data. The median is the fourth piece.

Arrange the prices in increasing order.
$7.50, $8.00, $8.50, $8.50, $9.00, $9.50, $10.00

The median price of a movie ticket is $8.50.

9. Find the median of each set.
   a) 3, 5, 1, 0, 8
   b) 12, 10, 9, 15, 10, 15, 14
   c) 33, 41, 87, 93, 54, 50, 22, 50
   d) 35, 96, 10, 64, 100, 56

10. Find the median of each set of data.
   a) Heights, in metres: 89, 105, 77, 95, 121
   b) Test scores, out of ten: 7, 9, 6, 5, 10, 7, 9, 8, 8, 10, 7, 8, 9, 6, 9
   c) Cost of one dozen eggs, in dollars: 1.95, 1.89, 1.90, 2.05, 1.97, 1.85, 1.98

11.1 Compare and Order Integers

Focus on...
- comparing integers
- ordering integers
- using a number line to represent integers

When a spacecraft is about to be launched, you may hear the ground crew say something like "T minus 10 minutes." What does this phrase mean? How can the phrase be represented by an **integer**?

integers
- build one of the numbers ..., −3, −2, −1, 0, +1, +2, +3, ...

Discover the Math

How can you compare integers?

The bar graph shows the mean high temperatures, during the month of December, for six Ontario cities.

1. **a)** Which of the six cities is the coldest in December?
   b) Which city is the warmest?

2. **a)** Order the cities from coldest to warmest.
   b) List the temperatures in increasing order.

3. Which city do you think is farthest north? Why?

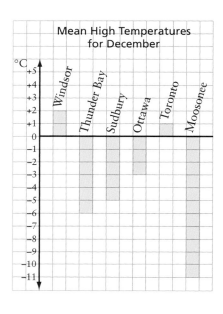

346 MHR • Chapter 11

4. a) Waterloo's mean high temperature in December is 0°C. Which of the six cities are colder than Waterloo?

b) The mean high temperature in Ottawa last December was −4°C. Was it colder or warmer than usual?

5. a) How many degrees warmer is Thunder Bay than Moosonee?

b) How many degrees colder is Ottawa than Toronto?

6. Reflect Think about the process you use to compare and order integers.

a) You are asked to compare three integers. One integer is negative, one is positive, and one is zero. How do you arrange the three integers in order from the least to the greatest?

b) Describe the steps you would use to arrange six different integers in increasing order.

Example 1: Use a Vertical Number Line

Use a number line to show the temperatures +4, −3, −4, +6, 0, and −5. List the temperatures from the coldest to the warmest.

Solution

Temperature is often shown on a vertical number line.

Positive integers are above 0.

Numbers are evenly spaced. The integers increase upward; they decrease downward.

Negative integers are below 0.

Listed from coldest to warmest, the temperatures are −5, −4, −3, 0, +4, +6.

positive integer
- one of the numbers +1, +2, +3, …

negative integer
- one of the numbers −1, −2, −3, …

Literacy Connections

Reading Vertical Number Lines
You see a vertical number line in some types of thermometer.

−4 and +4 are **opposite integers**. They are an equal distance to the left and right of zero on an integer number line.

opposite integers
- two integers with the same numeral but opposite signs

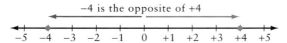

Example 2: Use a Horizontal Number Line

Compare the integers −6, −3, and +1 with the integer −4. Use a number line, words, and symbols.

Solution

Negative integers are to the left of 0.

Positive integers are to the right of 0.

−6 −5 −4 −3 −2 −1 0 +1 +2 +3 +4 +5 +6

Numbers are evenly spaced. The integers increase to the right and decrease to the left.

Literacy Connections

Reading > and <
The symbol > is wider on the left. So, a number to the left of > is larger than a number to the right. You read 10 > 8 as "ten is greater than eight."

Words	Symbols
−6 is less than −4	−6 < −4
−3 is greater than −4	−3 > −4
+1 is greater than −4	+1 > −4

Key Ideas

- A thermometer uses a vertical number line.

  positive temperatures
  0
  negative temperatures

- A horizontal number line is a useful tool for comparing and ordering integers. Numbers are evenly spaced to the left and right of 0.

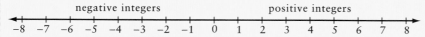

- Opposite integers are an equal distance to the left and right of 0.

- You can compare two different integers using the symbols > and <. For example, −3 < 0 means −3 is less than 0.

348 MHR • Chapter 11

Communicate the Ideas

1. On a horizontal integer number line, do the numbers increase or decrease from left to right? Explain.

2. No sign is used in front of the number 0. Why?

3. Use a number line to show one pair of opposite integers. Explain how you know that they are opposite integers.

4. Gus suggested a rule for comparing integers: "The number farthest from zero on the number line is always the greatest." Comment on the accuracy of his rule.

Check Your Understanding

Practise

For help with questions 5 to 7, refer to Example 1.

5. Four temperatures are shown.
   a) Which temperatures are greater than 0?
   b) Which temperatures are less than 0?
   c) List the temperatures in increasing order.
   d) Which temperatures are opposite integers?

6. True or false? Justify your answers.
   a) +5 > +1
   b) −2 > −5
   c) −2 > +1
   d) −5 < +5

7. Write the temperatures in each set in order, from coldest to warmest.
   a) 8°C, 0°C, −5°C, 20°C, 15°C
   b) −21°C, 12°C, 17°C, 8°C, −30°C, 0°C
   c) 32°C, 11°C, −2°C, −8°C, −1°C, 19°C

For help with questions 8 to 12, refer to Example 2.

8. Five integers are shown on a number line.

   a) Which integers are greater than 0?
   b) Which integers are less than 0?
   c) Write the integers in order from least to greatest.
   d) Which numbers are opposite integers?

9. a) Which of the following integers are greater than −4?
      −2, −5, +1, −8
   b) Which of the following integers are greater than −10?
      −9, 0, +5, +11

10. a) Which of the following integers are less than +2?

−2, 0, −5, +6

b) Which of the following integers are less than −6?

−5, −7, +5, +7

11. Which is farther left on a number line?
- **a)** −10 or −30
- **b)** +12 or −12
- **c)** 0 or −5
- **d)** −7 or −6
- **e)** −43 or 27
- **f)** −14 or −2

12. Use words to compare the integers in each pair. Then, use a < or > symbol.
- **a)** −10, −30
- **b)** +12, −12
- **c)** 0, −5
- **d)** −7, −6
- **e)** −43, 27
- **f)** −14, −2

Apply

13. Use integers to represent the following. Show the integers on a number line.
- **a)** 2 min before liftoff, 7 min after liftoff
- **b)** a loss of $1, a gain of $6
- **c)** 8 m above sea level, 3 m below sea level, sea level
- **d)** ground level, 6 floors up, 4 floors down

Did You Know?
Ancient civilizations used various symbols for counting numbers. Symbols for negative numbers were probably first used by the Chinese during the first century A.D. They used red rods to represent positive numbers and black rods to represent negative numbers.

14. a) If two integers have opposite signs, which one is greater? Explain.
b) If two integers are both negative, which one is greater? Explain.

15. "It is colder today than yesterday."
- **a)** Give an example, using negative temperatures, of this situation.
- **b)** Compare your temperatures using a < or > symbol.
- **c)** Show the temperatures on a number line.

16. When Sandy turned on the radio she heard the announcer say "T minus 15 seconds." Mario walked past the TV just as a commentator said "T minus 20 seconds." Who heard the report closer to liftoff time?

17. In hockey, players get +1 if they are on the ice when their team scores. They get −1 if they are on the ice when the other team scores. The sum of these +1 values and −1 values is a player's plus/minus rating. The scatterplot shows Jake's plus/minus rating for his first eight games.

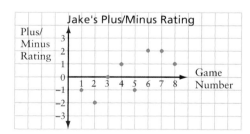

- **a)** What does the ordered pair (3, 0) represent?
- **b)** What does the ordered pair (1, −1) represent?
- **c)** Jake says he has been playing better in recent games. What might he say to support his claim?

Chapter Problem

18. One morning the air temperature was −10°C. The table shows how the wind chill value changes as the wind speed increases.

Wind Speed (km/h)	Wind Chill Value (for air temperature −10°C)
5	−13
10	−15
15	−17
20	−18
25	−19
30	−20
35	−20
40	−21
45	−21
50	−22
55	−22
60	−23

a) What is the wind chill value when the wind speed is 5 km/h? What if the wind speed is 10 km/h?

b) Make a scatterplot to show how the wind chill values change as the wind increases. Describe the pattern.

19. The mean temperature was recorded each day during the first week of February in Pembroke.

Day	1	2	3	4	5	6	7
Temperature (°C)	0	−2	−3	1	−4	−10	−8

a) Make a scatterplot of these data.
b) Which days were colder than February 2?
c) Which days were warmer than February 5?
d) How much colder was the temperature on February 6 than on February 5?
e) What was the median temperature? On which day was it?
f) Describe any trend in the temperature over the week.

Extend

20. On any given day, the temperature can vary greatly depending on where you are in Canada. One winter day, Ottawa experienced a record low temperature of −29°C. The median temperature of Vancouver, Ottawa, and Halifax was −12°C. What might the temperatures have been in Vancouver and in Halifax? Can you find more than one solution?

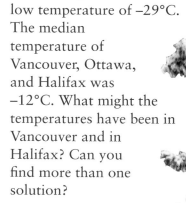

Did You Know?

The record wind chill in Ottawa is −48, on January 23, 1976. The air temperature was −31°C with a wind speed of 35 km/h. For Thunder Bay, the record is −58, on January 10, 1982. The air temperature was −36°C with a wind speed of 54 km/h.

11.2 Explore Integer Addition

Focus on...
- representing integers using manipulatives
- representing integer addition

In many situations colour is used to provide information. When booster cables are used to jump-start a car, the red cable ends must be attached to the positive terminal of the car battery. The convention is that red means positive and black means negative.

Discover the Math

Materials
- 20 chips (10 red and 10 blue)

How can you represent integers using coloured chips?

You can use coloured chips to represent integers.
One red chip represents +1.
One blue chip represents −1.

1. Look at the photo.

   a) What integer is Maria showing?
   b) What integer is Gabe showing?

2. Use coloured chips to model each integer.
   a) +2
   b) −2
   c) −5
   d) +7

3. **a)** Explain why one red chip and one blue chip together represent zero.
   b) Select 3 red chips and 3 blue chips. What integer is represented by each group of chips? What is the total value if you put the six chips together? Explain why.
   c) Select 5 blue chips and 5 red chips. What integer is represented by each group of chips? What is the total value if you put the ten chips together? Explain why.

4. **a)** Select 5 red and 3 blue chips. How many of the chips cancel each other out? How many chips remain? What colour are they?
   b) What integer sum have you shown?

5. **a)** Select 6 red and 4 blue chips. How many of each colour remain after cancelling zero pairs?
   b) What integer sum have you shown?

6. Consider the following chips. How many chips remain if you cancel out zero pairs? What integer sum have you shown?
   a) 7 red and 7 blue   **b)** 7 red and 2 blue   **c)** 5 red and 8 blue

   > **Strategies**
   > Which strategies are you using in steps 6 and 7?

7. **Reflect** Look for a pattern. Describe how you can model the sum of two integers using coloured chips.

Example 1: Use Integer Chips to Model Addition

Use integer chips to represent each sum. Give each result.
a) $(+6) + (-6)$ **b)** $(+4) + (-2)$ **c)** $(-5) + (+4)$

Solution

a)

$(+6) + (-6) = 0$

b)

$(+4) + (-2) = +2$

c)

$(-5) + (+4) = -1$

Example 2: Modelling Sums

Write an integer sum to represent each situation.
Model the sum using integer chips. Interpret the result.
a) Kahil earns $8 and then loses $2.
b) Etta owes her sister $6 and she owes her mother $5.

Solution
a) earns $8 → +8   loses $2 → –2   **b)** owes $6 → –6   owes $5 → –5

(+8) + (–2) = +6   (–6) + (–5) = –11
Kahil has gained $6.   Etta owes a total of $11.

Key Ideas

- Integer chips can be used to model integer addition. This textbook uses red for positive and blue for negative.

- Two chips of opposite colours cancel each other out. Each pair of cancelled chips equals zero. This is called the **zero principle**.

zero principle
• (+1) + (–1) = 0

Communicate the Ideas

1. What integers are represented by each group of chips? What sum do they represent?

2. Use integer chips to show that (+2) + (–5) has the same result as (–5) + (+2).

Check Your Understanding

Practise

For help with questions 3 to 9, refer to Example 1.

3. Interpret each group of integer chips. Write the addition statement for each.

a)    b)

4. What integer sum is modelled? Give each result.

a)

b)

5. What integer sum is modelled? Give each result.

   a)  b)

6. What integer sum is modelled? Give each result.

   a)  b)

7. What integer sum is modelled? Give each result.

   a)  b)

   c)

8. Use integer chips to model each sum. Give each result.

   a) (+2) + (−3)   b) (−2) + (+3)
   c) (−2) + (−3)   d) (−10) + (+8)
   e) (+8) + (−10) f) (−4) + (−3)

9. Use integer chips to represent each addition statement. Give each result. What do you notice about the results? Explain why this happens.

   a) (+5) + (−5)   b) (−4) + (+4)
   c) (+4) + (−4)   d) (−8) + (+8)

Apply

For help with question 10, refer to Example 2.

10. Use an integer sum to represent each situation. Model the sum using integer chips. Interpret the result.

    a) Ali was in a game show. He gained 6 points in the first round, but then he lost 9 points in the second round.
    b) Mae lost $2, but then she found $5.
    c) The temperature was 5°C and dropped 8°C over night.
    d) Brian trained hard and lost 4 kg in March. Then, he was injured and gained 7 kg in April.
    e) A dive to 3 m below sea level is followed by a further dive of 6 m down.

11. Make up an example, similar to those in question 10, for each addition expression. Use integer chips to model the expression and find the sum. Interpret the result.

    a) (+5) + (−8)
    b) (−2) + (+4)
    c) (−3) + (−7)
    d) (+10) + (−10)

Extend

12. On the stock market, prices regularly go up and down. The daily changes are recorded using positive and negative numbers. If a price changes +$0.89 on Monday and −$1.42 on Tuesday, is the net change for the two days positive or negative? Explain.

Making Connections

Shake and Spill

Work with a partner and take turns. Use a paper cup and six coins. Let heads represent +1 and tails represent −1. Shake the coins in the cup and spill them onto your desk. Record the integers that you see and their sum. After you have each had at least 10 turns, try to answer these questions:
- What are all the possible ways that the coins can land?
- How would the outcomes change if there were seven coins instead of six?

11.3 Adding Integers

Focus on...
- adding integers using a number line
- describing addition statements

Some sports use integers in their statistical data. In ice hockey, each goal scored by the team while a player is on the ice is recorded as +1 and each goal against is recorded as −1. The overall sum of these goals for and against is called the player's plus/minus rating.

While Inga was on the ice, her team scored 2 goals and the opposing team scored 1 goal. What was Inga's plus/minus rating for the game?

Discover the Math

How can you find the sum of two integers?

1. Paula used integer chips to model four integer sums. Greg says he prefers to show the sums on a number line. Compare the two methods. What sum is represented in each case?

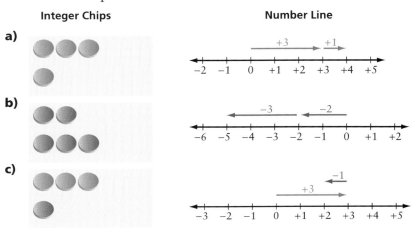

356 MHR • Chapter 11

d)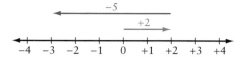

2. Look carefully at each number line model in step 1.
   a) How do you show the first integer in the sum? Where do you start? Which way do you go?
   b) How do you show the second integer in the sum?
   c) How do you know what the answer is?

3. **Reflect** Explore more sums, using integer chips or a number line. Look for patterns in the results. Copy and complete the following statements using negative, positive, or zero.
   a) The sum of two positive integers is always ▇▇▇ .
   b) The sum of two negative integers is always ▇▇▇ .
   c) The sum of two opposite integers is always ▇▇▇ .

Example 1: Add Integers Using a Number Line

Use a number line to add each pair of integers.
a) (+3) + (+5)   **b)** (−2) + (−3)
c) (+4) + (−4)   **d)** (+5) + (−8)

Solution

a)
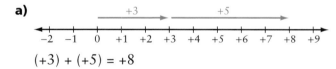
(+3) + (+5) = +8

b)
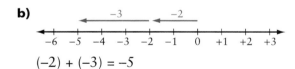
(−2) + (−3) = −5

c)
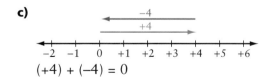
(+4) + (−4) = 0

d)
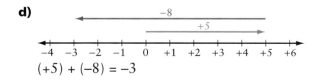
(+5) + (−8) = −3

Example 2: Describe Gains and Losses

a) Describe the addition statement (−5) + (+8) + (−12) in money terms.
b) Show the sum on an integer number line.
c) Find the sum numerically and interpret the result.

Solution

a) (−5) + (+8) + (−12) represents a loss of $5, followed by a gain of $8, and then a loss of $12.

b)
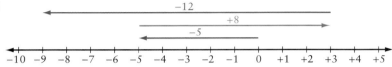

c) (−5) + (+8) + (−12)
= (+3) + (−12)
= −9
The result is an overall loss of $9.

Key Ideas

- You can use an integer number line to show the sum of integers.

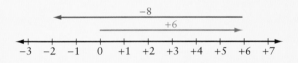

- You can also use integer chips to show the sum of integers.

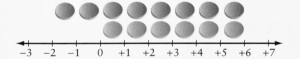

Communicate the Ideas

1. What integer sum is shown on each number line?

a)
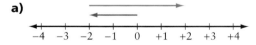

b)
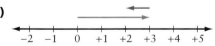

2. Tell whether each sum has a positive or a negative answer. Explain your reasoning.
   a) (+5) + (+5)   **b)** (−3) + (−4)
   c) (+8) + (−6)   **d)** (−7) + (+2)

3. On a quiz, Leanne provided the following solutions.
   a) (−10) + (+7) = −3   **b)** (+10) + (−7) = −3
   Are Leanne's solutions correct? If you feel that she has made an error, explain why and give a correct solution.

Check Your Understanding

Practise

For help with questions 4 to 8, refer to Example 1.

4. What integer sum is shown? Give each result.

a)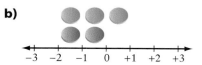

b)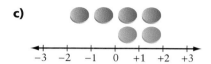

c)

5. What integer sum is shown? Give each result.

a)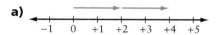

b)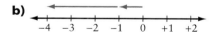

c)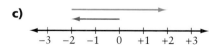

d)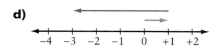

6. What integer sum is shown? What do you notice about the results? Explain why this happens.

a)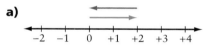

b)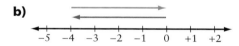

c)

7. Use a number line to model each sum.
 a) (+2) + (+7)
 b) (−3) + (−5)
 c) (−9) + (+3)
 d) (−4) + (+1)
 e) (−6) + (+6)
 f) (+8) + (−2)

8. Use integer chips or a number line to model each sum.
 a) (+3) + (−9)
 b) (−2) + (−4)
 c) (−5) + (+7)
 d) (+1) + (−9)
 e) (+3) + (+2)
 f) (−6) + (−4)

9. Decide whether each sum is positive, negative, or 0.
 a) Two positive integers are added.
 b) Two opposite integers are added.
 c) A large positive integer is added to an integer that is just less than 0.
 d) A small positive integer is added to an integer that is much less than 0.
 e) Two negative integers are added.

10. Use mental arithmetic to find each sum.
 a) (+6) + (−4)
 b) (+5) + (−8)
 c) (−5) + (−7)
 d) (−3) + (−4)
 e) (−7) + (+10)
 f) (−9) + (+5)

11. Find each sum.
 a) (+7) + (+9)
 b) (+18) + (−18)
 c) (−14) + (+14)
 d) (−53) + (+55)
 e) (−20) + (−20)
 f) (+100) + (−105)

For help with questions 12 to 14, refer to Example 2.

12. Find each sum. Use integer chips or a number line if you need them.
   a) (−2) + (−3) + (−5)
   b) (−6) + (+4) + (+7)
   c) (+8) + (+2) + (−5)
   d) (+7) + (+6) + (−13)
   e) (−10) + (−7) + (+5)
   f) (−21) + (+17) + (−16)
   g) (+15) + (−5) + (+18)
   h) (−3) + (−4) + (+29)

Apply

13. Represent each situation as the sum of two integers. Find the result.
   a) Kajan owed $20 and then he earned $15 cutting lawns.
   b) The Rockets basketball team scored 89 points but they had 95 points scored against them.
   c) Sylvie parked 3 floors below ground level and then took the elevator up 10 floors.
   d) The kite rose 83 m but then dropped 23 m.

14. Describe each addition statement in money terms. Find and interpret each sum.
   a) (+6) + (−5)
   b) (−10) + (−20)
   c) (−12) + (+8)
   d) (+50) + (−20)

15. The sum of two integers is −3. What are the integers? Give three possible answers.

16. The sum of two integers is −8. What are the integers? Give three possible answers. Compare your answers with a classmate.

17. Write an integer addition expression for each situation. Then, find the sum.
   a) The temperature is 12°C and then drops 15°C.
   b) Last season, when Jason was on the ice, 25 goals were scored by his team and 32 goals were scored by the other team.
   c) An elevator goes up 18 floors and then down 11 floors.

18. Copy and complete the pyramid. What number belongs in the top triangle?

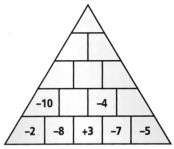

19. Stephanie made up a coding system using integers.

A	B	C	D	E	F	G	H	I
−12	−11	−10	−9	−8	−7	−6	−5	−4

J	K	L	M	N	O	P	Q	R
−3	−2	−1	0	+1	+2	+3	+4	+5

S	T	U	V	W	X	Y	Z
+6	+7	+8	+9	+10	+11	+12	+13

To code a word she adds the value of each letter. For example, ME = 0 + (−8), or −8. Use her code to find the value of each of the following names.
   a) VY
   b) DON
   c) PAT
   d) ZACK
   e) LENA
   f) *your* name

20. In a darts game, each ring is given a positive or a negative score.

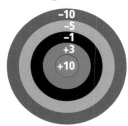

Ring	Colour	Value
Centre	Red	+10
First	Green	+3
Second	Black	−1
Third	Orange	−5
Outer	Blue	−10

Each player throws four darts. Find each person's score if their darts land as follows.

a) Jack: green, blue, black, green
b) Fiona: blue, black, black, red
c) Rohan: orange, orange, orange, blue
d) Anika: red, green, blue, black
e) Who won this round? Explain.

 21. Make up a situation or puzzle that involves integer addition. Provide a solution.

Extend

22. Refer to question 21. If Mona scored a total of 0 points, where did her four darts land? Provide three possible solutions.

23. In a magic square, each row, column, and diagonal has the same total.

a) Place the numbers −4, −3, −2, −1, 0, +1, +2, +3, and +4 in a three-by-three magic square so that the magic sum is zero.

b) Another three-by-three magic square has −9 as its magic sum. What nine numbers should be used, and where should they be placed?

Making Connections

Integer Game

Materials
- two different-coloured dice (one for positive numbers, one for negative numbers)
- position markers (e.g., coin, button, eraser)
- a number line clearly labelled from −10 to +10

- Each player places a marker on any integer between −6 and +6.
- Take turns rolling the dice and adding the two integers represented. For a positive sum, the player moves that many steps to the right. For a negative sum, the player moves that many steps to the left.
- The first player to reach either end of the number line wins the game.

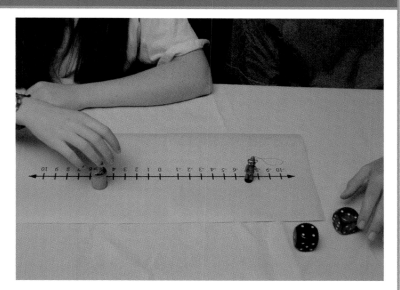

Play the game with a partner. Is there a best place to start from? Explain.

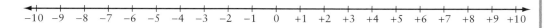

11.4 Explore Integer Subtraction

Focus on...
- identifying when to subtract integers
- using integer chips to model subtraction of integers

What is the difference in the temperatures shown on the thermometer? What is another situation where you need to subtract integers?

Discover the Math

Materials
- 20 chips (10 of one colour and 10 of another colour)

How can you model subtraction using integer chips?

1. Subtraction can be described as taking away. What integer subtraction is modelled by the steps shown?

2. Use integer chips to demonstrate each subtraction.
   a) (+4) − (+3)
   b)  (−4) − (−2)
   c) (−4) − (−4)

3. a) Try to demonstrate $(+2) - (+5)$. What happens?
  b) Remember that each pair of opposite chips is zero. Add zero pairs until you have enough red chips to be able to take away 5. These steps are shown below.

  How does adding three zero pairs help? Finish the subtraction. What is the result of $(+2) - (+5)$?
  c) Describe, in words, the strategy used to model $(+2) - (+5)$. Why do you need to add 3 red and 3 blue chips in the second step?

4. Use your strategy from step 3 to model each subtraction.
  a) $(+2) - (+6)$
  b) $(+3) - (+4)$
  c) $(+1) - (+3)$

5. a) Try to demonstrate $(-2) - (-5)$. What happens?
  b) The following steps show how to model $(-2) - (-5)$.

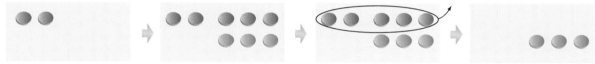

  c) Describe the strategy used.

6. Use your strategy from step 5 to subtract.
  a) $(-1) - (-3)$
  b) $(-3) - (-4)$
  c) $(-2) - (-3)$

7. a) Use integer chips to model $(-2) - (+5)$. What strategy did you use?
  b) Use integer chips to model $(+2) - (-5)$. What strategy did you use?

8. Use your strategies to subtract.
  a) $(-1) - (+6)$
  b) $(-3) - (+4)$
  c) $(+4) - (-3)$

9. Reflect Summarize how to model subtraction using integer chips. When do you need to add zero pairs? How do you determine how many zero pairs you need to use?

Example: Use Integer Chips to Subtract Integers

Use integer chips to find each difference.
a) $(-6) - (-4)$
b) $(-2) - (-7)$
c) $(+4) - (-2)$
d) $(-3) - (+1)$

Solution

a)

To subtract (–4), remove 4 blue chips.

$(-6) - (-4) = -2$

b)

There are not enough blue chips to take away 7 of them.

Add zero pairs until you can take away 7 blue chips. Add 5 blue and 5 red chips.

To subtract (–7), remove 7 blue chips.

$(-2) - (-7) = +5$

c)

There are no blue chips, so you cannot take away 2 of them.

Add two zero pairs.

To subtract (–2), remove 2 blue chips.

$(+4) - (-2) = +6$

d)

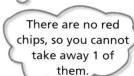

There are no red chips, so you cannot take away 1 of them.

Add one zero pair.

To subtract (+1), remove 1 red chip.

$(-3) - (+1) = -4$

Key Ideas

- When using integer chips to model subtraction of integers, you may need to add extra chips.
  ○ If so, add the same number of positive and negative chips. You can do this because you are adding zero pairs.
  ○ Then, take away the appropriate number of chips to represent the subtraction.
  ○ The chips remaining represent the final result.

Communicate the Ideas

1. To model $(+3) - (-3)$, Theo places 3 red chips on the desk. To subtract (-3), he knows he must take away 3 blue chips. What should his next step be?

2. Kelly used the steps shown to model $(-5) - (+3)$. What should her next step be?

Check Your Understanding

Practise

For help with questions 3 to 7, refer to the Example.

3. Write the subtraction statement that is modelled by each sequence.

a)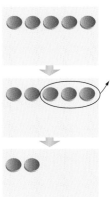

b)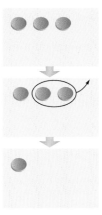

c)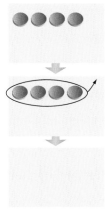

4. Write the subtraction statement that is modelled by each sequence.

a)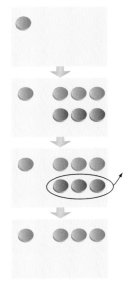

b)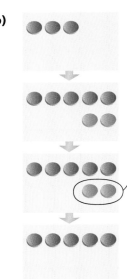

5. Use integer chips to model each subtraction statement.
   a) $(+9) - (+5)$
   b) $(-3) - (-1)$
   c) $(-4) - (+8)$
   d) $(+3) - (-2)$

6. Find the result of each subtraction.
   a) (+5) − (+10)
   b) (−5) − (+10)
   c) (+5) − (−10)
   d) (−5) − (−10)

7. Find each result. Use integer chips if necessary.
   a) (+6) − (+8)
   b) (+3) − (+7)
   c) (−2) − (+5)
   d) (−4) − (+4)
   e) (+6) − (−3)
   f) (−5) − (−2)

8. Find each result.
   a) (+4) − (−6)
   b) (−10) − (+4)
   c) (+1) − (−9)
   d) (−4) − (−12)
   e) (+10) − (−5)
   f) (+9) − (+11)

9. Which expressions can you simplify mentally? Find each result, using integer chips when necessary.
   a) (−2) − (−9)
   b) (+7) − (+3)
   c) (+4) − (−9)
   d) (−6) − (−2)
   e) (+3) − (+8)
   f) (−1) − (−1)

Apply

10. Write each situation as an integer subtraction statement. Then, model it using integer chips. Interpret the result.
   a) The temperature was 4°C, but it fell by 8°C.
   b) A slug climbed 2 m, but then fell 3 m.
   c) Waleed borrowed $7, but paid off a previous debt of $5.

11. Does (−3) − (−5) give the same result as (−5) − (−3)? Explain using words and diagrams.

12. Using any of the integers +2, −2, +7, and −7, two at a time, list all the subtraction statements that result in a difference of −5.

13. The table shows the high and low temperatures one day in six cities.

City	High (°C)	Low (°C)
Vancouver	+8	−1
Edmonton	−5	−10
Ottawa	+3	−5
Trois-Rivières	0	−7
Fredericton	−3	−5
Saint John	−2	−9

   a) Find the temperature difference for each city.
   b) Which city had the greatest change in temperatures?
   c) Which city had the least change in temperatures?

14. Which is greater, (−2) − (+5) or (+5) − (−2)? Explain using words and diagrams.

Extend

15. Use the four integers −3, −5, +4, and +7, and the addition or subtraction operation, to write an expression that has each result.
   a) +5
   b) −19

16. a) Evaluate each integer expression. Describe the pattern in the results.
   (−1) − (−2) = ■
   (−1) − (−2) − (−3) = ■
   (−1) − (−2) − (−3) − (−4) = ■
   (−1) − (−2) − (−3) − (−4) − (−5) = ■
   b) What is the result of
   (−1) − (−2) − (−3) − ... − (−15)?

11.5 Extension: Subtracting Integers

Focus on...
- subtracting integers
- comparing subtraction and addition of integers

Temperatures in space can be much hotter or colder than on Earth. The mean temperature on Earth is 15°C. The mean temperature on Mars is –63°C. What is the difference between the mean temperatures on Earth and on Mars?

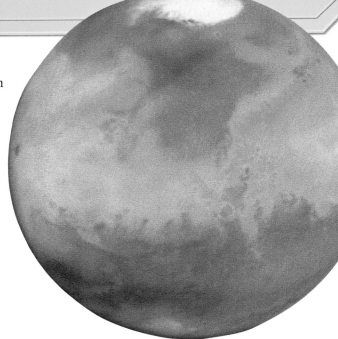

Discover the Math

Materials
- 20 chips (10 of one colour and 10 of another colour)

Is there another way to subtract integers?

1. You know that 6 – 2 = 4. You can show this difference on a number line.

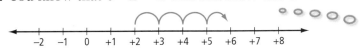

> From 2 to 6 is 4 steps to the right.
> 6 – 2 = +4.

You can also use a number line to find (+6) – (–2).

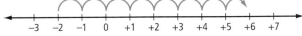

a) Why do you start counting steps from –2
b) How many steps is it from –2 to +6? Which direction are you moving?
c) What is the result of (+6) – (–2)?

2. The difference (–5) – (–1) is shown on the number line.

a) Why do you start counting steps at (–1) and end at (–5)?
b) How many steps is it? In which direction?
c) What is the result of (–5) – (1)? Check using integer chips.

368 MHR • Chapter 11

3. The difference (−2) − (+3) is shown on the number line.

   a) Why do you start counting steps at (+3) and end at (−2)?
   b) How many steps is it? In which direction?
   c) What is the result of (−2) − (+3)? Check using integer chips.

4. Use integer chips or a number line to find each difference and each sum. What patterns do you see?

	Difference	Sum
a)	(+7) − (+3)	(+7) + (−3)
b)	(−5) − (−5)	(−5) + (+5)
c)	(−4) − (+6)	(−4) + (−6)
d)	(−2) − (−3)	(−2) + (+3)
e)	(+6) − (−1)	(+6) + (+1)

5. **Reflect** Describe another way to subtract an integer, by using a related sum. Provide two examples of your own to show how the method works.

Example 1: Find Differences

Find each difference in two ways.
- using a number line
- by adding the opposite

a) (+3) − (+5) b) (+2) − (−4) c) (−2) − (+3)

Solution

Method 1: Use a number line

a) (+3) − (+5) = −2

b) (+2) − (−4) = +6

c) (−2) − (+3) = −5

Method 2: Add the opposite.

(+3) − (+5) = (+3) + (−5)
          = −2

(+2) − (−4) = (+2) + (+4)
          = +6

(−2) − (+3) = (−2) + (−3)
          = −5

Example 2: Find Temperature Differences

The table shows the minimum temperatures on Monday and Tuesday in five cities.
a) Find the change in temperature from Monday to Tuesday for each city.
b) Which city had the greatest increase in temperature?
c) Which city had the greatest decrease in temperature?

City	Monday Temperature (°C)	Tuesday Temperature (°C)
Halifax	−1	+8
Saint John	−10	−5
Ottawa	+5	−3
London	−7	0
Trois-Rivières	+7	+4

Solution

a) Visualize the difference between the temperatures on the thermometer. Or, rewrite each subtraction statement as a related addition statement.

City	Change in Temperature (°C) Tuesday − Monday
Halifax	(+8) − (−1) = (+8) + (+1) = +9
Saint John	(−5) − (−10) = −5 + (+10) = +5
Ottawa	(−3) − (+5) = (−3) + (−5) = −8
London	0 − (−7) = 0 + (+7) = +7
Trois-Rivières	(+4) − (+7) = (+4) + (−7) = −3

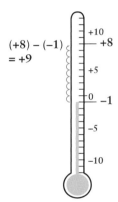

(+8) − (−1) = +9

b) Halifax had the greatest increase in temperature.
c) Ottawa had the greatest decrease in temperature.

Key Ideas

- Subtraction of integers can be modelled using integer chips or a number line.
- On a number line, the difference is the distance and direction from the second integer to the first. For example, (−5) − (+2) = −7.

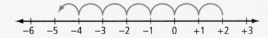

- Subtraction of integers can be expressed as addition of the opposite integer. For example, (−5) − (+2) = (−5) + (−2)
   = −7

Communicate the Ideas

1. The number line shows (−1) − (−7). What is the result? Rewrite the expression as an addition statement. Do you get the same result?

2. The number line shows (−7) − (−1). What is the result? Rewrite the expression as an addition statement. Do you get the same result?

3. Compare and discuss the different ways you can use to find the result of (+3) − (−4). Which method do you prefer? Why?

Check Your Understanding

Practise

For help with questions 4 to 7, refer to Example 1.

4. Write the opposite of each integer.
   a) −2
   b) +5
   c) −8
   d) −15
   e) +12
   f) −10

5. The number line shows (+5) − (−2).

   a) What is the result?
   b) Copy and complete the related sum:
   (+5) − (−2) = (+5) + ■
   = ■

6. The number line shows (−4) − (−8).

   a) What is the result?
   b) Copy and complete the related sum:
   (−4) − (−8) = (−4) + ■
   = ■

7. Find each difference.
   a) (+10) − (+6)   b) (+2) − (+9)
   c) (−3) − (+11)   d) (−15) − (+2)
   e) (+9) − (−3)   f) (+14) − (−6)
   g) (−3) − (−7)   h) −20 − (−15)

For help with question 8, refer to Example 2.

8. Copy and complete the table to find the change in temperature for each city one January 1.

City	Day's High (°C)	Day's Low (°C)	Change (°C) (Low − High)
Kingston	+3	−8	
Kenora	−10	−18	
Sudbury	−15	−22	
Oshawa	0	−11	

9. Add or subtract as indicated.
   a) (+2) + (−3)   b) (−9) − (+5)
   c) (+5) − (+9)   d) (+8) − (+9)
   e) (−1) + (−7)   f) (+2) − (+8)
   g) (−6) − (−5)   h) (+5) − (−4)

Apply

10. Each thermometer has two temperatures shown. Write each as a subtraction statement. Subtract the lower temperature from the higher temperature.

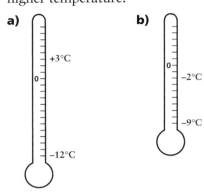

11. Find the missing integer in each.
   a) −6 − ■ = −9   **b)** −2 − ■ = 3
   c) 5 − ■ = −7   **d)** 7 − ■ = −9
   e) −4 − ■ = 0   **f)** ■ − 3 = −7
   g) ■ − (−5) = 6   **h)** ■ − (−10) = 5

12. The main part of a home refrigerator should be kept at +3°C. The freezer part should be −17°C. How much colder is it in the freezer compartment?

13. Write each of the following as an integer subtraction statement. Give the outcome of each statement.
   a) Tekesha deposited $8, and then wrote a cheque for $15.
   b) The temperature was −8°C, and then dropped by 15°C.
   c) A gain of 4 points was followed by a loss of 7 points.
   d) Yan dived to 5 m below sea level, then dove a further 4 m.
   e) Jo's mother lost $100 on the stock market, then lost another $600.
   f) Muhammad owed $20, but his friend reduced the debt by $5.

14. Write a situation describing each subtraction statement, similar to those in question 13. Use a different type of situation for each.
   a) (−40) − (+25)   **b)** (+10) − (+12)
   c) (+15) − (−8)   **d)** (−6) − (−2)

15. Evaluate.
   a) (+6) − (+13) − (−15)
   b) (−12) − (+12) − (−12)
   c) (+15) − (+20) − (+25)
   d) (−34) − (−24) − (+10)

16. Time zones are often listed as GMT + or − a number of hours. GMT stands for Greenwich Mean Time, which refers to Greenwich, in England. The table gives the time zone references for four cities.

City	Time Zone
Charlottetown, PE	−4
Rainy River, ON	−6
Beijing, China	+8
Tel Aviv, Israel	+2

a)

Copy and extend the number line to show the position of the four cities.
   b) If it is 6:00 p.m. in Tel Aviv, what time is it in Beijing?
   c) If it is 3:00 a.m. in Rainy River, what time is it in Tel Aviv?
   d) If it is 11:30 p.m. in Beijing, what time is it in Charlottetown?
   e) If it is 10:45 a.m. in Rainy River, what time is it in Beijing?

Did You Know?
China is almost as large as Canada and should have five time zones. However, all of China uses the same time.

Chapter Problem

17. One day the wind speed is light, at 10 km/h. However, the temperature is falling. The chart shows the change in wind chill values.

Wind Chill Chart

Wind Speed (km/h)	Air Temperatures (°C)						
	5	0	–5	–10	–15	–20	–25	–30
10	3	–3	–9	–15	–21	–27	–33	–39

a) When the wind speed is 10 km/h and the air temperature is –15°C, what is the wind chill value?

b) Find the change in the wind chill values when the air temperature falls from 0°C to –15°C.

c) Look across the row of wind chill values. Describe the pattern.

18. Canada is so large that it has several time zones. The map shows how time changes relative to Eastern Standard Time. For example, when it is noon in Toronto, it is one hour earlier, or 11 a.m., in Winnipeg, Manitoba.

a) How does the time in Vancouver compare to the time in Ontario? When it is noon in Toronto, what time is it in Vancouver?

b) How does the time in Halifax compare to the time in Ontario? When it is 9 a.m. in Niagara Falls, what time is it in Halifax?

c) How does the time in Edmonton compare to the time in Halifax?

19. Which is less (–7) – (+3) or (+3) – (–7)? Explain using diagrams.

Extend

20. Use the time zone map in question 18. Tina lives in Kingston, Ontario. It is noon on Thursday. Tina needs to talk to her uncle in Vancouver as soon as possible. She knows that he leaves for work at 8 a.m. and usually returns home at 6 p.m. What would be a good time for Tina to phone her uncle? Explain your reasoning.

21. You are in charge of keeping track of the financial records for your student council. You thought the student council had $50 in its account. When you checked your records, you found that the student council owed $50. This is recorded as –$50.

a) What is the difference in the two amounts?

b) Explain the error you could have made.

22. a) Write –14 as the sum of four consecutive numbers.

b) Write –14 as the difference of four consecutive numbers.

c) Are the numbers you used in part a) the same or different from those in part b)? Explain.

Did You Know?

Earth rotates once every 24 h. In 1 h, Earth turns through 360° ÷ 24 or 15°. Vancouver is 123°W and Halifax is 63°W, so the cities are 60° apart. When the sun rises in Halifax, it will not be seen until 4 h later in Vancouver. This is the idea behind time zones.

11.5 Extension: Subtracting Integers • MHR   373

11.6 Integers Using a Calculator

Focus on...
- adding and subtracting integers using a calculator

Mike Weir's scores for the three rounds in a tournament were 2 under par, 3 over par, and 5 under par. What was Mike's final score?

Discover the Math

How can you find integer sums using a calculator?

Your calculator may have a (−) key or a (+/−) key that you can use to enter negative numbers.

1. Mike Weir's score for the three rounds was (−2) + (+3) + (−5). Add these integers manually. Use integer chips or a number line to help if you need to.

2. Check the sum using a calculator.
   - If your calculator has a (+/−) key, use
     (C) 2 (+/−) (+) 3 (+) 5 (+/−) (=)
   - If your calculator has a (−) key, use
     (C) (−) 2 (+) 3 (+) (−) 5 (=)

3. **Reflect** Write brief notes to remind yourself how to enter negative integers on the calculator that you use.

Literacy Connections

Reading Positive Integers
Whenever you see a number with no sign in front of it, the number is positive. Similarly, a calculator assumes that a number entered is positive. You use the (+/−) or (−) key to tell the calculator that a number is negative.

Did You Know?

In golf, *par for the course* means the average number of strokes needed by an expert golfer to complete the round.

Example: Combine Profit and Loss

Zack operates a booth every Saturday at a flea market. For the first four weeks, he noted his profit or loss on a calendar.
a) Express Zack's overall profit and loss for the four weeks as an integer expression.
b) Use a calculator to find his overall profit or loss. Use estimation to check that your answer is reasonable.

iday	Saturday
	6 loss $85
	13 loss $122
	20 profit $64
	27 profit $193

Solution

a) (−85) + (−122) + 64 + 193

b) C 85 +/− + 122 +/− + 64 + 193 = 50
Estimate: (−85) + (−122) ≐ −200 and 64 + 193 ≐ 250
−200 + 250 = 50
The calculator answer is reasonable.

Zack's overall profit was $50.

Key Ideas

- Calculators can be used to help simplify integer expressions.
- Calculator keystokes vary. To check how to enter negative integers on your calculator, use a simple sum that you can answer in your head.

Communicate the Ideas

1. Write the calculator keystrokes you would use to evaluate (−45) + 17 − (−12).

2. Which of the following would you do in your head and which would you use a calculator for? Explain your choice.
   a) (−3) + (−2)
   b) −42 + 44
   c) 571 + (−363) − (−210)

3. To simplify longer integer sums, David rearranges them so that all the positives come first and then all the negatives. How would this method help? How would you apply his method to evaluate (−11) + 23 + 79 + (−18)?

Check Your Understanding

Practise

4. Use integer chips or a number line to model each sum. Use a calculator to check the results.
   a) (+3) + (−8)   b) (−2) + (−7)
   c) (−5) + 12   d) 16 + (−4)

5. Use integer chips or a number line to model each difference. Use a calculator to check the results.
   a) (−5) − (−6)   b) 11 − (−5)
   c) (−1) − 4   d) −7 − (−7)

6. Decide which of the following can be done without a calculator. Use a calculator where necessary to help evaluate each expression.
   a) 6 + 10 − 18
   b) (−3) − 7 + 4
   c) (−1) − (−6) + 6
   d) (−3) − 3 − 3 − 3 − 3 − 3
   e) 11 − 17 + 12 − 18
   f) 200 − 150 − 250 + 100 + 100
   g) 18 + (−20) − (−9) − 10
   h) 500 + 200 − 700 − (−300)

7. Evaluate each expression.
   a) 134 − 218 − (−317)
   b) 45 + (−47) − (−23) − 61
   c) (−52) − 41 − (−17) + 19
   d) (−122) + (−141) − 78 + 89
   e) (−49) − 49 − 49 − 49
   f) 32 − (−16) + 48 − (−136)

Apply

8. Write an integer expression to represent the overall results of each situation. Find each result using an appropriate tool.
   a) Sam's bank balance was $70. Then, he wrote a cheque for $13 and a cheque for $22.
   b) Barbara borrowed $4 from her mom and $9 from her dad. She repaid $10 of what she owed to her brother.
   c) Sarah gained 5 points in a trivia game. She gained another 15 points before losing 25 points.
   d) Wayne's plus/minus ratings in four games during a hockey tournament were +3, −4, +2, and −1.

9. Start with 1, subtract 2, add 3, subtract 4, add 5, and so on. Describe the pattern that results after each step.

10. Find the missing numbers. Explain your strategy.
    a) −2 + 7 − ■ = −10
    b) 3 + ■ + 4 = −6
    c) 5 − ■ − 8 = −15
    d) ■ + 12 − 8 = −9
    e) −14 − 3 − ■ = 17
    f) ■ − 10 + 15 = 0

11. Lori Kane is another famous Canadian golfer. Her scores in a tournament were 2 over par, 4 under par, 3 under par, and 5 under par. What was Lori's final score?

12. On a TV game show, contestants can win or lose money depending on whether they answer questions correctly or incorrectly. Tasneem played in a round against Yvonne and their results were as follows.

    Tasneem: −$100, −$200, +$50, +$100, −$300, +$400, −$500
    Yvonne: +$50, +$100, −$50, −$200, +$300, +$350, −$500, +250.
    Who finished with the higher amount?

13. The integer –6 can be expressed as the sum of three consecutive numbers, (–1) + (–2) + (–3) = –6. Write each of the following as the sum of three consecutive numbers.
a) –12
b) –30
c) 0

14. Lizo recorded his daily profit or loss at his booth at a collector's fair.

Day 1 $320 profit
Day 2 $210 loss
Day 3 $165 loss
Day 4 $412 profit
Day 5 $382 profit

What was Lizo's overall profit or loss for the five-day fair?

15. Place the numbers –6, –5, –4, –3, –2, –1, 0, +1, and +2 into a magic square so that the sum of each row, column, and diagonal is –6.

16. The boiling points of substances vary.

liquid oxygen   –183°C
liquid nitrogen   –196°C
water   100°C
helium   –269°C
nickel   3278°C

a) How much colder is the boiling point of liquid nitrogen than that of liquid oxygen?
b) How much hotter is the boiling point of nickel than that of liquid nitrogen?
c) How much hotter is the boiling point of nickel than that of liquid oxygen?
d) How much colder is the boiling point of helium than that of water?

17. The table shows the temperature records.

	Canada	World
High temperature	45°C at Midale, Saskatchewan	58°C at Al'azizyal, Libya
Low temperature	–63°C at Snag, Yukon	–89°C at Vostok, Antarctica

a) What is the difference between the coldest world temperature and the coldest Canadian temperature?
b) How much greater is the record high temperature than the record low temperature, for Canada?

18. The table shows the profit or the loss of six companies at the end of two different years.

Company	Year 1 ($)	Year 2 ($)	Change ($) (Year 2 – Year 1)
Integers & Co.	230 000	212 000	
Geometry, Inc.	4 500 000		–5 500 000
Algebra, Ltd.	–50 000	–140 000	
Measures R Us	–150 000		60 000
Fractions, Ltd.		–41 500	–10 500

a) Copy and complete the table.
b) Rank the companies from the one with the greatest profit to the one with the greatest loss in year 1.

Extend

19. How do the results of the following expressions compare? What rule can you make about combining addition and subtraction of integers?
a) 2 – 3 + 4
b) 2 + 4 – 3
c) 4 + 2 – 3
d) –3 + 2 + 4
e) 2 + (–3) + 4
f) 4 + 2 – (+3)

20. The nightly low temperatures one week in Fredericton, New Brunswick, were 4°C, –5°C, –2°C, 6°C, 3°C, –4°C, and –9°C. What was the mean nightly low temperature? Explain your solution.

CHAPTER 11 Review

Key Words

1. Match each term with an example.

Term	Example
a) positive integers	A −1 and −2
b) negative integers	B (+1) + (−1) = 0
c) opposite integers	C +2 and −2
d) zero principle	D +1 and +2

11.1 Compare and Order Integers, pages 346–351

2. Use words, then one of the symbols >, <, or =, to compare the integers in each pair.
   a) +3, −5
   b) −7, +7
   c) −6, −10
   d) 0, −18

3. a) Show the following integers on a number line.
   −5, +2, 0, +8, −10, −3, −2, −8
   b) List the integers in order, from least to greatest.

11.2 Explore Integer Addition, pages 352–355

4. What integer sum is modelled? Give each result.

   a)

   b)

   c)

5. Write an integer sum to represent each situation. Model the sum using integer chips. Interpret the result.
   a) A stock gained $9, then lost $3.
   b) The temperature was −8°C. Then, it increased by 6°C.
   c) Elizabeth's golf scores, for two rounds, were 3 over par and 4 under par.
   d) Keith earned $7, but owed $10 to his friend.
   e) Sunita counted 6 votes in favour and 8 votes against a candidate for student council.

11.3 Adding Integers, pages 356–361

6. Use a number line to find each sum.
   a) (−4) + (−6)
   b) (+5) + (−5)
   c) (−7) + (+9)
   d) (+2) + (−9)
   e) (−12) + (−3)

7. Ashraf has $150 in his bank account. He owes $45 to his sister, $25 to his brother, and $70 to his friend. How much money does he have to save, if he needs $400 to buy a bike? Justify your answer.

8. What is the missing integer in each pattern?
   a) −6, −3, ■, +3, +6, …
   b) −11, −8, −5, ■, +1, …
   c) −22, −17, −12, −7, ■, +3, …
   d) −10, −6, −2, ■, +6, …
   e) 0, −7, ■, −21, −28, −35, …

11.4 Explore Integer Subtraction, pages 362–367

9. Find each result. Use integer chips, if necessary.
   a) $(-5) - (-3)$
   b) $(-3) - (+7)$
   c) $(+4) - (+9)$
   d) $(+6) - (-2)$

10. Does $(-2) - (+5)$ give the same result as $(+5) - (-2)$? Explain why using words and pictures.

11.5 Extension: Subtracting Integers, pages 368–373

11. $(+7) - (-5)$ gives the same result as $(+7) + (+5)$. Explain why using words and pictures.

12. Find each difference. Use integer chips or a number line, if necessary.
    a) $(+8) - (+12)$
    b) $(-5) - (+9)$
    c) $(-6) - (+2)$
    d) $(+6) - (-7)$
    e) $(-20) - (-12)$
    f) $(+15) - (+25)$

13. Find the missing integer in each.
    a) $(+3) - \blacksquare = -7$
    b) $-8 - \blacksquare = +2$
    c) $\blacksquare - (+4) = +5$
    d) $\blacksquare - (-1) = -4$

14. Evaluate.
    a) $(+16) - (+10) - (+17)$
    b) $(-5) - (-2) - (+8)$

15. Jenna's plus/minus ratings for her first six games of the hockey season were +1, −3, −2, +2, +1, and +2. What was Jenna's overall plus/minus rating?

11.6 Integers Using a Calculator, pages 374–377

16. The table shows the elevations, relative to sea level, of six places.

Place	Elevation (m)
Death Valley, California	−86
Dead Sea, Israel/Jordan	−411
Valdez Peninsula, Argentina	−40
Lake Assal, Djibouti	−156
Mt. Everest, Tibet/Nepal	8863
Mt. Kilimanjaro, Tanzania	5895

a) What is the difference between the elevations of Death Valley and the Valdez Peninsula?
b) What is the difference between the elevations of Mt. Kilimanjaro and Lake Assal?
c) What is the difference between the highest and lowest elevations?

17. The zoo veterinarian suspects that a giraffe is ill. The mass of the giraffe is recorded each week. Over six weeks the following changes, in kilograms, were noted: −6, +2, +3, −4, −7, −4.

a) Write an integer expression for the change over the six weeks.
b) What was the giraffe's overall change in mass?

CHAPTER 11 Practice Test

| Strand Questions | NSN 1–12 | MEA | GSS | PA 1–10 | DMP 12 |

Multiple Choice

For questions 1 to 8, select the correct answer.

1. Which statement compares −3 and +2 correctly?
   A −3 > +2
   B +2 < −3
   C −3 < +2
   D None of these.

2.

   The number line models
   A a profit of $4 followed by a loss of $4
   B a profit of $4 followed by a loss of $7
   C a loss of $3 followed by a profit of $4
   D a loss of $7 followed by a loss of $3

3.

   The sum that is modelled is
   A (+7) + (−3)
   B (−7) + (−3)
   C (+3) + (+7)
   D (−7) + (+3)

4.

   The sum that is modelled is
   A (−3) + (+2)
   B (−3) + (−5)
   C (+2) + (+5)
   D (−3) + (+5)

5.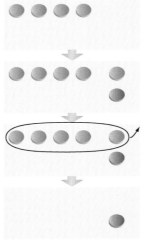

   The subtraction that is modelled is
   A (−4) − (−5)   B (−5) − (−4)
   C (+4) − (+5)   D (−4) + (−5)

6. The result of (−10) + (+3) is
   A −13   B −7
   C +3   D +7

7. Which of the following does not have a result of 0?
   A (−6) + (+6)   B (+6) − (−6)
   C (−6) − (−6)   D (+6) − (+6)

8. The result of (−2) − (−8) is
   A −10   B −6
   C +6   D none of the above

Short Answer

9. Evaluate. Use integer chips or a number line to help. Then, arrange the answers in order from least to greatest.
   a) (−3) + (−6)   b) (−2) + (+7)
   c) (+5) − (+12)   d) (+10) − (−6)
   e) (−5) + (−5)

380 MHR • Chapter 11

10. Jan went on a diving expedition in the St. Lawrence River. She went down 1 m every 2 s. The surface of the river is at an elevation of 8 m above sea level. What was Jan's depth, relative to sea level, after 30 s?

11.

Start: −6 → Add −3 → ○ → Subtract +8 → ○ → Subtract −2 → ○ → Add −4 → ○ → Add −1 → End

Short Cut: from after "Add −3" directly to before "Add −1"

a) Follow the path to find the number that belongs in the end square.
b) What instruction belongs beside the short cut?

Extended Response

12. The table shows the temperature taken every hour one night.

Time	Temperature, °C
10:00	6
11:00	4
12:00	2
1:00	0
2:00	−2
3:00	−6
4:00	−7
5:00	−4
6:00	−2

a) Find the drop in temperature from 11:00 p.m. to 3:00 a.m.
b) Which hour had the greatest drop in temperature? What was it?
c) Which hour had the greatest increase in temperature? What was it?
d) Plot a graph of the data.
e) Describe any trends in the graph.
f) Predict the change in temperature between 6:00 a.m. and 7:00 a.m. Justify your answer.

Chapter Problem Wrap-Up

In question 18 on page 351 and question 17 on page 373 you explored some parts of the wind chill chart.

What patterns or trends can you find in the wind chill chart? Describe them using
- integers
- words
- graphs, charts, or diagrams

Wind Chill Chart

Wind Speed (km/h)	Air Temperature (°C)							
	5	0	−5	−10	−15	−20	−25	−30
5	4	−2	−7	−13	−19	−24	−30	−36
10	3	−3	−9	−15	−21	−27	−33	−39
15	2	−4	−11	−17	−23	−29	−35	−41
20	1	−5	−12	−18	−24	−31	−37	−43
25	1	−6	−12	−19	−25	−32	−38	−45
30	0	−7	−13	−20	−26	−33	−39	−46
35	0	−7	−14	−20	−27	−33	−40	−47
40	−1	−7	−14	−21	−27	−34	−41	−48
45	−1	−8	−15	−21	−28	−35	−42	−48
50	−1	−8	−15	−22	−29	−35	−42	−49
55	−2	−9	−15	−22	−29	−36	−43	−50
60	−2	−9	−16	−23	−30	−37	−43	−50

Patterning and Algebra
- Relate whole numbers and variables.
- Evaluate expressions by substituting whole numbers.
- Translate simple statements into expressions or equations.
- Solve equations and problems giving rise to them, by inspection and systematic trial.
- Realize that a solution to an equation makes the equation true.
- Write statements to interpret simple formulas.
- Present and explain solutions to patterning problems.

Number Sense and Numeration
- Explain the problem solving process in mathematical language.

Measurement
- Describe measurement concepts in measurement language.

Key Words
variable
variable expression
solution

CHAPTER 12

Patterning and Equations

Geometric designs can give walls, floors, and furniture an attractive finish.

Some designs use circles, squares, or other geometric forms in patterns.

You can use patterning and equation skills to plan similar designs of different sizes. In this chapter, you will work with many patterns, including tile designs. You will describe patterns using variable expressions and determine the number of items in various stages of patterns. You will use equations and patterning to solve problems.

Chapter Problem

Vicki is making coffee tables. She is going to copy the floor pattern here for one of her designs.

Extend the pattern to show two more tile designs. Explain how the steps in your pattern are related.

Design 1

Design 2

Design 3

Get Ready

Use the Order of Operations

In math, the correct **order of operations** is

- B — Brackets, then
- O — Order:
- D, M — Division and Multiplication, from left to right
- A, S — Addition and Subtraction, from left to right

Making Connections

You used the order of operations in Chapter 4.

For example,

$12 - 4 \times 2$
$= 12 - 8$ — Multiply first.
$= 4$ — Then, subtract.

$2 \times (3 + 4)$
$= 2 \times 7$ — Do brackets first.
$= 14$

1. Evaluate each expression.
  a) $5 \times 2 + 3$
  b) $3 \times 12 - 5$
  c) $2 \times 6 + 2 \times 7$
  d) $9 \times 12 \div 4$

2. Evaluate each expression.
  a) $3 \times (5 + 2)$
  b) $(15 - 7) \div 4$
  c) $(3 \times 1.5 + 0.5) \times 13$

Translate Into Mathematics

To solve a problem, you sometimes need to translate words into operations. For example, "the *sum* of 3 and 8" translates as $3 + 8$.

For example, find the number that is 12 *more than* the *product* of 2 and 7:

$2 \times 7 + 12$
$= 14 + 12$
$= 26$

More than tells me to add. *Product* means to multiply. I multiplied first, because of BODMAS.

Literacy Connections

Which Operation?
- \+ sum, grows by, more than, increased
- − less than, difference, subtracted from, decreased
- × product, times, double
- ÷ share equally, the quotient

3. Translate each statement using the +, −, ×, and ÷ operations. Calculate the result.
  a) the sum of 3 and 13
  b) four less than 7
  c) double 6
  d) the difference of 13 and 8
  e) the product of 8 and 9

4. For each question, translate, and calculate the result.
  a) Share $20 equally among 4 people. How much does each person get?
  b) If 77 grows by 13, how big does it get?
  c) What is 6 less than double 17?
  d) What is 5 more than the product of 8 and 6?

Work With Formulas

When you work with formulas, substitute what you know, and evaluate using the order of operations.

To find the perimeter of this rectangle, apply the formula $P = 2(l + w)$.

$P = 2(l + w)$
$P = 2 \times (12 + 5)$
$P = 2 \times 17$
$P = 34$

$2(l + w)$ means add l and w, then multiply by 2.

The perimeter of the rectangle is 34 cm.

To find the rectangle's area, apply the formula $A = lw$.

$A = lw$
$A = 12 \times 5$
$A = 60$

lw means multiply $l \times w$.

The area of the rectangle is 60 cm².

5. Find the perimeter and the area of this rectangle.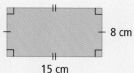

6. Find the area of this trapezoid. Use the formula $A = (a + b) \times h \div 2$.

Identify and Extend Patterns

When you work with patterns, ask yourself two questions:
• How does the pattern start?
• How does it change from one item to the next?

For example, the pattern 5, 7, 9, ... begins with 5 and increases by adding 2. The next two numbers are 11 and 13. You can describe this number pattern as follows:

$5 = 2 + 3$ or $5 = 5$
$7 = 2 + 2 + 3$ $7 = 5 + 2$
$9 = 2 + 2 + 2 + 3$ $9 = 5 + 2 + 2$

7. Describe each number pattern in words. Then, find the next two numbers.

a) 4, 8, 12, ...
b) 6, 10, 14, ...
c) 5, 9, 13, ...

8. a) Extend this pattern to two more diagrams.
b) How many dots are in the 6th diagram? Explain.
c) How many dots are in the 15th diagram? Explain.

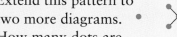
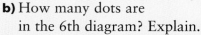

Get Ready • MHR

12.1 Variables and Expressions

Focus on...
- representing variables
- modelling expressions

You can use cups and counters to model math expressions. What could the cups in the photograph represent? Why do you think they both hold the same number of counters?

Materials
- 2 cups or containers
- supply of counters

Optional:
- BLM 12.1A Model Expressions Without Manipulatives

variable
- a letter that represents an unknown number

Discover the Math

How can you represent variables and model expressions?

1. Use a cup to represent an unknown number of counters. What would this diagram represent?

2. Use the **variable** C to represent the counters in the cup. Write an expression to model the total number of counters.

3. a) Suppose you put 10 counters in the cup. Instead of an unknown number, you now know the cup has a value. What value?
  b) How many counters will you have in total?
  c) If you let $C = 10$, what is the value of your expression in step 2?

4. a) Each cup in this diagram contains the same number of counters. Use C to show what the diagram represents.
  b) Use cups and counters. Try various numbers for the counters in the cups. Show each solution using the variable C and then using numbers.

5. Reflect Describe in words what each set of cups and counters models. Then, use a **variable expression** to model the diagram.

a)  **b)**

variable expression
- contains variables and operations with numbers
- $C + 3$ and $2C$ are variable expressions

Example 1: Model Number Phrases

Model each phrase with cups and counters. Then, write each as a variable expression.
a) 4 more than a number
b) 5 times a number plus 1

Solution

a)

+ ● ● ● ●

Let p represent the number.
$p + 4$

I could also use b for the variable. Part a) would be $b + 4$.

b)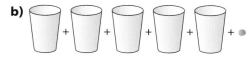

Let x represent the number.
$5x + 1$

Another way of showing this is $x + x + x + x + x + 1$. But $5x + 1$ is more efficient.

Example 2: Translate Models

a) Use cups and counters to model the variable expression $3x + 4$.
b) Evaluate the expression for these values of x: $x = 3$, $x = 4$, and $x = 5$.

Solution
a)

b) *Method 1: Use Cups and Counters*

I could also say
$3 \times 3 + 4 = 13$.

For $x = 3$: $3 + 3 + 3 + 4$
            $= 13$

I could say
$3 \times 4 + 4 = 16$.

For $x = 4$: $4 + 4 + 4 + 4$
            $= 16$

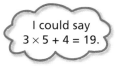

I could say
$3 \times 5 + 4 = 19$.

For $x = 5$: $5 + 5 + 5 + 4$
            $= 19$

Method 2: Substitute Into the Expression $3x + 4$

For $x = 3$:
 $3 \times 3 + 4$
 $= 9 + 4$
 $= 13$

For $x = 4$:
 $3 \times 4 + 4$
 $= 12 + 4$
 $= 16$

For $x = 5$:
 $3 \times 5 + 4$
 $= 15 + 4$
 $= 19$

Key Ideas

- A container, such as a cup, can be used to model an unknown number.
- Counters, such as integer chips or algebra tiles, can be used to represent known numbers.
- A variable represents an unknown number. For example, you could use C to represent the contents of the cups in the diagram.
- Variable expressions have a combination of variables, numbers, and operations. For example, another way of showing the expression in the diagram is 2C + 3.
- You can substitute a number for a variable, and then evaluate a variable expression. For example, evaluate the expression on the right if there are 7 counters in the cup.

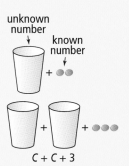

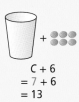

C + 6
= 7 + 6
= 13

Communicate the Ideas

1. a) Identify the number and the variable in the expression C + 6.
   b) Describe two advantages of using letters, instead of cups, for variables.

2. Develop a variable expression for the diagram. Describe how you did it.

3. The class was asked to model "2 more than 3 times a number." Look at the following solutions. Identify any errors. How would you correct them?

a) Mark's model:

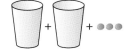

b) Rhiann's model:

c) Kayla's model:

4. a) Use cups and counters to model the variable expression 5a + 2.
   b) Create another variable expression that means the same thing but does not use multiplication. Justify your response.

Check Your Understanding

Practise

For help with questions 5 to 8, refer to Example 1.

5. Write each model as a variable expression.

   a)

   b)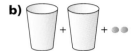

   c)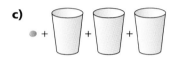

   d)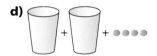

6. John uses a large square to represent an unknown quantity and buttons to represent counters. What variable expression is modelled by each diagram?

   a)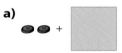

   b)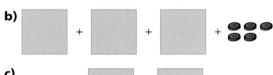

   c)

7. Model each phrase with cups and counters.
   a) a number plus 5
   b) double a number

8. Model each phrase with cups and counters.
   a) 2 more than 3 times a number
   b) 4 less than 5 times a number

For help with questions 9 to 12, refer to Example 2.

9. Use cups and counters to model each variable expression.
   a) $2a + 3$
   b) $4n + 5$
   c) $2 + 5x$
   d) $3w + 10$

10. a) For each expression in question 9, put 2 counters in each cup. Evaluate the expression.
    b) Repeat part a) using 3 and then 4 counters in each cup. Evaluate the expression.

11. Substitute $k = 5$ and evaluate each expression.
    a) $k + 6$
    b) $k - 3$
    c) $3k$
    d) $2k + 3$
    e) $12 - k$
    f) $4 + 3k$

12. For each expression in question 11, substitute $k = 8$ and evaluate.

Apply

13. Model each variable expression.
    a) $4 + 3t$
    b) $7t - 4$
    c) $5 + 6t$
    d) $10t + 3$

14. For each expression in question 13, substitute $t = 3$ and evaluate.

15. Show each phrase with a variable expression, using addition, subtraction, or multiplication.
    a) $10 *more than* an unknown price
    b) the *product* of 8 and Jessica's age
    c) the area *increased by* 10 cm^2
    d) *double* the length

16. Write a variable expression for each phrase.
   a) 10 cm *shorter* than you
   b) 15 *less than* your opponent's score in a card game
   c) *triple* the elephant's mass
   d) the *sum* of 12 and Kenneth's points

17. The expression $5h + 3$ describes Sonja's pay scale for baby-sitting.
   a) Model this expression with cups and counters.
   b) Explain what you think the expression means.
   c) If Sonja baby-sits from 6:30 P.M. to 11:30 P.M., how much does she earn?

18. You can use the expression $\frac{n}{20}$ to estimate your average walking speed, in kilometres per hour. The variable n is the number of steps you take in 1 min. Estimate your walking speed if you take 80 steps in 1 min.

19. a) Evaluate the expression $3x + 4$ for each value: $x = 1, 2, 3, 4,$ and 5.
   b) Describe the pattern you see in your answers. What could produce this pattern?

20. You can use the expression $\frac{m}{13}$ to estimate the total volume of blood in your body, in litres. The variable m is your mass, in kilograms.
   a) Estimate how much blood a 39-kg teenager has.
   b) Estimate how much blood a 70-kg adult has.
   c) Estimate the volume of blood in your body.

21. a) Model this phrase with a variable expression: double the area, increased by 5 cm².
   b) Evaluate your expression in part a) for each area: 5 cm², 10 cm², 15 cm², 20 cm².
   c) Describe a pattern rule for your expression values.
   d) Try to extend the pattern. Identify the 10th value.

Extend

22. The average January temperature can be estimated for any city in the world. The expression $33 - 0.75L$ estimates the temperature, in degrees Celsius, for a city with a latitude of L degrees. What is the estimated mean January temperature for
   a) Kingston, Jamaica, at 17°N?
   b) Yellowknife, Yukon, at 62°N?
   c) London, England, at 52°N?
   d) Is the estimate for part c) accurate? Explain.
   Go to www.mcgrawhill.ca/links/math7 and follow the links to find out.

Literacy Connections

Latitude
Latitude measures how far north or south you are from the equator. The equator is 0°. The North Pole is 90°N. The South Pole is 90°S.

23. a) Substitute $x = 5$ into each of the expressions $x + 4$ and $2x$. Evaluate each expression.
   b) Which expression gives the greater result? Will this be true for all values of x? Justify your answer.

12.2 Solve Equations by Inspection

Focus on...
- mental math
- substituting into equations

This balance is showing 230 g. The smaller mass is 30 g. What are each of the larger masses?

What does the equation $30 + 2m = 230$ have to do with this situation?

Discover the Math

Materials
- 5 cups or containers
- supply of counters

How can you find the solution to an equation using mental math?

1. Model each diagram with cups and counters. Determine the number of counters in each cup.

   a)

   b)

   c)

   d)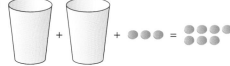

2. a) Describe how you came up with the answer in each part of step 1.
  b) How could you use mental math to answer step 1?
  c) How can you model step 1 using a balance, like the one in the picture?

3. a) Use one of your strategies from step 2 to find the value of the variable *n* in the equation $n + 4 = 5$.
  b) Is your answer to part a) the only **solution**? Explain.

solution (of an equation)
• a number that makes an equation true
• $y = 3$ is the solution to $y + 1 = 4$

4. How could you use mental math to find the value of the variable in each equation?
  a) $p - 5 = 2$
  b) $w + 5 = 10$
  c) $5 \times c = 10$
  d) $3 \times x = 6$

5. Reflect Explain the similarities between your method in step 1 and your method in step 4. Describe how you can use mental math to solve simple equations.

Literacy Connections

The pictures of balances are designed to help you visualize the idea of balancing an equation.

Example 1: Solve by Inspection

For each equation, use mental math to try a number as a solution.
  a) $3k = 6$
  b) $m + 3 = 12$
  c) $7 - x = 4$
  d) $\dfrac{a}{3} = 4$

Solution

a) $3 \times 2 = 6$
   So, the solution is $k = 2$.

b) $9 + 3 = 12$
   So, the solution is $m = 9$.

c) $7 - 3 = 4$
   The solution is $x = 3$.

d) $\dfrac{12}{3} = 4$
   The solution is $a = 12$.

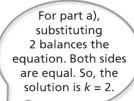
For part a), substituting 2 balances the equation. Both sides are equal. So, the solution is $k = 2$.

Example 2: Test for Solutions

Which of the numbers 5, 6, and 7 is a solution to the equation $4x + 3 = 27$?

Solution

Try each number.

$4 \times 5 + 3$
$= 20 + 3$
$= 23$
$x = 5$ does not work.

$4 \times 6 + 3$
$= 24 + 3$
$= 27$
With $x = 6$, the equation is true.

$4 \times 7 + 3$
$= 28 + 3$
$= 31$
$x = 7$ does not work.

The solution is $x = 6$.

> I know $x = 6$ is the solution. So, I think $x = 7$ won't work. I'll check.

Key Ideas

- A number is the solution to an equation if the two sides of the equation are equal after substituting.

- Simple equations can be solved by inspection, using mental math. For example, what mass would balance the scales in the top diagram?

- You can test possible solutions to an equation. For example, which of $b = 2$, $b = 3$, and $b = 4$ is the solution to $3b + 1 = 10$? Testing shows that $b = 3$ is the solution.
    $3 \times 3 + 1$
    $= 9 + 1$
    $= 10$

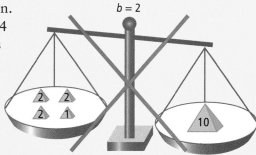

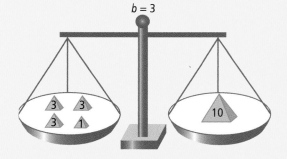

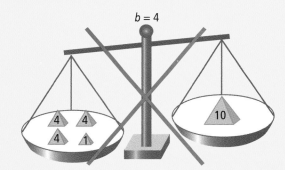

Communicate the Ideas

1. The diagram shows how Tom used modelling to solve an equation.
   a) What is being tested here?
   b) Do the cups show a possible solution? Explain.

2. Describe how to use inspection to solve the equation $9k = 72$.

3. To solve the equation $7 + 6y = 31$, you are given these values to test:
   $y = 3$, $y = 4$, $y = 5$, and $y = 6$. Will you have to test all four values?
   Explain why or why not.

Check Your Understanding

Practise

For help with questions 4 to 9, refer to Example 1.

4. What mass will keep the balance?

   a)

   b)

   c)

5. What number on each weight will keep the balance?

   a)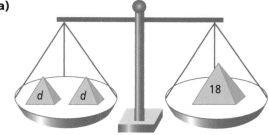

   b)

   c)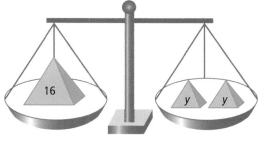

6. Find the value of each cup to make the sentence true.

a)

b)

c)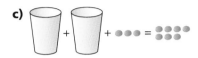

7. Find the value of each cup to make the sentence true.

a)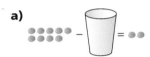

b)

c)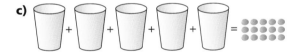

8. Solve by inspection or using manipulatives.
   a) $7b = 21$ b) $x + 15 = 25$
   c) $8v = 56$ d) $13 = 3 + n$
   e) $r - 10 = 0$ f) $19 - g = 13$

9. Solve by inspection.
   a) $n + 7 = 26$ b) $35 = y - 15$
   c) $63 = 7t$ d) $2a + 2 = 10$
   e) $25 - x = 16$ f) $30 = 5q + 5$

For help with questions 10 to 15, refer to Example 2.

10. How many counters should go in the cup? Test the possible solutions. Which number balances the equation?

Test $b = 11$, 12, and 13.

11. How many counters should go in the cup? Justify your answer.

Test $y = 21$, 22, and 23.

12. For each diagram, how many counters should go in each cup? Test the possible solutions.

a)

Test $c = 3$, 4, and 5.

b)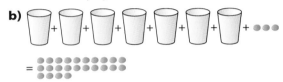

Test $k = 2$, 3, and 4.

c)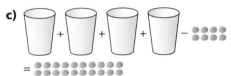

Test $t = 6$, 7, and 8.

13. Show whether or not $x = 4$ is the solution to each equation.
   a) $6x = 24$ b) $17 - x = 10$
   c) $36 \div x = 8$ d) $x + 27 = 33$
   e) $3x + 1 = 15$ f) $20 = x + 10$

14. Determine whether $k = 9$ or 10 is the solution to each equation.
   a) $k + 5 = 14$ b) $k - 8 = 1$
   c) $k + 6 = 16$ d) $2k = 18$
   e) $k \div 2 = 5$ f) $16 = 2x - 4$

15. Solve each equation.
   a) $h + 7 = 10$ b) $6m = 36$
   c) $y - 17 = 21$ d) $f \div 7 = 11$
   e) $50c = 250$ f) $3 + 20x = 83$

Apply

16. Draw a balance to show the equation $230 = 30 + 2m$.

   a) What total mass does the balance show?
   b) What does the equation tell you about the smaller mass? the two larger masses?
   c) Solve the equation to determine each larger mass.

17. a) Write two different equations that do *not* have 3 as a solution. Each equation should use a different operation (addition, subtraction, multiplication, division).
   b) Describe how you came up with your equations.

18. If Annie had $5 more in her pocket, she could afford a $14 CD. An equation modelling this is $x + 5 = 14$.

   a) Explain what the variable x represents. How do you know?
   b) Solve the equation to find how much money Annie has.

19. Kathy walks at a speed of 100 m per minute on her way to school. She lives 1500 m from school. Kathy's walk can be modelled by the equation $100t = 1500$. In this equation, t is the time, in minutes, it takes Kathy to get to school. Solve the equation to determine how long it takes Kathy to get to school.

20. The perimeter of this triangle is 70 cm. An equation modelling the perimeter is $x + 20 + 16 = 70$. Determine the length of side x.

21. a) Write two different equations that have 5 as a solution. Each equation should use a different operation (addition, subtraction, multiplication, division).
   b) Describe how you came up with your equations.

Extend

22. A competitor's total score in a diving competition can be calculated using the formula $T = Mn$.

   • M is the mean score.
   • T is the total score.
   • n is the number of judges.

   If Alex's mean score was 8 and his total score was 48, how many judges were there?

23. The sum of 3 and a number is –10.

   a) Write an equation to model this situation.
   b) Solve the equation to find the unknown number.

24. When you know your speed and travel time, the distance can be calculated using the formula $d = s \times t$.

   a) How fast are you driving if it takes 5 h to go 350 km?
   b) The Cassini space probe arrived at Saturn in 2004. How fast did the probe travel if it took 5 h to go 2400 km?

Did You Know?

Saturn's largest moon, Titan, has a thick methane atmosphere. Cassini will teach scientists a lot about this strange moon.

12.3 Model Patterns With Equations

Focus on...
- geometric and other patterns
- modelling with equations

In this honeycomb pattern, each row has one more cell than the row above it. What equation models the row of cells that uses 21 toothpicks?

Discover the Math

Materials
- supply of counters
- 4 cups or containers

Optional:
- BLM 12.3A Develop an Equation Step by Step

How can you use patterning to write equations?

1. Use counters to make this pattern of rectangles. Extend the pattern as far as Rectangle 5.

   Rectangle 1   Rectangle 2   Rectangle 3

2. **a)** Rectangle 1 can be modelled as in the photo

   Model Rectangles 2 and 3 using cups and counters. Hint: In your models, continue the pattern started in the photograph. Think about what part of the pattern each cup will represent.

   b) Use a table like the one below to record what you are doing in each model.

Rectangle number	Cups and Counters	Equation
1	2 cups with 1 counter each + 2 counters	$2 \times 1 + 2 = 4$
2	2 cups with ▊ counters each + 2 counters	$2 \times ▊ + 2 = ▊$
3		

3. **a)** Extend your table to predict what will happen in Rectangles 4 and 5.
   b) Check by modelling. Revise your equation, if necessary.

4. **Reflect** How did patterning help you develop your equations?

Example 1: Write an Equation for a Patterning Problem

Cosmic Toys makes coloured rods.
- Cubes are joined in rows to make the rods.
- The rods are then dipped in paint.
- Finally, a smiley face sticker is fixed to every exposed vertical face.

For example, the two-cube rod has 6 smiley faces.

How many cubes are needed to have 30 smiley faces?
Model this problem with an equation.

Solution

The equation has to model the number of smiley faces on a rod.
The expression for the number of smiley faces should equal 30.

1. Use a table to organize the pattern and create a set of equations.
2. Use the equations in the table to help you write an equation that models the rod with 30 smiley faces.

1.

Number of Cubes	Number of Smiley Faces	Equation
1	1 at the front + 1 at the back + 2 at the ends = 4	2 × 1 + 2 = 4
2	2 at the front + 2 at the back + 2 at the ends = 6	2 × 2 + 2 = 6
3	3 at the front + 3 at the back + 2 at the ends = 8	2 × 3 + 2 = 8
4	4 at the front + 4 at the back + 2 at the ends = 10	2 × 4 + 2 = 10

Strategies
Make a table or chart

2. Let n represent the number of cubes.
   The equation for the block pattern with 30 smiley faces is $2n + 2 = 30$.

If you double the number of cubes and add 2, you get the number of smiley faces.

Substitute the values 1, 2, 3, and 4 for n and check against the equations in the table. For example:
$2n + 2 = 2 \times 1 + 2$
$ = 4$

One cube has 4 smiley faces.

Example 2: Model Patterns With Equations

Study this pattern of marble diagrams.

Diagram 1   Diagram 2   Diagram 3

a) Use equations to model the first three diagrams.
b) Use an equation to model the diagram with 17 marbles. Explain what your equation means.

Solution

a)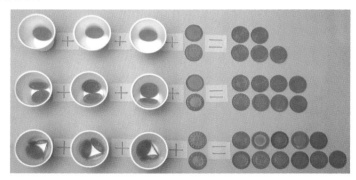

Strategies
Act it out

Let d stand for the number of the diagram.
Diagram 1: $3d + 2 = 5$
Diagram 2: $3d + 2 = 8$
Diagram 3: $3d + 2 = 11$

The letter d also stands for the number of counters in each cup. In the model for diagram 2, there are 3 cups with 2 counters in each cup. Plus there are 2 more counters. I can use the equation $3d + 2 = 8$.

b) The equation is

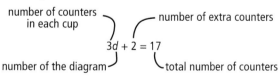

number of counters in each cup
number of extra counters
$3d + 2 = 17$
number of the diagram
total number of counters

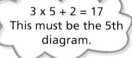

$3 \times 5 + 2 = 17$
This must be the 5th diagram.

I used the same variable expression for Diagrams 1, 2, and 3. I must use the same expression for the diagram with 17 marbles. The equation is $3d + 2 = 17$.

Key Ideas

- The items in a pattern can be translated into equations, using numbers or variables.
- An equation for a pattern can be explained in terms of the pattern. For example, look at the pattern to the right.

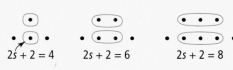

- You can write an equation in different ways. For example, the equations $2x + 3 = 12$, $12 = 2x + 3$, and $3 + 2x = 12$ all mean the same thing.

Communicate the Ideas

1. Helen wrote an equation as $2x + 3 = 7$. Kenneth wrote it as $7 = 2a + 3$. Jeeva wrote it as $7 = 3 + 2m$. Are all of these the same or different? Explain.

2. Which stage in this pattern has 19 dots?

Sameh started by writing
$2 + 1 + 2 = 5$
$2 + 2 + 2 = 6$
$2 + 3 + 2 = 7$

Rosie started by writing
An expression for the pattern is $4 + n$.

a) Complete each solution. What are the advantages of each method?
b) Can you think of any other methods?

3. a) Kim wrote the expression $n - 3$ for a dot pattern. Draw three stages of a dot pattern for this expression.
b) Make an equation to show the number of dots in the third stage of your dot pattern.
c) Will your equation be exactly the same as those developed by other students? Explain.

Literacy Connections

Choosing Variables
Any letter can be used for a variable. Try to use a letter that helps you remember what the variable represents. In the Key Ideas, s refers to the step number.

Check Your Understanding

Practise

For help with questions 4 and 5, refer to Example 1.

4. These rods have smiley stickers on all their square faces except the two end faces.

a) Copy and complete this table.

Number of Cubes	Number of Smiley Faces	Equation
1	1 (front) + 1 (back) + 1 (top) + 1 (base) = 4	$4 \times 1 = 4$
2		
3		

b) How long is the rod with 28 smiley faces? Write an equation for this rod.

5. a) Copy and complete this table for the toothpick pattern shown.

Number of Rows	Perimeter	Equation
1	1 + 1 + 1	$3 \times 1 = 3$
2	2 + 2 + 2	
3		

b) Write an equation for the number of rows of triangles needed to have a perimeter of 27 toothpicks.

For help with questions 6 and 7, refer to Example 2.

6. a) Use equations to model the first three diagrams in this pattern.

Diagram 1   Diagram 2   Diagram 3

b) One diagram has the equation $2 + 2d = 14$. How many dots does the diagram have? Explain.

7. Study this pattern of marbles.

a) Use equations to model the first 3 diagrams.

b) Use an equation to model the diagram with 17 marbles. Explain what your equation means.

8. Write each equation in two other ways.
a) $k + 7 = 13$ b) $6n + 3 = 21$

9. Write each equation in two other ways.
a) $37 = 4w + 5$ b) $6 + 2a = 26$

Apply

10. a) Use equations to model the first three diagrams.

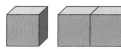 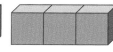

b) Use an equation to model the diagram with 75 model cars.

11. A pattern of blocks is shown. A certain number of blocks are needed to have 20 unexposed vertical faces. Write an equation for this problem.

Chapter Problem

12. Vicki makes coffee tables for sale. The visuals below show her first three designs.

Design 1

Design 2

Design 3

a) Develop a variable expression for the number of red tiles in each design. Hint: You can use tiles to model this.
b) Predict how many red tiles will be in Design 5. Explain why you are making this prediction.
c) Make a model of Design 5. If necessary, revise your ideas.

13. If c represents an unknown number of interlocking cubes, write a situation that could be modelled by each equation.

a) $c - 2 = 4$
b) $2c + 1 = 23$
c) $4c - 2 = 14$

14. a) Create a pattern that could be modelled by this expression.

$$8 + 3h$$

b) Write equations for the first 3 diagrams in the pattern.
c) Write an equation for the 7th diagram in the pattern. What does this equation model? Explain.

Extend

15. These four-sided and eight-sided dice have spot patterns on their faces. The patterns have 1, 2, 3, and 4 spots for the four-sided die, and 1, 2, 3, 4, 5, 6, 7, and 8 spots for the eight-sided die. Let m represent the number of four-sided dice. Let n represent the number of eight-sided dice.

Spot Patterns:
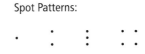

Spot Patterns:

a) Write an equation, using m and n, for a combination of dice with a total of 102 spots.
b) Investigate the possible total numbers of spots, using different combinations of dice.

16. Cosmic Toys makes a line of pyramid puzzles. The building block is a pyramid with equilateral triangles for its base and side faces. Pyramid puzzles with base lengths 3 and 4 look like this.

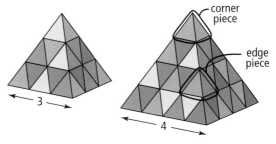

How many edge and corner pieces would a puzzle with base length n have?

12.3 Model Patterns With Equations • MHR   **403**

12.4 Solve Equations by Systematic Trial

Focus on...
- modelling with equations
- systematic trial

This geoboard shows a geometric pattern. If the pattern were extended, which item would have a perimeter of 32?

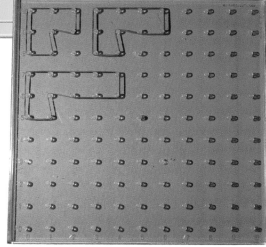

Discover the Math

How can you use systematic trial to solve an equation?

1. Each cup contains an unknown number of counters. When more than one cup is used, each contains an equal number of counters. What number of counters in the cup makes each picture true? For each picture, record each guess until you get the correct answer.

   a) ▢ + ▢ = ●●●●●●●●●●●

   b) ▢ − ●●●●●●●● = ●●

   c) ▢ + ▢ + ▢ + ▢ + ▢ + ▢ + ▢ − ●●●●●●●●● = ●●●●●

2. Describe your method of finding the numbers of counters in step 1.

What visual can I use to show this?

3. Use your method to answer this question: If you triple the number of counters and add 15 more counters, you have 63 counters. How many counters did you start with?

4. Make up two examples of your own. Share them with a partner. Ask your partner to describe his or her solving strategy.

5. **Reflect** Can you improve your method? If so, describe the improved method. If not, explain why your method is effective as it is.

Example 1: Model and Solve Pattern Problems

a) Write an equation for the diagram with 26 marbles.
b) Solve your equation. State what the solution means.

Diagram 1

Diagram 2

Diagram 3

Solution

a)

Diagram Number	Number of Marbles	Pattern
1	5	$3 \times 1 + 2$
2	8	$3 \times 2 + 2$
3	11	$3 \times 3 + 2$
d	26	$3 \times d + 2$

Strategies
Make a table or chart

The equation is $3d + 2 = 26$.

b) Use systematic trial.
Try $d = 5$.
 $3 \times 5 + 2$
$= 15 + 2$
$= 17$
Too small.

Use d to stand for the number of the diagram.

Try $d = 10$.
 $3 \times 10 + 2$
$= 30 + 2$
$= 32$
Too large.

I'll try a larger number for d.

Try $d = 8$.
 $3 \times 8 + 2$
$= 24 + 2$
$= 26$
Correct.

I'll try a number that's between 5 and 10.

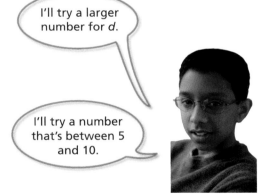

In Diagram 8, there will be 26 marbles.

Example 2: Practise Systematic Trial

Solve each equation.
a) $2x - 7 = 25$
b) $7a + 12 = 54$

Solution

a) Try $x = 20$:
$2 \times 20 - 7$
$= 40 - 7$
$= 33$
Too large.

I need a number that is more than 25 when I double it. I'll try 20.

Try $x = 15$:
$2 \times 15 - 7$
$= 30 - 7$
$= 23$
Slightly too small.

Try $x = 16$:
$2 \times 16 - 7$
$= 32 - 7$
$= 25$
Correct.

The solution is $x = 16$.

b) Try $a = 5$:
$7 \times 5 + 12$
$= 35 + 12$
$= 47$
Too small.

Try $a = 7$:
$7 \times 7 + 12$
$= 49 + 12$
$= 61$
Too large.

Try $a = 6$:
$7 \times 6 + 12$
$= 42 + 12$
$= 54$
Correct.

5 is too small and 7 is too large. The solution should be $a = 6$.

The solution is $a = 6$.

Example 3: Model and Solve Number Pattern Problems

a) Develop an expression to model the number pattern 2, 7, 12, 17, ...
b) Which step is the number 52?

Solution

a) $1 \times 5 - 3 = 2$
$2 \times 5 - 3 = 7$
$3 \times 5 - 3 = 12$
$4 \times 5 - 3 = 17$

The steps in the pattern repeatedly jump by 5. Show repeatedly adding 5 as multiplying the step number by 5.

The expression $n \times 5 - 3$ or $5n - 3$ describes the number pattern.

b) The equation $5n - 3 = 52$ models the step in the pattern that equals 52.

Try $n = 10$.
$5 \times 10 - 3$
$= 50 - 3$
$= 47$
Slightly too small.

Try $n = 11$.
$5 \times 11 - 3$
$= 55 - 3$
$= 52$
Correct.

Strategies
Use systematic trial

The 11th step in the pattern is 52.

Key Ideas

- To solve equations by systematic trial, substitute values for the variable until you get the correct answer.
- Use a reasonable guess for the first value you substitute. Think about the relationship between the numbers in the equation.
- For each value that you substitute, think about the previous values. Were they too large? too small? For example, solve $3n - 8 = 31$.

"We need a number that's bigger than 31 when you multiply it by 3. Start with 12."

$3(12) - 8$
$= 36 - 8$
$= 28$

"12 is a bit too small. What about 13?"

$3(13) - 8$
$= 39 - 8$
$= 31$

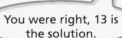

"You were right, 13 is the solution."

Communicate the Ideas

1. A question asks you to solve the equation $4w - 1 = 15$. What does the word "solve" mean?

2. When solving $5k + 15 = 60$, Mario chose 10 for his first value of k. Jenna chose 20. Which number is the better choice, and why?

3. Kajan has one more than twice as many candies as Lena.
   a) Model this situation visually or with cups and counters.
   b) Explain whether each equation could model the situation.
   - $2L + 1 = 15$, where L is the number of candies Lena has
   - $2K + 1 = 15$, where K is the number of candies Kajan has
   c) In each equation in part b), what does the number 15 mean?

Check Your Understanding

Practise

For help with questions 4 to 7, refer to Example 1.

4. Study the pattern of marbles.

Diagram 1   Diagram 2   Diagram 3

   a) Copy and complete this table.

Diagram Number	Number of Dots	Pattern
1	1 on the left, 2 on the right = 3	$1 + 1 \times 2$
2	1 on the left, 4 on the right = 5	$1 + 2 \times 2$
3		
d		

   b) Write an equation for the diagram with 15 dots.
   c) Solve your equation. State what the solution means.

5. Study the pattern of dots.

   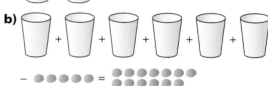

   a) Copy and complete this table.

Diagram Number	Number of Dots	Pattern
1	2 on the left, 1 in the middle, 2 on the right = 5	$2 + 1 \times 1 + 2$
2	2 on the left, 2 in the middle, 2 on the right = 6	$2 + 2 \times 1 + 2$
3		
d		

   b) Write an equation for the diagram with 11 dots.
   c) Solve your equation. State what the solution means.

6. What number makes each situation true? Solve by inspection or by systematic trial.

7. Solve by inspection or by systematic trial.

   a) [cup] + [cup] + ●●●● = ●●●●●●●●●
   b) [cup] + [cup] + [cup] + [cup] + [cup] − ●●●●● = ●●●●●●●●●●●●●

For help with questions 8 and 9, refer to Example 2.

8. Solve each equation by systematic trial.
   a) $2x + 3 = 17$ b) $3q + 2 = 20$
   c) $5 + 7z = 19$ d) $20 = 6 + 2a$

9. Solve each equation by systematic trial.
   a) $2m + 12 = 56$ b) $8w + 5 = 37$
   c) $34 = 9 + 5p$ d) $5n - 2 = 33$

For help with questions 10 to 12, refer to Example 3.

10. a) Develop an expression to model the number pattern 7, 10, 13, 16, ….
    b) Which step is the number 34?

11. a) Develop an expression to model the number pattern 10, 17, 24, 31, ….
    b) Which step is the number 45?

12. a) Develop an expression to model the number pattern 19, 18, 17, 16, ….
    b) Which step is the number 7?

Apply

13. Solve by systematic trial.
   a) $7 + 3c = 91$ b) $123 - a = 97$
   c) $10x - 6 = 84$ d) $48 = 9c - 6$

14. Solve each equation.
   a) $30 - 2m = 18$ b) $64 - 3k = 34$
   c) $100 - 5c = 35$ d) $89 = 45 + 2b$

15. a) Five times a number plus 13 gives 48. Model this with an equation.
   b) What is the number?
   c) Explain your solution method.

16. a) Develop an expression to model the shrinking number pattern 49, 45, 41, 37, …. Hint: What is the jump in this pattern? What number, minus the jump, gives the first step of 49?
   b) Which step is the number 13?

17. Study the toothpick pattern. Use equations to determine the number of squares that give

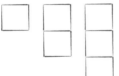

   a) a total of 52 toothpicks
   b) a perimeter of 48 toothpicks

18. a) It cost Jenna $50 to set up her business selling smiley pins. She charges $2 a pin. In her first week, Jenna made a profit of $80. The equation modelling her profit is $2n - 50 = 80$, where n is the number of pins sold. How many pins did Jenna sell?
   b) Explain the strategy you used. How else might you have solved the problem?

19. a) Write an equation, containing multiplication and addition, that has 8 as a solution.
   b) Write an equation, containing division and subtraction, that has 8 as a solution.

Chapter Problem

20. The diagrams show the first three of Vicki's coffee table designs.

Design 1   Design 2   Design 3

   a) Develop a variable expression for the number of red tiles in each design.
   b) Which design will need 24 red tiles? Show and explain your solution.

 21. Suppose this geoboard pattern were extended. Write and solve an equation for the item in the pattern with a perimeter of 32.

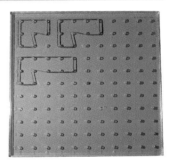

Extend

22. Solve each equation.
   a) $1.4y + 15 = 29$
   b) $150 = 2.5w + 50$

23. For a new style of rod, Cosmic Toys decides to put a smiley face sticker on every exposed square face.

The number of exposed smiley stickers in a rod of length n can be calculated using an equation. Is there a rod with 49 exposed smiley faces? 106 exposed smiley faces? Justify your answers.

12.5 Model With Equations

Focus on...
- modelling real-world problems
- applying equation skills

Banks and stores need quick ways of counting their coins. Rolls of $2 coins contain 25 coins, or $50. What would be a quick way of counting these coins? How could you model their value using an equation?

Discover the Math

How can you model and solve a problem using an equation?

1. Barbara is counting the money in her piggy bank. She has a $10 bill and a number of $2 coins.

   a) Barbara modelled the number of $2 coins with the expression $2T$. What does T represent?
   b) Explain why the value of Barbara's $2 coins can be modelled with the expression $2T$.

2. Explain why the total value of Barbara's coins can be modelled with the expression $2T + 10$.

3. Barbara has a total of $18.
   a) Write an equation modelling this total.
   b) Describe the equation in words.
   c) Solve the equation. Explain what the solution means.

4. **Reflect** How would you change the equation if Barbara had a total of $30? Explain.

Example 1: Systematic Trial in Formulas

The perimeter of a rectangle is 32 cm. The length is 9 cm. What is the width of the rectangle?

Solution

Method 1: Work With the Formula
$P = 2l + 2w$
$32 = 2(9) + 2w$
$32 = 18 + 2w$
$2w$ is unknown
So, $32 = 18 + 2w$. Solve for the width.

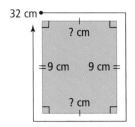

Strategies
Make a picture or diagram

Try $w = 8$:
    $18 + 2 \times 8$
$= 18 + 16$
$= 34$
The right side is 34. 8 is slightly too high.

Try $w = 7$:
    $18 + 2 \times 7$
$= 18 + 14$
$= 32$
7 is correct.

2w has to be 18 less than 32. w should be less than 10.

The width of the rectangle is 7 cm.

Method 2: Work Backward
Two of the sides are 9 cm. So, the leftover perimeter is
    $32 - 2 \times 9$
$= 32 - 18$
$= 14$

Strategies
Work backward

The two widths total 14 cm. So,
$2w = 14$
$w = 7$

Strategies
What other strategy might you use?

The width of the rectangle is 7 cm.

Example 2: Model With an Equation

At a garage sale, paperback books are priced at $2 each.
One customer buys $12 worth of books.
a) Solve this problem using a method of your choice.
b) Use an equation to solve this problem.
c) Compare your solutions from a) and b). What conclusion can you make?

Solution
a) $2 \times 6 = 12$
The solution is $b = 6$. This means that the customer bought 6 books for $12.

b) Let b represent the number of books.
The equation is $2b = 12$.
price of books ⌋ ⌊ total cost
number of books

I used cups and counters to model the situation.
2 cups = 12 red counters

c) There are different ways to solve this problem. The three solutions here all work.

Key Ideas

- Situations can be modelled with an equation by translating the situation into numbers and variables. Use operations to show what happens to the numbers.
- Solve the equation using an appropriate method.

$32 = 2 \times 9 + 2w$
$P = 32$ cm

Solving methods:
- work backward
- systematic trial
- make a diagram

Communicate the Ideas

1. When Jane's age is doubled and 3 is added, the result is 17. Which of the following equations are appropriate models? For each equation, explain your answer.
   a) $2a + 3 = 17$
   b) $17 = 3 + 2a$
   c) $3a + 2 = 17$

2. The perimeter of an isosceles triangle is 30 cm. The two equal sides are each 12 cm long. What clues in this statement help you develop an equation?

3. As a math puzzle, Ted wrote the expressions $3f - 6 = 21$ and $2f = 18$ to describe the number of people in his family. Is it possible to have two equations modelling the same situation? Explain.

Check Your Understanding

Practise

4. Write each sentence as an equation. Then, solve.
   a) Four less than a number is 3.
   b) The sum of a number and 5 is 12.
   c) Ten more than the product of a number and 3 is 31.
   d) Double a number, decreased by 10, is 15.

5. Write each sentence as an equation. Then, solve.
   a) A cost shared by 4 people amounts to $10 each.
   b) There are 42 oranges. This is 14 more than double the number of apples.
   c) The number of students increased by 15 to 32.
   d) 70 cm is 10 cm less than half of Bill's height.

For help with questions 6 to 9, refer to Example 1.

6. The perimeter of a rectangle is 39 cm. Its width is 5 cm.
   a) Model this situation with an equation.
   b) Solve the equation to find the length of the rectangle.

7. The perimeter of a rectangle is 42 cm. The width is 4 cm. What is the length of the rectangle?

8. The perimeter of an isosceles triangle is 20 cm. The two equal sides each measure 6 cm. How long is the third side?

9. The formula for the area of a triangle is $A = b \times h \div 2$. Find h for a triangle with base 4 cm and area 10 cm².

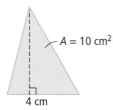

For help with questions 10 and 11, refer to Example 2.

10. The variable d represents an unknown number of DVDs. Write a situation that could be modelled by the equation $20d = 60$.

11. The variable b represents an unknown number of books. Write a situation that could be modelled by each equation.
    a) $b + 5 = 7$
    b) $7b = 28$
    c) $b + 14 = 19$

Apply

12. a) Find b for a triangle with area 40 cm² and height 10 cm.

b) Find h for a triangle with area 36 cm² and base 4 cm.

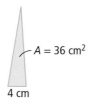

13. Khalid is 1 year older than 3 times his son's age. Khalid is 31 years old. Explain why both $31 = 3s + 1$ and $1 + 3s = 31$ can be used to model this situation.

14. Describe a situation involving money that can be modelled by the equation $20T + 12 = 72$.

15. The Student Council is planning a year-end party. It costs $300 for the D.J. for the evening. They plan to spend $5 per student on food.
    a) What is the total cost for s students?
    b) The school budgeted $1000 for the party. How many students can attend?
    c) How did you solve this problem? What other strategy might you use?

16.
    a) The rectangular room shown has a perimeter of 26 m. Write an equation to model the perimeter.
    b) The room has area 40 m². Write an equation to model the area.

17. An experiment found that you can predict the approximate area of the palm of your hand using the formula $A = 20L - 67$. In this equation,
    - A is the area, in square centimetres, of your palm
    - L is the length, in centimetres, of your index finger
    a) Measure your index finger. Use the formula to estimate the area of the palm of your hand.
    b) Approximately how long would a person's index finger be if his or her hand had an area of 153 cm²?

18. A volleyball court is 9 m wide and has a perimeter of 54 m. What is its length?

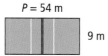

19. a) A basketball court is 14 m wide with an area of 364 m². What is its perimeter?

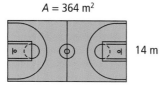

b) How did you solve this problem? What other strategy might you have used? Which strategy do you prefer, and why?

20. In a science experiment, various masses were suspended from a spring. The stretched length of the spring is recorded in the table.

Mass (g)	5	10	15	20
Stretch Distance (cm)	12	22	32	42

What mass would stretch the spring to a length of 122 cm?

21. Sandi is delivering flyers. She earns $5 plus $3 per bundle of 100 flyers.

a) How much is Sandi paid to deliver 1 bundle of flyers? 2 bundles of flyers? x bundles of flyers?

b) Sandi earned $29. How many bundles of flyers did she deliver?

Extend

22. The perimeter of a rectangular playing field is 300 m. The length is double the width.

a) Model this situation with an equation. Use one variable only.

b) How could you simplify the equation?

23. The formula $T = 33 - 0.75L$ estimates the mean January temperature, in degrees Celsius, for a city with a latitude of L degrees. At what latitude would Moscow, Russia, be with a mean January temperature of $-9°C$?

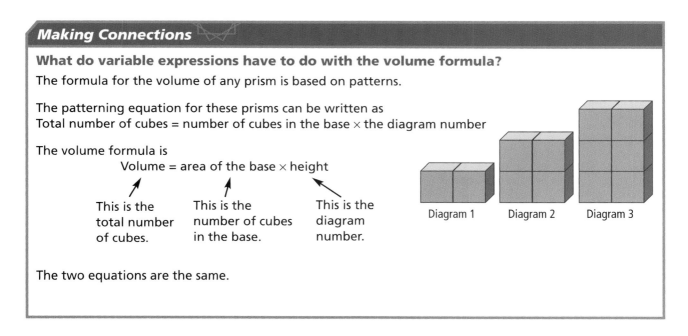

CHAPTER 12 Review

Key Words

Use these words to copy and complete the statements in questions 1 and 2.

variable   variable expression
solution   equation

1. In the 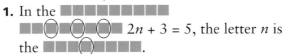 $2n + 3 = 5$, the letter n is the .

2. A ▨▨▨▨▨▨▨▨ ▨▨▨▨▨▨▨▨ⓞⓞ uses numbers, ▨▨▨▨▨▨ⓞs, and operations.

3. Use the circled letters from questions 1 and 2 to finish this sentence: The value that makes an equation true is the 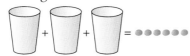.

12.1 Variables and Expressions, pages 386–391

4. Model each phrase with cups and counters.
   a) a number, less 3
   b) 3 times a number

5. **a)** Describe, in words, the variable expression that is modelled by this diagram.

   ⊔ + ⊔ + ⊔ = ●●●●●●

   b) Write the variable expression.

6. A car is travelling at an average speed of 60 km/h. The expression $60t$ tells you the distance driven, in kilometres. The variable t stands for the time, in hours.
   a) How far would this car travel in 3 h?
   b) How far would this car travel in 7 h?

12.2 Solve Equations by Inspection, pages 392–397

7. What number on each mass will keep the balance?

   a)
   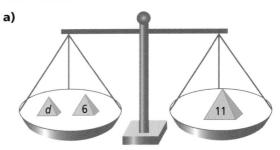

   b)
   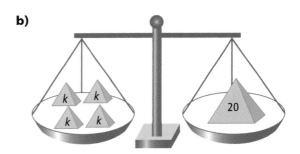

8. Solve each equation by inspection.
   a) $c - 12 = 25$
   b) $9x = 81$
   c) $n - 4 = 10$

9. Ken's parents asked him his average mark on his report card. Ken said that the sum of his 8 marks was 560. What was Ken's average mark?

12.3 Model Patterns With Equations, pages 398–403

10. a) Write equations for the first three shapes in this pattern of marbles.

b) Write an equation for the shape that uses 27 marbles.

11. Study this toothpick pattern. Write an equation for the diagram that uses 46 toothpicks.

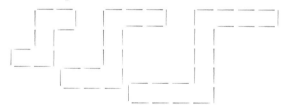

12.4 Solve Equations by Systematic Trial, pages 404–409

12. Model each situation using an equation.

a) A number increased by 9 is 15.
b) A number, when doubled, gives 24.
c) The product of 4 and a number, minus 5, is 27.
d) Two more than three times a number is 8.

13. Solve each equation in question 12.

14. Solve each equation by systematic trial.

a) $7y - 8 = 55$
b) $70 - 4n = 22$
c) $75 = 40 + 5w$
d) $4y - 6 = 58$
e) $11x + 15 = 202$
f) $25 = 0.5q + 18$

15. Study the pattern of toothpick triangles.

a) Develop an expression for the number of toothpicks needed for a diagram that has n triangles.
b) Write an equation modelling this pattern when 51 toothpicks are used.
c) Solve the equation. How many triangles will be formed with 51 toothpicks?

12.5 Model With Equations, pages 410–415

16. Model each situation using an equation. Solve each equation.

a) Double a length, decreased by 4, is 13.
b) The cost of a computer is 6 times the cost of a printer. The computer costs $1200.
c) Chan's mass is 30 kg less than double Juan's mass. Juan's mass is 45 kg.
d) The regular price of a hardcover novel has been lowered by $6 to $35.
e) Double a number, less 7, is 23.
f) The difference between 180 cm and a person's height is 30 cm.

17. Saturn has 31 moons. Jupiter has 1 fewer than double the number of moons that Saturn has.

a) Model this situation with an equation.
b) How many moons does Jupiter have?

Did You Know?

In 1979, the Voyager 1 space probe discovered rings around Jupiter. Unlike Saturn's rings, they are too faint to be seen by a telescope.

CHAPTER 12 Practice Test

| Strand Questions | NSN 2, 9, 11–13 | MEA | GSS | PA 1–13 | DMP |

Multiple Choice

For questions 1 to 5, select the correct answer.

1. Which expression does the illustration model?

   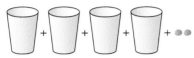

   A $4x + 2$ **B** $2x + 4$
   C $4x - 2$ **D** $4 + 2x$

2. The solution to the equation $k - 12 = 15$ is
   A 3   **B** 12
   C 27   **D** none of the above

3. "Double a number, increased by 5, is 17" can be modelled as
   A $n + 2 \times 5 = 17$ **B** $2n + 5 = 17$
   C $17 = 2n - 5$ **D** $5n + 17 = 2$

4. Which dot pattern matches the equation $3n + 1 = 22$?

   A, **B**, **C**, **D** (dot patterns)

5. Which equation could model a number in the pattern 39, 35, 31, …?
   A $4n - 5 = 39$
   B $39 - 4n = 17$
   C $43 - 4n = 11$
   D $42 - 3n = 12$

Short Answer

6. Model each expression using diagrams of cups and counters.
   a) $2C + 6$
   b) $3C - 2$
   c) $7 - C$

7. Model each situation using an equation.
   a) Your mother's job takes the same number of hours each day for 5 days. She works a total of 35 h.
   b) Your father's job takes the same number of hours each day for 3 days. He works a total of 42 h.
   c) $2 more than 4 times your allowance is $38.
   d) Triple your allowance increased by $10 gives $55.

8. Evaluate.
   a) $n + 5$ for $n = 7$
   b) $3x$ for $x = 10$
   c) $8k - 3$ for $k = 2$
   d) $12 - b \div 3$ for $b = 30$

9. Solve each equation. Use a method of your choice.
   a) $m + 5 = 17$
   b) $6k = 18$
   c) $4w = 92$
   d) $2r - 5 = 19$
   e) $15x + 12 = 177$

10. Study this pattern of marbles. Write an equation for the shape that uses 27 marbles.

11. A Mats Sundin collector card sells for $10 more than a Vince Carter card. A Mats Sundin card sells for $75.

   a) Write an equation modelling this situation. Explain how you developed the equation.

   b) Solve the equation to find the selling price of a Vince Carter collector card.

12. A roll of $2 coins contains 25 coins, or $50. Write and solve an equation to model each situation below. Explain what your variable represents for each situation.

   a) The total number of toonies is 175.
   b) The total value of the toonies is $350.

Extended Response

13. Acme Toy Company makes rods from cubes. The cubes are joined end to end as shown and then dipped in paint.

Develop and solve an equation to find the total number of cubes used when 86 faces are painted.

Chapter Problem Wrap-Up

Vicki is planning a set of rectangular tables. Here are her first two designs.

1. Write a variable expression to model the number of red tiles in this pattern. Explain your method.

2. Design your own table pattern using red and white tiles. Use pictures, words, numbers, and equations to show and discuss your pattern.

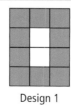
Design 1

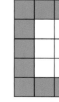
Design 2

Making Connections

Equation Puzzle

Copy this three-by-three grid onto a piece of paper or thin cardboard. Make sure each section is square. Then, cut out the nine squares.

Rearrange the squares so that each equation is next to its solution. There will be extra equations and numbers around the outside edge that do not match up.

Hint: The piece from the top left corner slides, with no turn, to become the centre piece.

Materials
- scissors

Optional:
- BLM 11/12 Task A Equation Puzzle

7 / $4k+3=7$ / $4m+9=10$ / $6m+7=37$	$m-6=7$ / 13 / 7 / 2	4 / $3c-4=5$ / $2b-4=10$ / 8
$c+6=10$ / 1 / 3 / $7w+3=24$	$2k+1=11$ / 2 / 2 / $7x-3=4$	$4k+1=9$ / $y-5=3$ / 6 / 4
9 / $5x+4=24$ / $2x+5=9$ / 1	5 / 3 / $8n=24$ / $5x=15$	$2m-6=8$ / $3k-1=5$ / 13 / 8

Magic Squares

TASK

1. Neeta and Sam found this card in the class math centre. They were not sure how to solve it. Help them find the missing numbers.

   Magic Square

2		
	5	1
		8

   What numbers belong in the empty squares? Remember that the rows, columns, and diagonals of a magic square all add to the same total.

2. Neeta made up this magic square. Sam says that it is not a magic square. Who is correct? Explain why.

−3	+2	+1
+4	0	−4
−1	−2	+3

3. Sam found another card with these two magic squares. He thinks they must be related. Solve each magic square and explain how they are related.

		5 + 1
5 + 4	5	
5 − 1		

	$x + 2$	
$x + 4$	x	
	$x − 2$	

4. Create at least one original magic square. Use some positive and some negative integers. Try to develop another using a variable.

Task: Magic Squares • MHR 421

Chapters 9–12 Review

Chapter 9 Data Management: Collection and Display

1. Ariel counted the number of coloured candies in a package.

Colour	Tally	Frequency
Red	I	
Green	III	
Blue	IIII	
Yellow	IIII	
Orange	III	

a) Copy and complete the frequency table.
b) Draw a bar graph to show the data.
c) Draw a pictograph to show the data.
d) Identify one advantage of each type of graph.

2. a) Organize the following test scores using a stem-and-leaf plot.
82 77 54 66 82 69 71 90
84 62 70 55 74 83 94 96
b) How many scores are in the 70s?
c) Which was the most common score?

3. Four friends worked together on a presentation. Kelly worked 4 h, Hai worked 3 h, Julio worked 3 h, and Annapola worked 2 h. Draw a circle graph that shows how the work was divided.

4. a) What is a database? Give an example.
b) What is a spreadsheet? Describe a situation in which you would use one.

Chapter 10 Data Management: Analysis and Evaluation

5. a) Describe each sales trend.
b) What product had the greatest sales in 2000?
c) Predict CD sales in 2005. Explain your prediction.

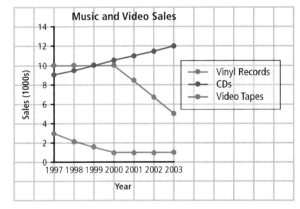

6. Shelley records the number of friends that attend her pool party each year: 45, 36, 13, 40, and 36.

a) Find the mean, median, and mode.
b) Which measure of central tendency best describes this set of data? Explain.

7. The results of a taste test are used in an advertisement.

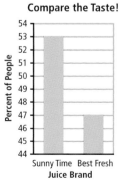

a) Which company do you think created this graph? Explain.
b) Explain why this graph is misleading.
c) Draw a new graph for the data that is not misleading.
d) Compare the two graphs. Describe what you notice.

Chapter 11 Integers

8. Temperatures can be converted between degrees Celsius (°C) and degrees Fahrenheit (°F) using the table.

Degrees Celsius	Degrees Fahrenheit
−20	−4
−15	5
−10	14
−5	23
0	32
5	41
10	50
15	59
20	68

   a) Describe the pattern using words and integers.
   b) Convert −25°C to degrees Fahrenheit. Explain your strategy.
   c) Convert −40°F to degrees Celsius. Explain your strategy.

9. Model each statement using integer chips or a number line. State the result of each.
   a) (−5) + (−3)   b) (+4) + (−9)
   c) (−8) − (−3)   d) (+7) − (+13)

10. Evaluate.
    a) (+4) + (−10)   b) (−12) − (−5)
    c) (−8) + (+17)   d) (+22) − (−21)

11. Use a calculator to help evaluate each expression.
    a) 30 − 45 − 15   b) 74 − (−16) − 50
    c) 33 + 44 − 87
    d) (−100) − 200 + 300
    e) (−250) + (−350) − (−450)

12. A scuba diver is at an elevation of 6 m above sea level. She dives to a depth of 8 m below the surface and then a further 7 m. How far below sea level did the diver end up?

Chapter 12 Patterning and Equations

13. The formula for the area of a trapezoid is $A = \frac{1}{2}(a + b)h$. Use the formula to find the area of the trapezoid shown.

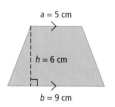

14. For what whole number values of k is $2k + 6$ greater than $3k$?

15. Find the number that makes each situation true.
    a)
    b)

16. Solve.
    a) $2a = 10$
    b) $m − 5 = 9$
    c) $y + 6 = 30$
    d) $3x + 1 = 16$
    e) $11w − 13 = 64$

17. The formula for the perimeter of an equilateral triangle is $P = 3s$. What side length is needed to make an equilateral triangle with a perimeter of 36 cm?

18. The diagram shows the posts and rails of a fence. How many posts are needed for a fence with 72 rails?

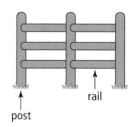

Geometry and Spatial Sense

- Explore transformations of geometric figures.
- Understand, apply, and analyse key concepts in transformational geometry.
- Recognize the image of a two-dimensional shape under a translation, a reflection, and a rotation.
- Create, analyse, and describe designs that include congruent, translated, rotated, and reflected two-dimensional images.
- Identify whether a figure will tile a plane.
- Construct and analyse tiling patterns with congruent tiles.

Key Words

frieze pattern

transformation

translation

rotation

reflection

image

tiling pattern

tessellation

tiling the plane

CHAPTER 13

Geometry of Transformations

A **tessellation** is another name for a mosaic. A mosaic is a picture or design made of small shapes of different colours. Ancient Romans made the world's most famous mosaics from small tiles. The mosaics covered and decorated surfaces, such as floors and walls. The tiles in a Roman mosaic were quadrilaterals, most often squares.

A tile used to make a mosaic is called a "tessera." This word comes from the ancient Greek word *tessares*, which means four. The tiles used to make ancient mosaics had four corners.

Think of some examples of tessellations or mosaics that you have seen in your home, school, or community. What shape are the tiles that are used?

In this chapter, you will explore transformations. You will use transformations to create, analyse, and describe designs.

Chapter Problem

Square tiles can be combined to make different shapes.

domino   triomino   tetromino   pentomino

What other shapes can be made from two or more square tiles? How can you create tessellations using shapes made up of 2, 3, 4, or 5 squares?

Get Ready

Congruent Figures

The equal angles and equal sides in these **congruent figures** are called corresponding parts.

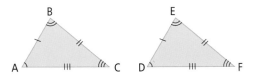

The corresponding parts are
∠A = ∠D   AB = DE
∠B = ∠E   BC = EF
∠C = ∠F   AC = DF

1. Are the figures in each pair congruent? Explain your reasoning.

   a)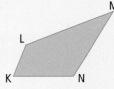

   b)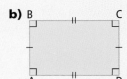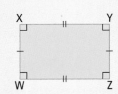

   c)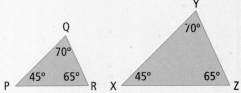

2. For any congruent figures you found in question 1, list the corresponding parts.

Regular and Irregular Polygons

A polygon with all sides equal and all angles equal is a **regular polygon**. An equilateral triangle is an example of a regular polygon. A polygon that is not regular is called an **irregular polygon**. An isosceles triangle is an example of an irregular polygon.

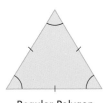 

Regular Polygon   Irregular Polygon

3. Decide if each polygon is regular or irregular. Give reasons for your decisions.

   a) 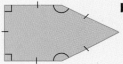 b)

   c)   d)

Transformations

A **transformation** moves one geometric figure onto another. Three common types of transformations are **translations**, **rotations**, and **reflections**.

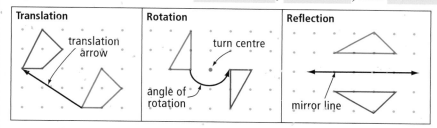

4. Name a transformation that moves the red figure onto the blue figure in each pair. Explain your reasoning.

   a)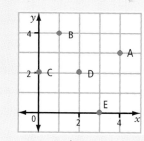

   b)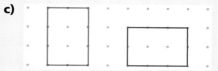

   c)

5. In each part of question 4, are the red and blue figures congruent? Explain.

Graphing Skills

The points A(1, 3), B(5, 2), C(6, 4), D(4, 0), and E(0, 1) can be plotted on a **coordinate grid** as shown.

Each point is named with an **ordered pair**.

A(1, 3)
↑ ↑
x-coordinate y-coordinate

6. State an ordered pair for each point on the grid.

7. Plot these points on a coordinate grid. Join the points in alphabetical order. What letter shape is formed?

   a) A(2, 6)   b) B(3, 2)   c) C(4, 5)
   d) D(5, 2)   e) E(6, 6)

13.1 Explore Transformations

Focus on...
- exploring transformations
- creating and analysing designs

frieze pattern
- a design pattern that repeats in one direction
- the word "frieze" sounds the same as the word "freeze"

A strip pattern or **frieze pattern** repeats in one direction. Frieze patterns are often used as decoration in arts, crafts, and architecture. You may have seen frieze patterns on textiles, pottery, and buildings. Describe the frieze pattern you see in this section of stone architecture.

Discover the Math

Materials
- grid paper

How can you use frieze patterns to explore transformations?

You can create a frieze pattern by transforming a design. Here is an example of a design.

transformation
- a change in a figure that results in a different position, orientation, or size

1. The diagrams show ways of using the design to create frieze patterns. Describe how **transformations** are used in each case.

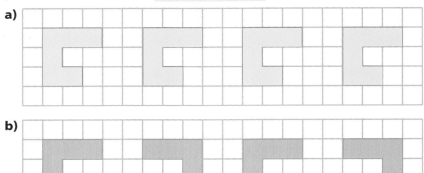

c)

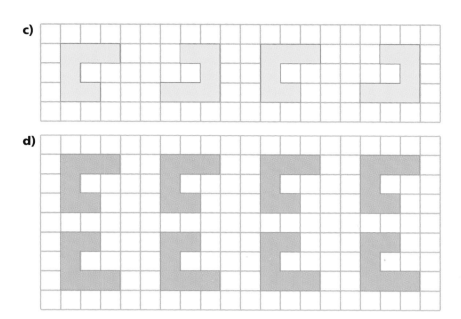

d)

2. Create your own frieze pattern on grid paper. Describe how you created it.

3. Compare your pattern with a classmate's pattern. Describe how you think your classmate's pattern was created.

4. **Reflect** Explain why transformations are often used in art and design.

Key Ideas

- Three common types of transformations are **translations**, **rotations**, and **reflections**.

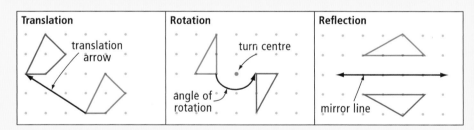

- A translation, rotation, or reflection **image** is congruent to the original figure.

translation
- a slide along a straight line

rotation
- a turn about a fixed point called the turn centre

reflection
- a flip over a mirror line

image
- a figure resulting from a transformation

Communicate the Ideas

1. Name the transformation in each picture.

   a)

   b)

2. List examples of translations, reflections, and rotations you see in your classroom. Discuss your list with your classmates.

3. The blue figure is the translation image of the red figure. Describe the translation.

   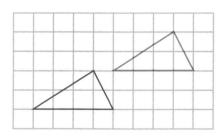

Literacy Connections

Describing Translations
Describe the movement left or right first. Then, describe the movement up or down.

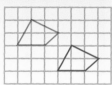

The red figure is translated 3 units right and 2 units down.

Check Your Understanding

Practise

4. Name the type of transformation that relates each pair of figures.

   a)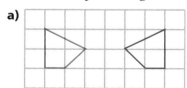

   b)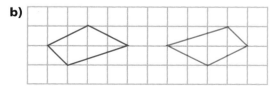

   c)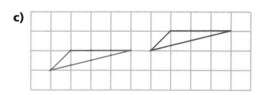

5. Create a frieze pattern by translating an irregular shape.

6. The blue figure is the translation image of the red figure. Describe the translation in words.

a)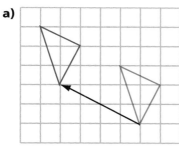

b)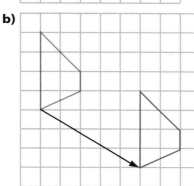

7. The blue figure is the reflection image of the red figure. Copy the diagram. Show the location of the mirror line.

a)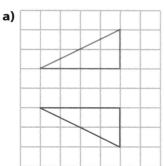

b)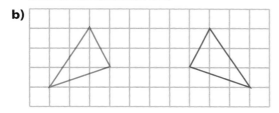

8. The blue figure is the rotation image of the red figure. Copy the diagram. Show the turn centre and the angle of rotation.

a)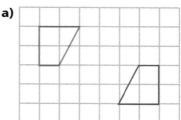

b)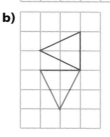

Apply

9. Explain why translations, rotations, and reflections can be called "congruency transformations."

10. Frieze patterns are used for decoration in many cultures. To find out more, go to **www.mcgrawhill.ca/links/math7** and follow the links. Choose a frieze pattern you like. Describe how and where it is used.

11. When you use a combination lock, you use rotations and translations. Describe them.

12. How do you use transformations when you ride a bicycle?

13. a) Describe a type of window that is opened and closed using translations.
    b) Describe a type of window that is opened and closed using rotations.

14. Can you change the shape or size of a figure by translating, rotating, or reflecting it? Use diagrams to show how you know.

15. A domino is a figure made by joining two congruent squares along whole sides.

    a) Where could you place a turn centre to rotate one square onto the other? Explain your reasoning.
    b) What would the angle of rotation be?

Chapter Problem

16. You can make or draw a tetromino by joining 4 congruent squares along whole edges.

    This is a tetromino:

    This is the same tetromino:

    This is not a tetromino:

    Two figures related by a transformation count as one tetromino. How many different tetrominoes can you make? Draw all the possible different tetrominoes on grid paper.

17. Some quilts are made from square blocks of material that are stitched together. The diagrams show two possible designs for a square block.

    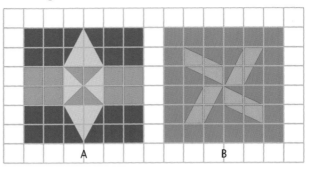

    a) Describe the transformations that relate the congruent parts in each design. Is there more than one possible answer in some cases? Explain.
    b) Describe reflections or rotations that would move each whole block onto itself.

18. a) Design your own square block for a quilt.
    b) Describe the transformations you used to create the block.

19. The diagram shows how a triangle can be translated across a parallelogram.

    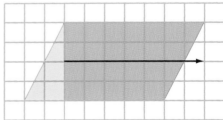

    a) Name the type of figure that results.
    b) How do the areas of the parallelogram and the resulting figure compare? Explain.
    c) How do the perimeters of the parallelogram and the resulting figure compare? Explain.
    d) Explain why the translation does not result in a figure that is congruent to the parallelogram.

Extend

20. Figures with the same shape but different sizes are known as similar figures.

a) Draw a figure on grid paper that is similar to the red figure by doubling the length of each side.

b) Draw another figure on grid paper that is similar to the same red figure by tripling the length of each side.

c) The figures you have drawn are known as enlargements of the original figure. What does "enlargement" mean?

d) Draw a figure on grid paper that is similar to the red figure, but with sides that are half as long.

e) Explain why the figure you drew in part e) is called a "reduction" of the red figure.

21. A figure is rotated twice in the same direction about the same turn centre. The angle of rotation is 180° for each rotation. How do the locations of the original figure and the final image compare? Draw a diagram to show why.

22. The blue figure is the translation image of the red figure.

You can also transform the red figure onto the blue figure by using two reflections, one after the other. Copy the diagram onto grid paper. Draw the two mirror lines on grid lines between the two figures. Explain your reasoning.

Making Connections

What does math have to do with snow?

Everyone who lives in Canada knows something about snow. When you walk or ski across a thick blanket of snow, you stand on huge numbers of snowflakes packed together. But have you ever looked at the shape of just one snowflake?

You may have heard that no two snowflakes are exactly alike. This means that no snowflake shape is the translation, reflection, or rotation image of another snowflake shape.

Look at the pictures of the snowflakes. Can you reflect or rotate each shape onto *itself*? If so, show how you can do this for each snowflake.

Use your research skills to find out if you could transform all snowflake shapes onto themselves in the same ways. Explain why or why not.

13.2 Investigate Frieze Patterns With *The Geometer's Sketchpad®*

Focus on...
- creating designs using *The Geometer's Sketchpad®*

Discover the Math

How do you construct frieze patterns with *The Geometer's Sketchpad®*?

1. Open *The Geometer's Sketchpad®* and begin a new sketch.

2. **a)** Near the top left corner of the screen, construct an irregular polygon and its interior.

   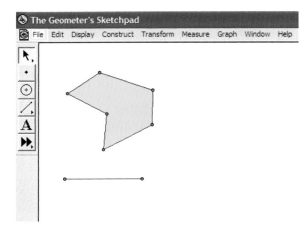

   b) To colour the interior, from the **Display** menu, choose **Color**. Click on a colour.

   c) Beneath the polygon, construct a horizontal segment.

   d) Select the left endpoint and the right endpoint of the segment in that order. From the **Transform** menu, choose **Mark Vector**.

   e) Select the polygon interior. From the **Transform** menu, choose **Translate**. Describe what happens.

   f) From the **Transform** menu, choose **Translate**. Repeat a few more times. Describe what happens.

   g) Drag points on the original polygon and the line segment to change the pattern. When you like the result, save it as a Sketchpad document using the filename **frieze1.gsp**.

   h) Describe the frieze pattern you made and the transformation you used.

3. **a)** Work with the pattern you saved in step 2g). Select the segment. From the **Transform** menu, choose **Mark Mirror**.

   b) Select the interiors of all the polygons in your pattern. From the **Transform** menu, choose **Reflect**. Describe what happens.

Materials
- computer
- *The Geometer's Sketchpad®* software (GSP 4)

Alternatives:
- TECH 13.2A Construct a Frieze Pattern (GSP 4)
- TECH 13.2B Construct a Frieze Pattern (GSP 3)

Technology Tip
- Another way to mark a segment as a mirror is to double click on the segment.

c) Drag points on the original polygon and the line segment to change the pattern. When you like the result, save it using the filename **frieze2.gsp**.

d) In the first row of your pattern, select the interiors of the second, fourth, and sixth polygons, and so on. In the second row of your pattern, select the interiors of the first, third, and fifth polygons, and so on. From the **Display** menu, choose **Hide Polygon Interiors**. Describe the result. What combination of transformations produces this frieze pattern?

e) Save your pattern from step 3d) as a Sketchpad document using the filename **frieze3.gsp**. Experiment with the pattern by dragging points. Describe what you see.

4. a) Reopen the document **frieze1.gsp**. Construct a point near the centre of the pattern.

b) From the **Transform** menu, choose **Mark Center**.

c) Select all the polygon interiors. From the **Transform** menu, choose **Rotate**. In the dialogue box, specify an angle of 180°. Then, click *Rotate*. Describe what happens.

d) Drag the point you constructed in step 4a) until you like the frieze pattern. What combination of transformations produces this frieze pattern?

e) Save the pattern as **frieze4.gsp**.

> **Technology Tip**
> - Another way to mark a point as a turn centre is to double click on the point.

5. Experiment by creating more frieze patterns. Save the results and share them with your classmates.

6. Reflect Describe how you created your patterns in step 5.

I will use one of my frieze patterns to decorate my binder. How will you use yours?

Making Connections

How can you animate a pattern using *The Geometer's Sketchpad*®?

Reopen any frieze pattern file you saved in the above activity. Select the interior of the original polygon you constructed. From the **Display** menu, choose **Animate Polygon Interior**. Describe what happens.

 "Dueling Pinwheels" is an activity that uses *The Geometer's Sketchpad*®. The activity involves transformations and animation. To try the activity, go to **www.mcgrawhill.ca/links/math7** and follow the links. Write a summary of what you learned from the activity.

13.3 Extension: Translations on a Coordinate Grid

Focus on...
- translating shapes on a coordinate grid

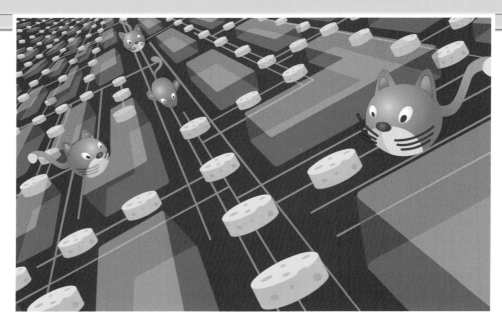

Many video games involve the movement of geometric shapes on a screen. The movements are programmed into the game as transformations on a coordinate grid. What transformations do you see in this screen display?

Discover the Math

Materials
- grid paper

Literacy Connections

Reading Figure Names
The image of a point can be named using prime symbols.
Read A′ as "A prime," B′ as "B prime," and so on.

How can you translate shapes on a coordinate grid?

1. Each diagram shows a translation. Describe in words how far the figure moves horizontally and vertically.

   a) △ABC and its image, △A′B′C′

   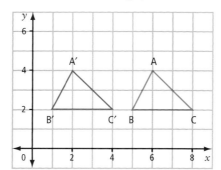

436 MHR • Chapter 13

b) quadrilateral KLMN and its image, quadrilateral K'L'M'N'

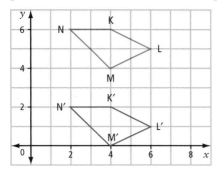

c) △PQR and its image, △P'Q'R'

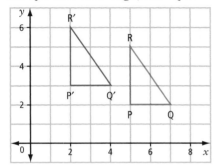

d) quadrilateral DEFG and its image, quadrilateral D'E'F'G'

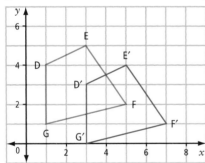

2. Reflect Describe how you found your answers in step 1.

Literacy Connections

Describing a Translation

A translation arrow can be used to describe the translation of a figure on a coordinate grid. The figure shown in the diagram has been translated 4 units left and 3 units up.

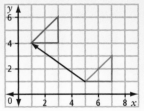

Describe the movement along the x-axis (left or right) first. Then, describe the movement along the y-axis (up or down).

Example: Translations on a Coordinate Grid

△ABC is translated 3 units right and 2 units down. Draw its translation image, △A′B′C′.

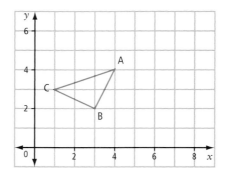

Solution

Method 1: Make a Model

Copy the diagram onto centimetre grid paper.

Trace △ABC on a separate sheet of paper.

Cut out the triangle you traced.

Place your cut-out △ABC on top of △ABC on the grid paper.

Move vertex B three grid squares right.

Strategies
Make a model

Now move vertex B two units down. The location you end up at is the image after both parts of the translation. Trace around your cut-out triangle. Label the image vertices A′, B′, and C′.

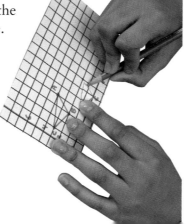

Method 2: Draw a Diagram

Draw the diagram on grid paper.

Count squares to move each vertex 3 units right and 2 units down.

Label the image vertices A′, B′, and C′.

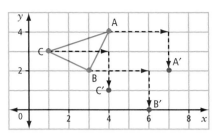

Strategies
Draw a picture or diagram

Join A′, B′, and C′ to form the translation image.

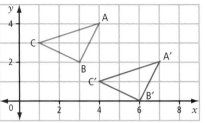

Key Ideas

- Translations can be carried out on a coordinate grid.
- A translation arrow can describe the translation of a figure on a coordinate grid. The translation arrow in the diagram shows a translation of 4 units left and 2 units up.

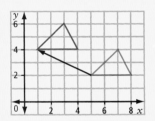

Communicate the Ideas

1. In the diagram shown, describe the translation that moves the red figure onto the blue figure.

2. A point and its translation image have the same first coordinate. Describe the direction of the translation. Show how you know.

3. The point P(4, 1) is translated 2 units left and 4 units up. How can you find the coordinates of the image without using a diagram?

4. A translation moves all the points on a figure to new positions. Why is only one translation arrow needed to show the distance and direction of a translation?

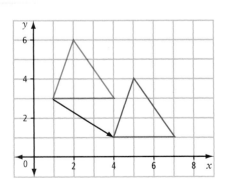

Check Your Understanding

Practise

5. Describe the translation that moves each figure onto its image.

  a) △ABC and its image, △A′B′C′

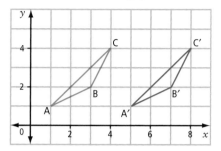

  b) quadrilateral WXYZ and its image, quadrilateral W′X′Y′Z′

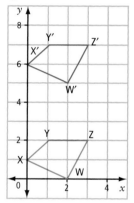

6. Describe the translation that moves each figure onto its image.

  a) quadrilateral KLMN and its image, quadrilateral K′L′M′N′

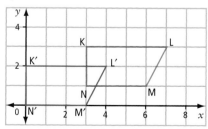

  b) pentagon PQRST and its image, pentagon P′Q′R′S′T′

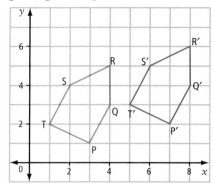

For help with questions 7 to 10, refer to the Example.

7. Draw the image of each point after the given translation. State the coordinates of the image point.

  a) A(3, 4); 2 units up
  b) B(2, 1); 1 unit left

8. Draw the image of each point after the given translation. State the coordinates of the image point.

  a) C(4, 6); 4 units left and 6 units down
  b) D(0, 2); 3 units right and 2 units up

9. Square WXYZ is translated 2 units left and 3 units up. Draw its translation image, square W′X′Y′Z′.

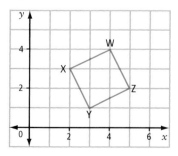

10. Parallelogram ABCD is translated 1 unit right and 2 units down. Draw its translation image, parallelogram A'B'C'D'.

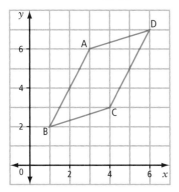

Apply

11. Fareeha drew the translation image of a figure on a coordinate grid. She described the translation to her friend, Michel. Michel could not see the figure or the image. But he told her how she could translate the image back onto the original figure.

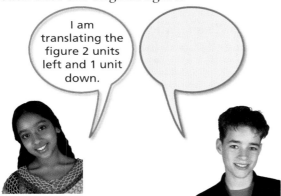

a) What did Michel say?
b) How did Michel know?

12. △XYZ with vertices X(1, 1), Y(3, 5), and Z(4, 3) is translated 4 units right and 1 unit down. The image is then translated 5 units left and 3 units up.

a) Find the coordinates of the vertices of the final image.
b) Describe a single translation that would move the original figure onto the final image.

13. △ABC has vertices A(4, 1), B(1, 1), and C(0, 4). △DEF has vertices D(8, 2), E(5, 2), and F(4, 4).

a) Draw the two triangles on a coordinate grid.
b) Is one triangle the translation image of the other? Explain how you know.

Extend

14. △ABC and △DEF are shown on a grid. You read A, B, and C in alphabetical order clockwise around the figure. We say that the "sense" of this triangle is clockwise. For △DEF, the sense is counterclockwise.

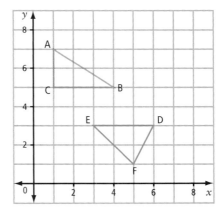

Maya thinks that translations do not change the sense of a triangle. Do you agree or disagree with her? Explain your reasoning.

Making Connections

In high school, you will learn more about transformations on a coordinate grid.

13.4 Identify Tiling Patterns and Tessellations

Focus on...
- identifying whether a figure will tile the plane
- constructing and analysing tiling patterns

In Section 13.1, you learned that a frieze pattern repeats in one direction. A wallpaper pattern repeats in two directions. Wallpaper patterns are not just found on wallpaper. They are common on carpets, fabrics, and baskets, for example. Describe the repeating pattern in the picture.

Many wallpaper patterns are made from shapes that cover the plane without overlapping or leaving gaps. These patterns are called **tiling patterns** or **tessellations**. Covering the plane in this way is called **tiling the plane**.

tiling pattern
- a pattern that covers a plane without overlapping or leaving gaps
- also called a **tessellation**

tiling the plane
- using repeated congruent shapes to cover a region completely

Materials
- set of pattern blocks or cardboard cutouts of pattern block shapes
- cardboard cutouts of a regular pentagon and a regular octagon
- cardboard
- scissors
- ruler

Alternative:
- BLM 13.4A Figures for Tiling the Plane

Discover the Math

Which regular and irregular figures can you use to tile the plane?

1. Use a pattern block hexagon or make a cardboard cutout of a regular hexagon.

2. **a)** Draw around the regular hexagon. Move the hexagon to a new position, so that the two hexagons share a common side. Draw around the hexagon again. Continue to see if a regular hexagon tiles the plane.
   b) Use the same method to find out if a regular (equilateral) triangle tiles the plane.
   c) How could you predict the answer to part b) without drawing any triangles?
   d) Does a square tile the plane? How do you know?

3. Try to tile the plane with each of the following. Describe your findings.
   a) a regular pentagon
   b) a regular octagon

4. Find out if the following irregular shapes in a set of pattern blocks will tile the plane. Explain how you decided.
   a) trapezoid
   b) rhombus

5. Cut out the shape of any irregular quadrilateral.
   a) Predict whether the shape will tile the plane. Justify your prediction.
   b) Try to tile the plane with the shape. What did you find?
   c) Share your findings with your classmates.
   d) Use the class results to write a rule.

> **Strategies**
> What strategy are you using here?

6. Repeat step 5, but use an irregular triangle instead of an irregular quadrilateral.

7. Try to tile the plane with the following shapes. What did you find?
   a) an irregular pentagon
   b) an irregular hexagon

8. **Reflect**
   a) What regular figures tile the plane? Explain why some regular shapes tile the plane but others do not. Hint: Look at the angles inside each shape. Can you find a pattern?
   b) Explain why some irregular figures tile the plane but others do not.

Key Ideas

- A tiling pattern or tessellation is a pattern that covers a plane without overlapping or leaving gaps.
- Only three types of regular figures tile the plane.
- Some types of irregular figures tile the plane.

Communicate the Ideas

1. Draw the three types of regular figures that tile the plane. Justify your choices.

2. Identify two types of irregular figures that tile the plane. Explain why they do.

3. Ana has to choose paving stones to pave her driveway. Should she choose paving stones in the shape of a regular octagon? Show how you know.

13.4 Identify Tiling Patterns and Tessellations • MHR 443

Check Your Understanding

Practise

4. Use this shape to tile the plane. Show and colour the result on grid paper.

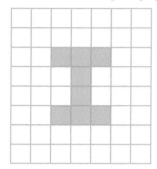

5. Tile the plane with an isosceles triangle. Use colours or shading to create an interesting design.

6. List tiling patterns you see at home or at school. Describe the shapes used to tile the plane. Share your descriptions with your classmates.

Apply

7. Rectangular hardwood strips are often used to tile floors. Aligning the strips in different ways can create attractive designs. An example is shown.

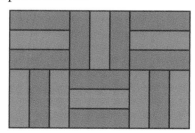

On grid paper, create a different floor design from congruent rectangular hardwood strips.

Chapter Problem

8. a) In question 16 on page 432, you made all the different tetrominoes. The simplest one is shown.

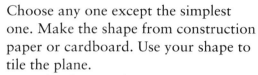

Choose any one except the simplest one. Make the shape from construction paper or cardboard. Use your shape to tile the plane.
b) Describe the transformations you used to tile the plane.
c) Colour your design and create a display to explain what you did.

9. There are many Web sites that deal with tessellations. To learn more, go to www.mcgrawhill.ca/links/math7 and follow the links. Find a tessellation you like and find out how it was created.

10. A wealth of information is available on historical quilts. The art and craft of quilt making continues today. So, you can also find information about modern quilts. To learn more about quilts, go to www.mcgrawhill.ca/links/math7 and follow the links. Choose a design that you like and explain why you like it. Describe how it uses transformations.

11. a) In question 18 on page 432, you designed a square block for a quilt. Explain why square blocks are often used to make quilts.
  b) Predict if blocks of other shapes could be used to make quilts. Justify your prediction.
  c) Design a quilt block that is not a square. Explain why you chose this shape.
  d) Can your block be used to make a quilt? Explain why or why not.
  e) If your block can be used to make a quilt, does the shape of the block have any disadvantages? Explain.

Extend

12. The diagram shows a tessellation of squares. A point has been added in the centre of each square. The points are joined by dashed segments perpendicular to common sides. The result is another tessellation. It is called the "dual" of the original tessellation.

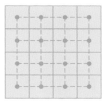

  a) Describe the dual of the original square tessellation.
  b) Draw a tessellation of regular hexagons. Draw and describe its dual.
  c) Draw a tessellation of equilateral triangles. Draw and describe its dual.

Making Connections

What does math have to do with patios?

Some patios are made from interlocking bricks. In many cases, the bricks are all the same shape, such as squares or rectangles. Sometimes two or more different shapes are used.

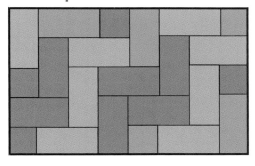

Materials
- set of pattern blocks or cardboard cutouts of the shapes
- cardboard cutout of a regular octagon
- cardboard or construction paper
- scissors
- ruler
- pencil crayons

Alternative:
- BLM 13.4A Figures for Tiling the Plane

1. a) Draw the result when you try to tile the plane with a regular octagon.
  b) Identify another regular shape you could use with a regular octagon to tile the plane.
  c) Design a patio using bricks in the shape of a regular octagon and the shape you found in part b).

2. Design patios using combinations of bricks with the following regular shapes. Use colours to create attractive designs.
  a) regular hexagons and equilateral triangles
  b) squares and equilateral triangles
  c) regular hexagons, squares, and equilateral triangles

Use Technology

This is a way to create tiling patterns using *The Geometer's Sketchpad®*.

Materials
- computer
- *The Geometer's Sketchpad®* software (GSP 4)

Alternatives:
- TECH 13.4A Create a Tiling Pattern (GSP 4)
- TECH 13.4B Create a Tiling Pattern (GSP 3)

Create Tiling Patterns Using *The Geometer's Sketchpad®*

Create a Tiling Pattern

In this activity, you will construct an irregular triangle. You will use transformations to create a tiling pattern.

1. Open *The Geometer's Sketchpad®* and begin a new sketch.

2. a) Near the top left corner of the screen, construct an irregular △ABC, as shown.

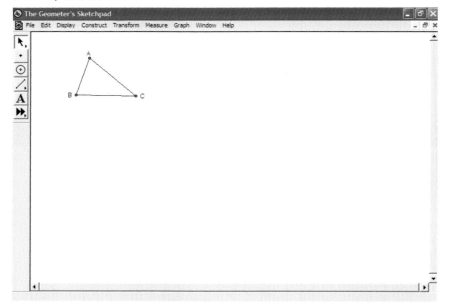

b) Select point B and point A in that order. From the **Transform** menu, choose **Mark Vector**.

c) Select point C. From the **Transform** menu, choose **Translate**. Name the new point as point D.

d) Construct segments AD and CD.

e) Construct the interior of △ABC.

f) Construct the interior of △ACD.

g) How do △ABC and △ACD compare? Hint: If you are not sure, measure side lengths and angles.

h) Select one of the triangles. From the **Display** menu, choose **Color**. Click on a colour. Then, select the other triangle and give it a different colour.

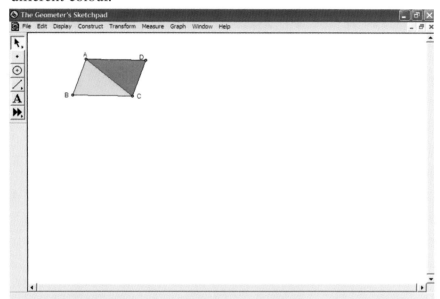

i) Select point B and point C in that order. From the **Transform** menu, choose **Mark Vector**.

j) Select the interiors of both triangles. From the **Transform** menu, choose **Translate**. Describe what happens.

k) From the **Transform** menu, choose **Translate**. Repeat a few more times. Describe what happens.

l) Select point A and point B in that order. From the **Transform** menu, choose **Mark Vector**.

m) Select all the triangles. From the **Transform** menu, choose **Translate**. Describe what happens.

n) From the **Transform** menu, choose **Translate**. Repeat a few more times. Describe what happens.

o) Drag points on △ABC. When you like the result, save it as a Sketchpad document.

p) Select the interior of △ABC. From the **Display** menu, choose **Animate Triangle**. Describe what happens.

3. Reflect Does any triangle tile the plane? Explain how you know.

13.5 Construct Translational Tessellations

Focus on...
- constructing tiling patterns
- creating designs using translated images

Did you know that tessellations have been used to create some beautiful works of art? The Dutch artist M.C. Escher (1898–1972) is famous for his use of tessellations. His work is in art galleries around the world. It also appears on T-shirts, posters, and other consumer products.

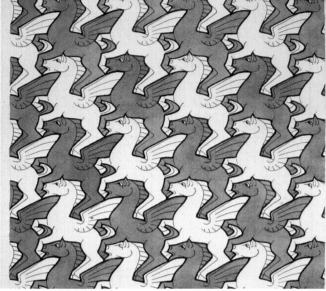

M.C. Escher's "Symmetry Drawing E105" © 2004 The M.C. Escher Company – Baarn – Holland. All rights reserved.

The picture shows one of Escher's works. How did he use translations to create it?

Discover the Math

Materials
- paper
- tracing paper
- glue stick
- adhesive tape
- cardboard or construction paper
- scissors

How can you tessellate using translations?

1. a) Draw a square with a side length of about 3 cm on a piece of paper. Cut out the square and glue it to a sheet of cardboard or construction paper. Cut out the square again.

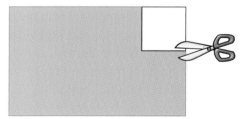

b) Inside the square, draw a curve that connects two adjacent vertices. Cut along the curve to remove a piece from one side of the square.

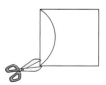

448 MHR • Chapter 13

c) Translate the piece to the opposite side of the square. Tape the piece in place.

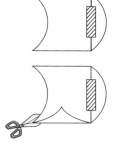

d) Now, draw a different curve on the third side of the square. Cut along this curve to remove another piece.

e) Translate the piece from part d) to the opposite side of the square. Tape the piece in place to complete your tile.

f) To tessellate the plane, draw around the tile on a piece of paper.

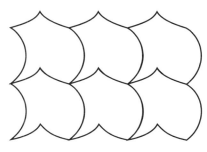

g) Add colour and designs to the tessellation to make a piece of art.

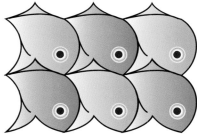

2. Start with another square. Use it to create your own work of art.

3. Experiment by starting with a rectangle, a parallelogram, or a rhombus, instead of a square.

4. Experiment by starting with a regular hexagon. Create an interesting tile by using the three pairs of opposite sides. Use the tile to create a work of art.

5. Reflect Describe how to use translations to create tessellations.

Did You Know?

The leading geometer of the 20th century was a professor at the University of Toronto. Donald Coxeter (1907–2003) was a friend of Escher's and gave him some ideas for his art. Professor Coxeter wrote some complex explanations of Escher's work. But Escher was not trained in math, so he could not understand them.

Use Technology

This is a way to tessellate with translations using *The Geometer's Sketchpad®*.

Materials
- computer
- *The Geometer's Sketchpad®* software (GSP 4)

Alternatives:
- TECH 13.5A Tessellate by Translation (GSP 4)
- TECH 13.5B Tessellate by Translation (GSP 3)

Tessellate by Translation Using *The Geometer's Sketchpad®*

Tessellate by Translation

In this activity, you will construct a rectangle. You will use translations to change it into a different irregular shape. You will then perform translations on this shape to tile the plane.

1. Open *The Geometer's Sketchpad®* and begin a new sketch.

2. Follow these steps to construct a rectangle ABCD.

   a) Near the top left corner of the screen, construct two points, A and B. Construct the segment that joins these points.
   b) Select point A and the segment. From the **Construct** menu, choose **Perpendicular Line**.
   c) Select point B and the segment. From the **Construct** menu, choose **Perpendicular Line**.
   d) Construct point C on the line that passes through B.
   e) Select point C and the segment AB. From the **Construct** menu, choose **Parallel Line**.
   f) Construct point D at the fourth vertex of the rectangle.
   g) Select the three lines, but not the segment AB. From the **Display** menu, choose **Hide Lines**.
   h) Construct three segments joining B and C, C and D, and D and A. Rectangle ABCD is now complete.
   i) Why does this method produce a rectangle?

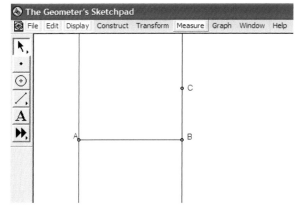

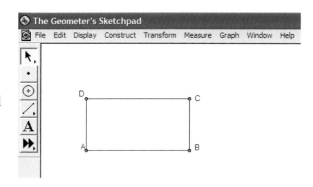

3. a) Construct a point close to side AD. Construct the segments from A and D to the new point.

   b) Select point A and point B in order. From the **Transform** menu, choose **Mark Vector**.

   c) Select the point and the two segments you constructed in step 3a). From the **Transform** menu, choose **Translate**. Describe what happens.

   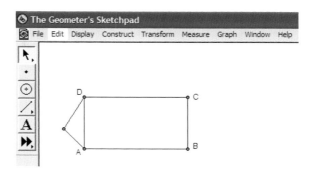

   d) Construct a point close to side CD. Construct the segments from C and D to the new point.

   e) Select point D and point A in order. From the **Transform** menu, choose **Mark Vector**.

   f) Select the point and the two segments you constructed in step 3d). From the **Transform** menu, choose **Translate**. Describe what happens.

   g) Select all the vertices of the irregular polygon you have created. From the **Construct** menu, choose **Polygon Interior**.

   h) From the **Transform** menu, choose **Translate**. Repeat at least twice more. Describe what happens.

   i) Select point A and point B in order. From the **Transform** menu, choose **Mark Vector**.

   j) Select the interiors of all the polygons. From the **Transform** menu, choose **Translate**. Repeat at least twice more. Describe what happens.

   k) Select individual polygons and use different colours to show the tessellation. Save the tessellation.

   l) Drag points on the original polygon and describe what you see.

   m) Reopen the file you saved in step 3k). Select the interior of the original irregular polygon. From the **Display** menu, choose **Animate Polygon Interior**. Describe what happens.

4. Investigate how to construct a parallelogram that is not a rectangle. Use translations to transform the parallelogram into another irregular polygon. Translate this polygon to tile the plane.

5. Experiment to produce some Escher-type art using translations. Share your art with your classmates.

6. **Reflect** Describe how to use translations to create tessellations.

The Art teacher really liked my Escher-type art. I think I'll try another one.

13.6 Construct Rotational Tessellations

Focus on...
- constructing tiling patterns
- creating designs using rotated images

The picture shows a piece of art created by the Dutch artist M.C. Escher. How did he use rotations to create it?

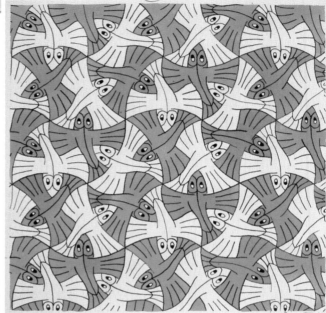

M.C. Escher's "Symmetry Drawing E99"
© 2004 The M.C. Escher Company –
Baarn – Holland. All rights reserved.

Discover the Math

Materials
- paper
- tracing paper
- glue stick
- adhesive tape
- cardboard or construction paper
- scissors

How can you tessellate using rotations about vertices?

1. a) Trace the equilateral triangle shown or draw an equilateral triangle with a side length of about 3 cm on a piece of paper. Cut out the triangle you drew and glue it to a sheet of cardboard or construction paper. Cut out the triangle again.

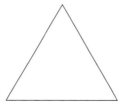
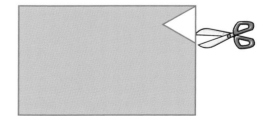

b) Inside the triangle, draw a curve that connects two adjacent vertices. Cut along the curve to remove a piece from the triangle.

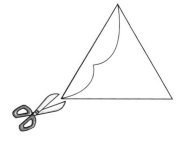

452 MHR • Chapter 13

c) Rotate the piece you removed 60° about a vertex at one end of the curve. This rotation moves the piece to another side of the triangle. Tape the piece in place to complete your tile.

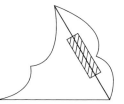

Did You Know?

The angles in an equilateral triangle are all 60°.

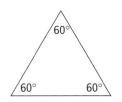

d) To tessellate the plane, draw around the tile on a piece of paper. Then, rotate and draw around the tile over and over until you have a design you like.

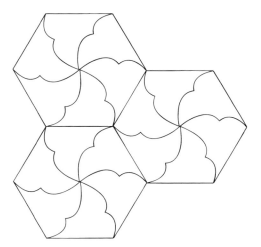

e) Add colour and designs to the tessellation to make a piece of art.

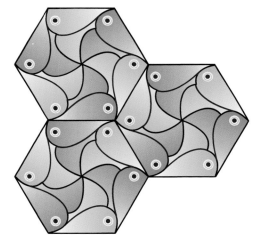

Making Connections

Escher Art

Internet Connect There are many Web sites that describe Escher's life and work. To learn more, go to www.mcgrawhill.ca/links/math7 and follow the links. Find a piece of Escher's art that you like. Think about the transformations he used to create it. Share your findings with your classmates.

2. Experiment by starting with a square or a regular hexagon. Use rotations about vertices to create a tile that will tessellate the plane. Use the tile to create a piece of art using rotations.

3. Reflect How can you tessellate using rotations about vertices?

13.6 Construct Rotational Tessellations • MHR   **453**

Use Technology

This is a way to tessellate with rotations using *The Geometer's Sketchpad®*.

Materials
- computer
- *The Geometer's Sketchpad®* software (GSP 4)

Alternatives:
- TECH 13.6A Tessellate by Rotation (GSP 4)
- TECH 13.6B Tessellate by Rotation (GSP 3)

Tessellate by Rotation Using *The Geometer's Sketchpad®*

Tessellate by Rotation

In this investigation, you will construct an equilateral triangle. You will use a rotation about a vertex to change the triangle into an irregular shape. You will then perform rotations on this shape to tile the plane.

1. Open *The Geometer's Sketchpad®* and begin a new sketch.

2. The first step is to construct an equilateral △ABC.
   a) Near the centre of the screen, construct two points, A and B. Construct the segment that connects them.
   b) Select point A. From the **Transform** menu, choose **Mark Center**.
   c) Select point B and the segment AB. From the **Transform** menu, choose **Rotate**. In the dialogue box, set the angle of rotation at 60°. Then, click on *Rotate*.
   d) Rename the new point, B′, as point C.
   e) Select point C. From the **Transform** menu, choose **Mark Center**.
   f) Select segment AC. From the **Transform** menu, choose **Rotate**. In the dialogue box, keep the angle of rotation at 60°. Then, click on *Rotate*.
   g) Explain why this method produces an equilateral triangle.

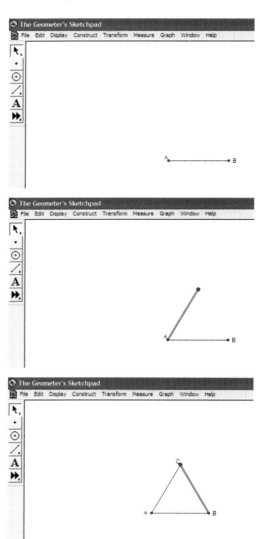

3. a) Construct a point close to side AC. Construct the segments from A and C to the new point.
  b) Select point C. From the **Transform** menu, choose **Mark Center**.
  c) Select the point and the two segments you constructed in step 3a). From the **Transform** menu, choose **Rotate**. In the dialogue box, set the angle of rotation at 60°. Then, click on *Rotate*. Describe what happens.
  d) Select all the vertices of the irregular pentagon you have created. From the **Construct** menu, choose **Pentagon Interior**.
  e) From the **Transform** menu, choose **Rotate**. Repeat four more times. Describe what happens.
  f) Select point A. From the **Transform** menu, choose **Mark Center**.
  g) Select the interiors of all the polygons. From the **Transform** menu, choose **Rotate**. In the dialogue box, set the angle of rotation at 120°. Then, click on *Rotate*. Rotate once more by the same angle. Describe what happens.
  h) Select point B. From the **Transform** menu, choose **Mark Center**.
  i) Select the interiors of all the polygons. Rotate by 120° about point B.
  j) Select individual polygons and use different colours to show the tessellation. Save the tessellation.
  k) Drag points on the original polygon and describe what you see.
  l) Reopen the file you saved in step 3j). Select the interior of the original irregular polygon. From the **Display** menu, choose **Animate Pentagon**. Describe what happens.

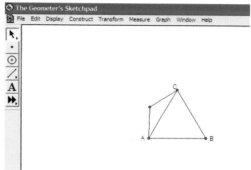

4. Investigate how to construct a square. Experiment by rotating shapes about the vertices to produce an irregular polygon. Create a tessellation pattern. Describe your results.

5. Experiment to produce some Escher-type art using rotations about vertices. Share your art with your classmates.

6. Reflect Describe how to use rotations about vertices to create tessellations.

We could use these designs to decorate the gym for the end of the year concert. Let's see how many different designs we can make?

CHAPTER 13 Review

Key Words

Use the clues to help you solve the puzzle.

Across
1. a figure resulting from a transformation
2. shows the distance and direction of a translation (2 words)

Down
3. the angle through which a figure turns (3 words)
4. moves one geometric figure onto another
5. a tiling pattern that covers a plane without overlapping
6. a slide along a straight line
7. a flip over a mirror line
8. a turn about a fixed point called the turn centre

13.1 Explore Transformations, pages 428–433

9. a) Design a frieze pattern that involves the translation and rotation of an irregular figure.
   b) Describe how you created your design.

10. The blue figure is the rotation image of the red figure. Copy the diagram and show the turn centre and the angle of rotation.

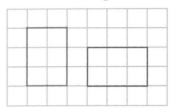

11. Johan drew a square on a piece of paper. He then used a transparent mirror to reflect the square onto itself. Where could he have placed the mirror? Explain.

12. A figure is flipped so that its image is horizontally beside it. Describe the mirror line.

13. The diagram shows a design for a stained-glass window.

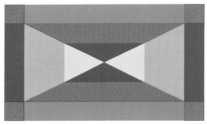

   a) Describe the transformations that relate the congruent parts in the design.
   b) Describe three ways to transform the whole window so that the original and the image are in the same position.
   c) If a classmate told you that the window was "upside down," what would you say? Explain.

14. a) Design your own stained-glass window on grid paper.
    b) Describe the transformations you used to create the window.

13.3 Translations on a Coordinate Grid, pages 436–441

15. Describe the translation that moves △ABC onto its translation image, △A′B′C′.

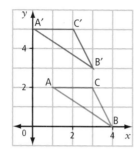

16. Kite KLMN is translated 3 units right and 2 units down. Draw the translation image, kite K′L′M′N′.

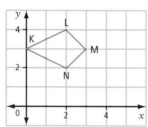

17. A point and its translation image have the same second coordinate. Describe the direction of the translation.

18. △ABC is translated 3 units right and 1 unit down to give its image, △A′B′C′. The image is shown in the diagram. What are the coordinates of points A, B, and C?

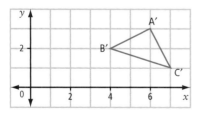

19. Describe the effect of the following transformation on any figure drawn on a coordinate grid.

Subtract 2 from the first coordinate of every point, and add 3 to the second coordinate of every point.

13.4 Identify Tiling Patterns and Tessellations, pages 442–445

20. Does a regular hexagon tile the plane? Justify your response.

21. Does a parallelogram tile the plane? Justify your response.

22. A triomino is a figure made by joining three congruent squares along whole sides.
   a) How many different triominoes are there? Draw them.
   b) Does each triomino tile the plane?

23. The diagram shows a garden path made from irregular 12-sided bricks.

   a) Explain why the 12-sided brick will tile the plane.
   b) Design an irregular 10-sided brick that could be used to make a path.
   c) Explain why your 10-sided brick will tile the plane.
   d) Design an irregular 6-sided brick that could be used to make a path.
   e) Explain why your 6-sided brick will tile the plane.

24. Create two different tiling patterns using rectangles.

CHAPTER 13 Practice Test

Multiple Choice

For questions 1 to 4, select the correct answer.

1. The transformation that relates the figures is

   A a translation
   B a reflection
   C a rotation
   D a translation and a reflection

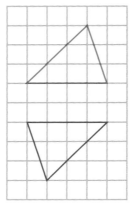

2. The transformation that relates the figures is

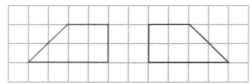

   A a translation
   B a reflection
   C a rotation
   D a translation and a rotation

3. Point P(3, 4) is translated 1 unit left and 2 units down. The coordinates of its translation image, point P′, are

   A (2, 2)
   B (1, 3)
   C (4, 2)
   D (5, 5)

4. You can tile the plane using

   A any figure
   B any irregular figure
   C any regular figure
   D some figures, but not others

Short Answer

5. a) Design a frieze pattern that involves the translation and reflection of an irregular figure.
   b) Describe how you used the transformations to create your design.

6. What effect does the translation, reflection, or rotation of a figure have on the side lengths and angle measures of the figure? Explain why.

7. Describe the translation that moves parallelogram ABCD onto its image, parallelogram A′B′C′D′.

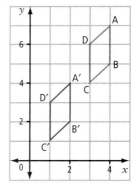

8. △PQR is translated 2 units left and 3 units up. Draw the translation image, △P′Q′R′.

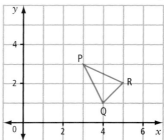

9. Does a regular pentagon tile the plane? Explain why or why not.

10. Does a scalene triangle tile the plane? Explain why or why not.

11. Create a tiling pattern using parallelograms that are not rectangles or squares. Colour or shade the pattern to create a design.

12. Seven square tiles have been arranged in a C-shape. Use grid paper to find out if this shape will tile the plane.

Extended Response

13. A tangram is an ancient Chinese puzzle. It includes 7 geometric pieces.

    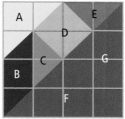

    a) Can you find pairs of pieces that are related by a translation? If so, describe the translation that relates the pieces in each pair.

    b) Can you find pairs of pieces that are related by a rotation? If so, describe the turn centre and turn angle for each pair.

    c) Can you find pairs of pieces that are related by a reflection? If so, describe the mirror line for each pair.

Chapter Problem Wrap-Up

In question 16 on page 432, you used identical squares to make all the possible tetrominoes. In question 8 on page 444, you chose a tetromino and used it as a tessellation tile to make a design.

Now, make or draw a pentomino by joining 5 congruent squares along whole edges. The simplest one is shown.

1. How many different pentominoes are there? Draw them on grid paper.

2. Choose any pentomino except the simplest one. Make the shape from construction paper or cardboard. Use your shape to tile the plane.

3. Describe the transformations you used to tile the plane.

4. Colour your design and create a display to explain what you did.

GET READY FOR GRADE 7

1 Fractions, Metric Units, Estimation, pages 2–3
1. a) $\frac{1}{3}$ **b)** $\frac{5}{4}$ **c)** $\frac{7}{9}$ **d)** $\frac{10}{3}$ **e)** $\frac{4}{5}$ **f)** $\frac{2}{1}$
3. B
5. C
7. B
9. B
11. a) Answers may vary. It gets smaller by half every time.
13. 128
15. Answers may vary. A fraction represents part of a whole object or a share of a group of objects.
17. Answers will vary. You might tell the classmate to picture a similar object that they know more about.

2 Multiplying and Dividing Decimals, Estimation, pages 4–5
1. a) 320 **b)** 3200 **c)** 3.2 **d)** 0.32
3. Using 32: 1a): move decimal one place to the right; 1b): move decimal two places to the right; 1c): move decimal one place to the left; 1d): move decimal two places to the left; 2a): move decimal one place to the left; 1b): move decimal two places to the left; 1c): move decimal one place to the right; 1d): move decimal two places to the right
5. a) 64.1 **b)** 64.1 **c)** 6.41 **d)** 6.41
7. C
9. C
11. C
13. Carriff: 7.8 km, Jeremy: 6.4 km, Len: 7.2 km, Meghan: 7.8 km, Amy: 6.3 km
15. Answers may vary. Organize the data for each event from best to worst.
17. Multiplying: count the total number of decimal places in the two numbers you start with. There will be this many decimal places in the answer. Dividing: there is no distinct relationship.
19. Try to estimate first to see roughly what your answer will be.

3 Patterns With Natural Numbers, Fractions, and Decimals, pages 6–7
1. a) 9, 11, 13 **b)** 10, 13, 16 **c)** 33, 43, 53 **d)** 25, 36, 49 **e)** 35, 46, 57 **f)** 32, 47, 65
3. a) increased by 3 **b)** multiplied by 2 and then increased by 3 **c)** multiplied by 3 **d)** multiplied by 2 and then increased by 1
5. Answers may vary. **a)** A = 0, B = 100, C = 120, D = 180, E = 250 **b)** Just after point C, but before point D. **c)** 180 **d)** It should be greater than 100 since it is the same distance from point A to 60 as from 60 to point C (C = 120).
7. Answers may vary.
9. Answers may vary. Multiples of 5.
11. They are prime numbers.
13. They are multiples of 2, 3, 6, and 9.
14.–18. Answers may vary.

CHAPTER 1

Get Ready, pages 10–11
1. a) 40 mm **b)** 24 m
3. a) 9 km **b)** 18 km **c)** 1.2 km **d)** 0.7 km
5. a) 16 cm^2 **b)** 8 cm^2 **c)** 15 cm^2

1.1 Perimeters of Two-Dimensional Shapes, pages 15–17
5. a) 22 cm **b)** 400 cm
7. 220 cm
9. a) 96 cm **b)** 9 m **c)** 6.9 cm **d)** 90 mm
11. Answers will vary.
13. Convert to the same units. 450 cm
15. a) 1200 m **b)** 24 km
17. Sasha; Anders forgot to change the measurements to the same units.
19. A loonie is a regular polygon with 11 sides, 7.7 cm
21. Answers may vary slightly. **a)** longer sides 2.4 cm, shorter sides 1.2 cm **b)** 7.2 cm **c)** Use a formula; $P = 2 \times (l + w)$.
23. No. Each side of the octagon measures 2.25 m.

1.2 Area of a Parallelogram, pages 20–21
3. a) 12 cm^2 **b)** 6 cm^2 **c)** 6 cm^2 **d)** 3 cm^2
5. Answers may vary slightly. **a)** 1.92 cm^2 **b)** 1 cm^2
7. a) No; the height is not given. **b)** Measure the distance between equal sides.
9. Joel measured a side instead of the height. He has to measure the perpendicular distance between two equal sides.
11. a) Monica is correct. **b)** Answers will vary. A rectangle is a special kind of parallelogram.

1.3 Area of a Triangle, pages 24–25
5. a) 6 cm^2 **b)** 24.5 mm^2 **c)** 13 m^2 **d)** 2.5 m^2
7. a) 60 cm^2 **b)** 18 m^2 **c)** 30 km^2 **d)** 7.5 mm^2
9. a) $b \times h$ comes from the formula for the area of a parallelogram. **b)** The area of a parallelogram divided by 2 is the area of a triangle. **c)** The area of a triangle is half the area of a parallelogram with the same base and height.
11. no
13. a) all equal areas **b)** Answers may vary slightly. Each base is 1.3 cm, each height is 1.2 cm. **c)** 1.56 cm^2

15. a) Diagrams may vary. Any triangle with base 4.0 m and height 2.5 m. **b)** No, the perimeters do not remain constant. The lengths of the sides change as the height changes.

1.4 Apply the Order of Operations, pages 28–29

5. 31.75
7. a) 11 **b)** 6 **c)** 15 **d)** 14
9. a) multiplication **b)** subtraction inside brackets **c)** division
11. a) In line 2, Vanya divided 16 by 4 first, she should have done 64 ÷ 16 first. In line 4, she subtracted 3 − 2, she should have done the multiplication 2 × 2 first. **b)** 0
13. a) 11.4 **b)** 2.1 **c)** 4.33 **d)** 8 **e)** 5
17. a) $A = 2 \times (4 \times 10) + 5 \times 6 \div 2$; 95 cm² **b)** Assume that the shape is symmetric.

1.5 Area of a Trapezoid, pages 32–33

5. a) $a = 8$ cm, $b = 18$ cm, $h = 6$ cm **b)** $a = 8$ mm, $b = 10$ mm, $h = 6$ mm **c)** $a = 2$ m, $b = 6.5$ m, $h = 3$ m **d)** $a = 5$ cm, $b = 6.4$ cm, $h = 1.4$ cm
7. a) $a = 1$ cm, $b = 2.1$ cm, $h = 1.3$ cm **b)** $a = 1.4$ cm, $b = 2.8$ cm, $h = 0.8$ cm
9. Answers may vary slightly. **6. a)** 2.4 cm² **b)** 1.8 cm² **7. a)** 2.0 cm² **b)** 1.7 cm²
11. Answers may vary.
13. 75 375 cm² (or about 7.54 m²)
15. Both trapezoids with height 3 m have equal area. The sum of their parallel sides is the same too, 12 m.
19. a) and **b)** Answers may vary. **c)** Use the formula for the area of a parallelogram and then divide by 2.

1.6 Draw Trapezoids, page 36

1.–8. Check your measurements with a good ruler.

1.7 Composite Shapes, pages 43–45

5. a) 6 m, 8 m **b)** 8 cm, 16 cm, 4 cm, 4 cm
7. 48 m
9. Answers may vary.
11. 13.6 m
13. a) Answers may vary. Split the logo into smaller shapes to calculate the area. **b)** $1275 **c)** Answers may vary. Find the area of the rectangle and subtract the four triangles.
15. Answers may vary.
17. a) $A = (17 \times 8) − (4 \times 5)$ **b)** 116 m² **c)** The answers are the same.
21. a) $825; area per sign estimated at 0.25 m²
b) Answers may vary.

Review, pages 46–47

1. D **3.** E **5.** A
7. 28 m
9. Diagrams may vary. The parallelogram should have base 6 and height 3.
11. 0.56 m²
13. a) Performed addition before division. **b)** Divided from right to left.
15. a) trapezoid, has one pair of parallel sides **b)** 850 cm²
17. a) no **b)** no
19. Answers may vary. An example could have $a = 10$ cm, $b = 14$ cm, and $h = 4$ cm.
21. Answers may vary slightly. **a)** 6.21 cm² **b)** 16.0 cm

Practice Test, pages 48–49

1. D **3.** B **5.** B
7. a) 8 **b)** 9 **c)** 21
9. Answers may vary. Use a ruler to check that the perimeter is 26 cm.
11. a) Use a ruler, measure the sides and the height. Make sure $a = 15$ cm, $b = 9$ cm, and $h = 4$ cm.
b) Answers may vary.
13. a) 122 m **b)** Answers may vary. **c)** 628 m², both areas same. Total area does not change if shape is split differently.

CHAPTER 2

Get Ready, pages 52–53

1. Answers may vary. Name two points that are joined by a line segment. AB, GH, BE, AD
3. Answers may vary. AB = BC
5. Answers may vary. **a)** 77° **b)** 100° **c)** 80°
7. a) acute angle **b)** right angle **c)** obtuse angle
9.

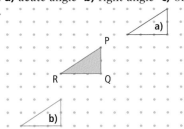

11.

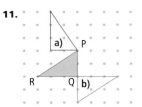

2.1 Classify Triangles, pages 57–59

5. a) isosceles **b)** scalene
7. a) right **b)** acute
9. a) right, scalene **b)** obtuse, scalene
11. a) △PQR, △PSR, △QSR
b) △PQR: right, isosceles, △PSR: acute, scalene, △RSQ: obtuse, scalene
13. a) PR = 7.3 cm, PQ = 5.0 cm, RQ = 4.4 cm
b) ∠Q = 110°, ∠R = 40°, ∠P = 30° **c)** obtuse, scalene

15. a) acute, isosceles   **b)** acute, isosceles

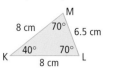

17. a) XY and YZ are of equal length, so △XYZ is an isosceles triangle.
b) Yes, ∠X = ∠Z.
19. a) ∠S = 60°, RS = 5 cm, TS = 5 cm   **b)** △RST is equilateral and acute
21. a) There are right, obtuse, isosceles, and scalene triangles.   **b)** Answers will vary.
23. Equilateral triangles are used in bridges because they are more rigid than squares, rectangles or other shapes. This makes bridges using them able to support heavier loads and stronger winds.

2.2 Classify Quadrilaterals, pages 63–65

5. a) parallelogram   **b)** rhombus
7. JKLO is a trapezoid, OLNM is a rectangle.
9. a) C rhombus **b) D** trapezoid **c) B** kite **d) A** square
11. a) Piece 4 is a square, 6 is a parallelogram **b)** Pieces 3 and 4, 4 and 5, or 5 and 6 form trapezoids.
13. a) trapezoid   **b)** rhombus
17.

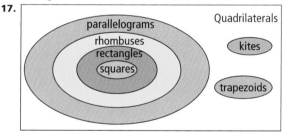

2.3 Congruent Figures, pages 68–69

3. a) Yes, they are each hexagons of the same size. **b)** No, although they are all parallelograms, they are all different sizes. **c)** Yes, they are all identical trapezoids.
5. △DEF and △GHI are congruent.
7. ∠A = ∠D, ∠B = ∠E, ∠C = ∠F, AC = DF, BC = EF, AB = DE
9. a) △AJG and △AJD **b)** △AIJ and △ABJ, △JIF and △JBE, △CBE and △HIF
c) △DCE and △GHF, △DBJ and △GIJ
11. No. For example, two squares with different side lengths. They are the same shape but different sizes.
13. No. For example, a square with side 2 cm has area 4 cm². A rectangle 4 cm by 1 cm also has area 4 cm².

2.4 Congruent and Similar Figures, pages 73–74

3. a) yes **b)** no
5. KLMN and WXYZ
7. yes

9. a) C **b)** E; each side of A is twice the length of the corresponding side of E.
11. a) Yes; pieces 1 and 2 are congruent, pieces 3 and 5 are congruent. **b)** Yes; pieces 1, 2, 3, 5, and 7 are all similar right isosceles triangles.
13. The diagonal cuts the rectangle into two sets of similar triangles (if the rectangles are similar).
17. It is a square.

Review, pages 76–77

1. a)   **b)**

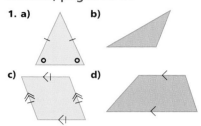

c)   **d)**

3. a) acute and equilateral **b)** scalene and obtuse
5. a) acute, isosceles **b)** obtuse, scalene
7. Squares, rectangles, trapezoids, rhombuses, and irregular quadrilaterals.
9. Answers may vary.
11. Not always.

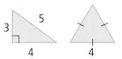

13. DEFG is similar to HIJK and LMNO is similar to PQRS. There are no congruent figures.
15. a) No, △ABC is taller than △DEF so they are not the same size. **b)** No, they are not the same shape either. FE is about as long as CB but DF is much shorter than AC.

Practice Test, pages 78–79

1. B   **3.** D   **5.** B
7. a)   **b)**

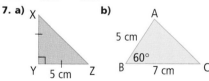

9. In order for triangles to be similar their angles must be identical. In a right triangle the other two angles are both acute. Because of this a right triangle can never be similar to an obtuse triangle.
11. ABCD: side length 4 cm, perimeter 16 cm, EFGH: side length 8 cm, perimeter 32 cm. These figures are similar.

CHAPTER 3

Get Ready, pages 84–85

1. a) $\dfrac{3}{2}$, $1\dfrac{1}{2}$ **b)** $\dfrac{7}{4}$, $1\dfrac{3}{4}$ **c)** $\dfrac{16}{6}$, $2\dfrac{4}{6}$

3. Diagrams may vary.
a) $\dfrac{1}{4} > \dfrac{1}{8}$ b) $\dfrac{1}{3} > \dfrac{1}{4}$ c) $\dfrac{2}{3} > \dfrac{5}{8}$
5. a) 2, 4, 6, 8, 10 **b)** 4, 8, 12, 16, 20 **c)** 5, 10, 15, 20, 25
7. a) $\dfrac{4}{12} = \dfrac{1}{3}$ **b)** $\dfrac{2}{8} = \dfrac{1}{4}$

3.1 Add Fractions Using Manipulatives, pages 88–89

5. a) $\dfrac{5}{6}$ **b)** $\dfrac{1}{2}$

7. a) $\dfrac{1}{6} + \dfrac{1}{3} = \dfrac{3}{6}$ or $\dfrac{1}{2}$ **b)** $\dfrac{1}{2} + \dfrac{1}{3} = \dfrac{5}{6}$

c) $\dfrac{1}{2} + \dfrac{1}{6} = \dfrac{4}{6}$ or $\dfrac{2}{3}$ **d)** $\dfrac{1}{2} + \dfrac{1}{3} + \dfrac{1}{6} = \dfrac{6}{6}$ or 1

9. a) $\dfrac{6}{6}$ or 1 **b)** $\dfrac{6}{6}$ or 1 **c)** $\dfrac{5}{6}$

11. a) $\dfrac{1}{2} + \dfrac{1}{2} + \dfrac{1}{3} = \dfrac{4}{3}$

b) $\dfrac{1}{3} + \dfrac{1}{3} + \dfrac{1}{3} + \dfrac{1}{6} + \dfrac{1}{6} + \dfrac{1}{6} + \dfrac{1}{3} = \dfrac{5}{3}$

c) $\dfrac{1}{6} + \dfrac{1}{6} + \dfrac{1}{6} + \dfrac{1}{6} + \dfrac{1}{6} + \dfrac{1}{6} + \dfrac{1}{3} + \dfrac{1}{3} + \dfrac{1}{2} = \dfrac{5}{2}$

13. a) $\dfrac{1}{2} + \dfrac{1}{2}$ **b)** $\dfrac{1}{2} + \dfrac{1}{2}$ **c)** $\dfrac{1}{2} + \dfrac{1}{4} + \dfrac{1}{4}$

15. Answers may vary.

17. Diagrams may vary. $\dfrac{1}{3} + \dfrac{1}{3} + \dfrac{1}{6} = \dfrac{5}{6}$

19. $1\dfrac{5}{6}$

3.2 Subtract Fractions Using Manipulatives, pages 92–93

3. a) $1 - \dfrac{1}{3} = \dfrac{2}{3}$ **b)** $\dfrac{1}{3} - \dfrac{1}{6} = \dfrac{1}{6}$ **c)** $\dfrac{5}{6} - \dfrac{1}{3} = \dfrac{1}{2}$

5. a) $\dfrac{1}{2} - \dfrac{1}{6} = \dfrac{1}{3}$ **b)** $\dfrac{2}{3} - \dfrac{1}{6} = \dfrac{1}{2}$ **c)** $\dfrac{1}{2} - \dfrac{1}{3} = \dfrac{1}{6}$

7. Diagrams may vary.
a) 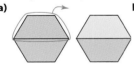   b)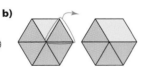
c) 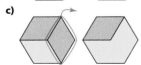   d)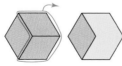

9. Diagrams may vary.
a)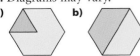

11. Diagrams may vary.
a) 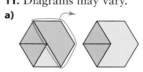   b)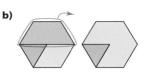

13. a) $1 - \dfrac{1}{2} = \dfrac{1}{2}$ **b)** $1 - \dfrac{1}{4} = \dfrac{3}{4}$

17. a) Diagrams may vary.

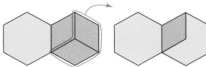

b) In this representation, each fraction is represented by pattern block pieces that are twice the size as they are when 1 hexagon = 1 whole. $\dfrac{1}{2}$ is represented by a whole hexagon and $\dfrac{1}{3}$ is represented by two blue rhombuses.

c) The numerical answer to part a) of $\dfrac{1}{6}$ is the same, but it is represented by pattern blocks twice the size as those that represent the answer when 1 hexagon = 1 whole.

3.3 Find Common Denominators, page 97

5. Answers may vary. **a)** 15 **b)** 21 **c)** 20 **d)** 24
7. Answers may vary. **a)** 6 **b)** 24 **c)** 15 **d)** 12
9. Answers may vary. **a)** 15, 30 **b)** 12, 24
11. 12, 24, 36

13. Answers may vary. **a)** 10, $\dfrac{3}{5} = \dfrac{6}{10}, \dfrac{1}{2} = \dfrac{5}{10}$

b) 8, $\dfrac{5}{8}, \dfrac{1}{4} = \dfrac{2}{8}$

15. Answers may vary. For example, you could find common multiples of 2, 3, and 4. **a)** 12 or 24 **b)** 60

3.4 Add and Subtract Fractions Using a Common Denominator, pages 101–103

5. a) $\dfrac{2}{6} + \dfrac{3}{6} = \dfrac{5}{6}$ **b)** $\dfrac{1}{4} + \dfrac{2}{4} = \dfrac{3}{4}$ **c)** $\dfrac{2}{8} + \dfrac{3}{8} = \dfrac{5}{8}$

d) $\dfrac{1}{6} + \dfrac{3}{6} = \dfrac{4}{6}$

7. a) $\dfrac{3}{4}$ **b)** $\dfrac{4}{3}$ or $1\dfrac{1}{3}$ **c)** $\dfrac{4}{5}$

9. Answers may vary. **a)** $\dfrac{2}{10} + \dfrac{5}{10} = \dfrac{7}{10}$

b) $\dfrac{3}{12} + \dfrac{10}{12} = \dfrac{13}{12}$ or $1\dfrac{1}{12}$

c) $\dfrac{6}{15} + \dfrac{10}{15} = \dfrac{16}{15}$ or $1\dfrac{1}{15}$

d) $\dfrac{1}{6} + \dfrac{3}{6} = \dfrac{4}{6}$ or $\dfrac{2}{3}$

11. a) $\dfrac{2}{2} = 1, \dfrac{3}{3} = 1, \dfrac{4}{4} = 1, \dfrac{5}{5} = 1$

b) They all equal one. **c)** 1

13. Answers may vary. Total number of pieces: $8 \times \frac{1}{8} = 1$; Number of pieces that did not fall out: $3 \times \frac{1}{8} = \frac{3}{8}$

15. a) Answers may vary. The sections being added are not the same size.

b)

17. $\frac{2}{5} + \frac{1}{2}$ is greater. $\frac{2}{5} + \frac{1}{2} = \frac{9}{10}$, $\frac{2}{3} + \frac{1}{6} = \frac{5}{6}$, $\frac{9}{10} > \frac{5}{6}$

19. a) $\frac{13}{12}$ or $1\frac{1}{12}$ **b)** $\frac{19}{20}$

21. Diagrams may vary. $1\frac{3}{4}$

23. $\frac{2}{5} + \frac{1}{4} + \frac{3}{10} = \frac{19}{20}$; Since $\frac{19}{20}$ is less than 1, the addition shows that together they did not clean all the windows. They should not be paid the full amount.

3.5 More Fraction Problems, pages 106–107

5. a) $\frac{17}{4}$ or $4\frac{1}{4}$ **b)** $\frac{17}{6}$ or $2\frac{5}{6}$ **c)** $\frac{27}{8}$ or $3\frac{3}{8}$

7. $\frac{5}{4}$ or $1\frac{1}{4}$

9. $\frac{7}{2}$ or $3\frac{1}{2}$

11. $\frac{3}{2} = 1\frac{1}{2}$, $\frac{8}{6} = 1\frac{2}{6}$ or $1\frac{1}{3}$

13. a) blue: $\frac{5}{16}$, purple: $\frac{3}{16}$, white: $\frac{2}{16}$ or $\frac{1}{8}$, green: $\frac{6}{16}$ or $\frac{3}{8}$ **b)** $\frac{11}{16}$ **c)** $\frac{13}{16}$

15. a) $\frac{1}{5}$ **b)** $\frac{1}{3}$ or $\frac{1}{7}$

19. 6

Review, pages 108–109

1. D **3.** H **5.** A **7.** G

9. Diagrams may vary.

a) **b)** **c)**

11. a) **b)** $\frac{1}{2} + \frac{1}{3} + \frac{1}{6} = 1$

13. a) $2 \times \frac{1}{3}$ **b)** $5 \times \frac{1}{6}$

15. a)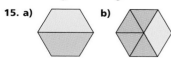

17. a) **b)**

19. Answers may vary. **a)** 12 **b)** 20
21. Answers may vary. **a)** 12 **b)** 10
23. a) $\frac{1}{2} - \frac{1}{3} = \frac{1}{6}$ **b)** $\frac{1}{3} + \frac{1}{6} + \frac{1}{6} = \frac{4}{6}$ or $\frac{2}{3}$
25. a) $\frac{7}{30}$ **b)** $\frac{13}{12}$ or $1\frac{1}{12}$
27. a) red: $\frac{6}{24}$ or $\frac{1}{4}$, blue: $\frac{4}{24}$ or $\frac{1}{6}$, white: $\frac{9}{24}$ or $\frac{3}{8}$, grey: $\frac{5}{24}$ **b)** $\frac{10}{24}$ or $\frac{5}{12}$ **c)** $\frac{15}{24}$ or $\frac{5}{8}$

Practice Test, pages 110–111

1. B **3.** B **5.** A
7. a) $4 \times \frac{1}{5} = \frac{4}{5}$ **b)** $5 \times \frac{2}{7} = \frac{10}{7}$ or $1\frac{3}{7}$
9. 12, 24
11. a) $1 - \frac{3}{10}$ **b)** Strategies may vary. You could subtract and compare your answers.
13. 16

CHAPTER 4

Get Ready, pages 114–115

1. a) 3 **b)** 165.6 cm **c)** 53.5 kg **d)** 15.2 mm **e)** 12.4 m **f)** 34.6 jellybeans
3. a, d
5. a) $\frac{8}{24}$ and $\frac{9}{24}$; $\frac{3}{8}$ **b)** $\frac{4}{10}$ and $\frac{5}{10}$; $\frac{1}{2}$
c) $\frac{20}{30}$ and $\frac{21}{30}$; $\frac{7}{10}$
7. a) yellow: $\frac{4}{8} = \frac{1}{2}$, blue: $\frac{3}{8}$, red: $\frac{1}{8}$
b) red, blue, yellow
9. a) 0.5 **b)** 0.375 **c)** 0.4

4.1 Introducing Probability, pages 118–120

5. Frequency: 13, 16, 11; Total trials: 40
7. a) $\frac{2}{7}$ **b)** $\frac{4}{400} = \frac{1}{100}$ **c)** $\frac{3}{40}$
9. a) It will stay the same. **b)** zippy zingers: $\frac{26}{50} = \frac{13}{25}$; stomach stirrers: $\frac{6}{50} = \frac{3}{25}$; tongue twisters: $\frac{4}{50} = \frac{2}{25}$; face freezers: $\frac{14}{50} = \frac{7}{25}$
11. a) $\frac{4}{13}$ **b)** tan; grey; The colour with the greatest number of pairs is tan. The colour with the least number of pairs is grey. **c)** The probabilities change because the number of favourable outcomes for each colour is different.

13. Answers may vary. She should conduct more trials or examine the spinner.
15. a) before: red = $\frac{7}{30}$, black: $\frac{6}{30} = \frac{1}{5}$, yellow: $\frac{4}{30} = \frac{2}{15}$, orange: $\frac{5}{30}, \frac{1}{6}$, green: $\frac{8}{30}, \frac{4}{15}$; after: red: $\frac{5}{20}, \frac{1}{4}$, black: $\frac{6}{20}, \frac{3}{10}$, yellow: $\frac{2}{20}, \frac{1}{10}$, orange: $\frac{4}{20}, \frac{1}{5}$, green: $\frac{3}{20}$ **b)** Answers may vary. She probably picked the colours she liked.
17. Answers will vary.

4.2 Organize Outcomes, pages 124–125

3. a) spinning red: $\frac{1}{3}$, spinning blue: $\frac{1}{3}$, spinning yellow: $\frac{1}{3}$
b) spinning red: $\frac{1}{6}$, spinning blue: $\frac{1}{6}$, spinning yellow: $\frac{1}{6}$, spinning pink: $\frac{1}{6}$, spinning orange: $\frac{1}{6}$, spinning brown: $\frac{1}{6}$
5. a) 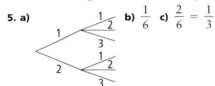 **b)** $\frac{1}{6}$ **c)** $\frac{2}{6} = \frac{1}{3}$

7. Answers will vary.
9. $\frac{4}{6} = \frac{2}{3}$
13. a) Answers will vary. **b)** Answers may vary. Player 1 = $\frac{1}{2}$, Player 2 = $\frac{1}{2}$ **c)** Both players are equally likely to win because their number of favourable outcomes is equal.

4.3 Use Outcomes to Predict Probabilities, pages 128–130

5. a) $\frac{1}{40}$ **b)** $\frac{1}{40}$ **c)** $\frac{2}{40} = \frac{1}{20}$ **d)** $\frac{8}{40} = \frac{1}{5}$ **e)** $\frac{8}{40} = \frac{1}{5}$
7. a) 1, 2, 3, 4, 5, or 6 **b)** A, B, or C **c)** red, black, green, or yellow
9. H/white, H/yellow, H/green, T/white, T/yellow, T/green
11. a) $\frac{3}{6} = \frac{1}{2}$; 3 favourable outcomes, 6 possible outcomes
b) $\frac{2}{6} = \frac{1}{3}$; 2 favourable outcomes, 6 possible outcomes
13. A, because it occurs most often.
15. a) Each side has a probability of $\frac{1}{8}$.
b) 4, because it occurs most often.
19. $\frac{4}{8} = \frac{1}{2}$

4.4 Extension: Simulations, pages 132–133

3. a) 12 **b)** O, since it was only chosen once.
5. Answers may vary. Each item should have a choice for every possible outcome. **a)** a coin **b)** an eight-sided die

c) a five-section spinner
7. Answers will vary.
9. Answers may vary. **a)** chossing at random from five different types of pop **b)** choosing at random from one black, one green, and two red jellybeans. **c)** choosing at random from two hot dogs, two hamburgers, and two steaks
11. a) Answers may vary. You could collect the different letters in "WINNER" (need only one N) or you could spell "WINNER" (need two Ns). 1st way: $\frac{1}{5}$, 2nd way: $\frac{2}{6} = \frac{1}{3}$
b) Each side has a letter from "WINNER." This fits the second way. **c)** Answers will vary.

4.5 Apply Probability in Sports and Games, pages 137–139

3. $\frac{2}{6} = \frac{1}{3}$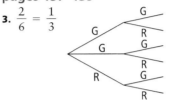

5. $\frac{6}{12} = \frac{1}{2}$

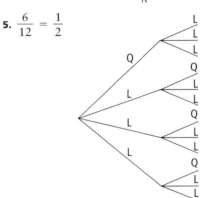

7. a)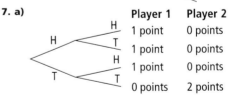

b) No. *Heads* has an advantage.
9. a) $\frac{250}{1000} = \frac{1}{4}$ **b)** $\frac{750}{1000} = \frac{3}{4}$; 0.75 **d)** 6
11. a) $\frac{4}{10} = \frac{2}{5}$ **b)** $\frac{6}{10} = \frac{3}{5}$ **c)** 1, represents the probability of every possible outcome.
13. Answers will vary.
15. a) $\frac{6}{36} = \frac{1}{6}$ **b)** $\frac{30}{36} = \frac{5}{6}$ **c)** It is difficult to get out of jail by rolling doubles.
19. a) $\frac{1}{2}$ **b)** no **c)** $\frac{7}{8}$

Review, pages 140–141
1. probab(i)lity, ta(l)ly chart, freque(n)cy table
3. (t)ree d(i)agra(m)
5. a) no **b)** $\frac{4}{16} = \frac{1}{4}$
7. a) most likely = orange, the most common colour; least likely = white, the least common colour **b)** There is a different number of marbles for each colour.
9. a)

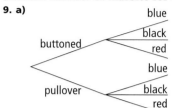

b) 6 **c)** No, it depends on Connie's preference.
11. a) $\frac{3}{6} = \frac{1}{2}$; 3 favourable outcomes, 6 possible outcomes
b) $\frac{3}{12} = \frac{1}{4}$; 3 favourable outcomes, 12 possible outcomes
13. 6; probability of $\frac{1}{10}$
15. a) $\frac{9}{100}, \frac{42}{100} = \frac{21}{50}, \frac{49}{100}$ **b)** coloured; coloured squares have the highest probability

Practice Test, pages 142–143
1. A **3.** B **5.** C
7. a)

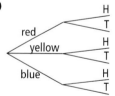

b) red/H, red/T, blue/H, blue/T, yellow/H, yellow/T
c) $\frac{1}{6}$ **d)** $\frac{2}{6} = \frac{1}{3}$
9. a)

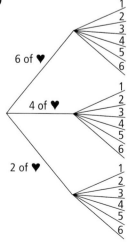

b) 6 of ♥/1, 6 of ♥/2, 6 of ♥/3, 6 of ♥/4, 6 of ♥/5, 6 of ♥/6, 4 of ♥/1, 4 of ♥/2, 4 of ♥/3, 4 of ♥/4, 4 of ♥/5, 4 of ♥/6, 2 of ♥/1, 2 of ♥/2, 2 of ♥/3, 2 of ♥/4, 2 of ♥/5, 2 of ♥/6
11. a) He should get two correct. **b)** Answers will vary.
c) 0 right = $\frac{1}{16}$; 1 right = $\frac{4}{16}$; 2 right = $\frac{6}{16}$;
3 right = $\frac{4}{16}$, 4 right = $\frac{1}{16}$ **d)** The number of trials is too small.

Chapters 1–4 Review, pages 146–147
1. a) 8 cm **b)** 6.8 cm
3. a) 23 **b)** 8 **c)** 9 **d)** 27 **e)** 15
5. Answers may vary. An example could have $a = 3$ cm, $b = 6$ cm, and $h = 4$ cm.
7. Diagrams may vary.
a) scalene triangle **b)** isosceles triangle
c) square or rhombus **d)** rectangle or parallelogram
9. a) $\frac{1}{2} + \frac{1}{6} = \frac{2}{3}$ **b)** $\frac{1}{3} + \frac{1}{6} = \frac{1}{2}$
11. a) 10 **b)** 12
13. a) $\frac{8}{24} = \frac{1}{3}$ **b)** $\frac{6}{24} = \frac{1}{4}$ **c)** 5 h, $\frac{5}{24}$
15. a) $\frac{1}{18}$ **b)** $\frac{2}{18} = \frac{1}{9}$ **c)** $\frac{3}{18} = \frac{1}{6}$ **d)** $\frac{6}{18} = \frac{1}{3}$

CHAPTER 5

Get Ready, pages 150–151
1. a) 0.3 **b)** 0.7 **c)** 0.01 **d)** 0.23
3. a) 0.2, 0.225, 0.25 **b)** 1.334, 1.34, 1.43
5. a)   **b)**

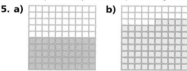

7. a) 43% **b)** 60% **c)** 5% **d)** 2%

5.1 Fractions and Decimals, pages 156–157
5. a)   **b)**   **c)**

d)

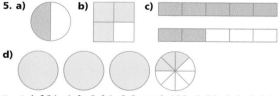

7. a) 1.301, 1.3, 0.34, 0.3 **b)** 0.489, 0.29, 0.2, 0.06
9. a) $\frac{1}{3}, \frac{4}{9}, \frac{1}{2}$ **b)** $1\frac{3}{8}, 1\frac{2}{3}, 1\frac{3}{4}, 1\frac{5}{6}$
11. a) $0.1\overline{6}$ **b)** $1.\overline{3}$ **c)** $3.\overline{6}$ **d)** $2.8\overline{3}$
13. $\frac{250}{1000}$ or $\frac{1}{4}$
15. a) $\frac{21}{27}, \frac{17}{20}, 0.87, \frac{23}{25}, 1.04, \frac{6}{5}$

b) $\frac{319}{25}$, 12.84, $12\frac{5}{6}$

17. a) $\frac{8}{8}$, $\frac{14}{16}$, $\frac{9}{12}$ **b)** Answers will vary.

19. a) 6.7, $6\frac{9}{20}$, $6\frac{2}{5}$, $6\frac{3}{8}$, $6\frac{1}{3}$, 6.05

b) Express all the numbers in decimal form or write them all in fraction form.

21. a) $\frac{3}{4}$ **b)** $\frac{4}{5}$ **c)** $\frac{1}{2}$ **d)** $\frac{13}{20}$

23. a) Yes, the digits begin to repeat on the calculator.
b) More digits of the number would aid in a decision.
c) Answers will vary, numbers like $\frac{4}{7}$ have sequences of numbers that repeat.

5.2 Calculate Percents, pages 160–161

5. a) 70% **b)** 85% **c)** 80%
7. Diagrams may vary. Use number lines, hundred charts, or circles.
9. 10
11. a) 75% **b)** Answers will vary. Think of a clock—45 min is $\frac{3}{4}$ of an hour (60 min). $\frac{3}{4}$ is 75%.
13. No, Amir mixed a percent with a score out of 100. He should say "I got 65 out of 100."
15. a) No, the decimal point should move 2 places to the right. **b)** Answers will vary.
17. a) Answers will vary. A minority government will have less than half of the 300 seats. For example, Liberal 140, Conservative 110, NDP 40, Bloc Quebecois 10.
b) Answers will vary. The party with the most seats will be the governing party. **c)** and **d)** Answers will vary.

5.3 Fractions, Decimals, and Percents, pages 164–165

3. a) $\frac{1}{4}$ **b)** $\frac{1}{2}$ **c)** $\frac{1}{10}$ **d)** $\frac{1}{5}$ **e)** $\frac{3}{4}$ **f)** 1
5. a) 0.32 **b)** 0.64 **c)** 0.7 **d)** 0.83 **e)** 0.05
7. $\frac{1}{2}$, $\frac{3}{4}$, $\frac{1}{10}$, $\frac{1}{4}$

9.

Fraction	Percent	Decimal
$\frac{3}{20}$	15%	0.15
$\frac{13}{20}$	65%	0.65
$\frac{1}{50}$	2%	0.02

11. A: 20, B: 17, C: 14
13. a) CD $\frac{3}{5}$, DVD $\frac{1}{5}$, Cassettes $\frac{3}{20}$, Other $\frac{1}{20}$
b) CD 0.60, DVD 0.20, Cassettes 0.15, Other 0.05
15. a) "Nearly two-thirds of children ..." **b)** Answers will vary. You would need information related to the water supply in the developing world and census statistics for the same areas where you conducted your research.

17. a) 0.53 **b)** 424 **c)** 376
19. Answers will vary. For example, Win/Loss percent = $\frac{\text{number of wins 2 number of losses}}{\text{number of total games}}$

5.4 Apply Fractions, Decimals, and Percents, pages 169–171

5. a) 18.75% **b)** 65% **c)** 83.33% **d)** 68%
7. a) $\frac{2}{5}$ **b)** 0.4 **c)** 40%
9. a) 20% **b)** 19%
11. 40%
13. a) $\frac{39}{50}$ **b)** $\frac{21}{100}$ **c)** $\frac{1}{100}$
15. Percent of Total Sales: Koala Cola 20.3%, Lizard Lime 7.8%, Lemur Lemon 12.5%, Gorilla Grape 4.7%, Roary Root Beer 14.1%, Oliphant Orange 15.6%, Jumping Ginger 3.1%, Canary Cream Soda 21.9% Visuals may vary. A vertical bar graph with bottle shapes coloured appropriately to match each flavour would be good. The height of each bottle would show the sales of that flavour.
17. a) Yes. Cut at least 11.5 min and at most 25 min to get between 50% and 65% of 90 min.
19. a) $105 **b)** In Ontario the jacket costs $120.75.

Review, pages 172–173

1. st(a)tistic
3. (p)ercent
5. a) ▨▨▨□□ **b)** ▨▨▨▨□
c) ▨▨▨□□

7. a) $0.58\overline{3}$ **b)** $0.\overline{5}$ **c)** 0.548
9. a) 65% **b)** 40% **c)** 23.7% **d)** 0.8%
11. a) 0.33 **b)** 0.06 **c)** $0.\overline{7}$ **d)** $0.1\overline{2}$
13. a) 30% **b)** .300 **c)** 54
15. 88%, 0.88, $\frac{22}{25}$; 36%, 0.36, $\frac{9}{25}$; 20%, 0.2, $\frac{1}{5}$; 42%, 0.42, $\frac{21}{50}$; 4%, 0.04, $\frac{1}{25}$
17. a) Answers may vary. Engineering on shelf 1, Math on shelf 2, and Science and other on shelf 3. **b)** Between 54 (30%) and 63 (35%) books will be on each shelf. Assume that each book is about the same thickness and weight.

Practice Test, pages 174–175

1. A **3.** C **5.** A
7. $\frac{7}{25}$, 0.28, 28%; $\frac{3}{8}$, 0.375, 37.5%; $\frac{9}{20}$, 0.45, 45%; $\frac{1}{20}$, 0.05, 5%; $\frac{24}{240}$, 0.1, 10%
9. a) Lab, Quiz, Test **b)** Answers will vary. Convert all the results into decimal form and then compare them.
11. a) Answers will vary. Blue is in the middle and may be easiest to hit. **b)** blue $\frac{9}{25}$, yellow $\frac{12}{25}$, red $\frac{4}{25}$
c) yellow, blue, red **d)** blue 36%, yellow 48%, red 16%

CHAPTER 6

Get Ready, pages 178–179

1. a) Pattern adds a vertical paper clip, followed by a horizontal paper clip, alternating at the top and bottom. **b)** Increasing by 2 each term. 10, 12, 14 **c)** Decreasing by 5 each term. 80, 75, 70 **d)** Each term is double the previous term. 48, 96, 19 **e)** Add alternating yellow and green right triangles, placed to form connecting squares. **f)** Each denominator is double the previous denominator. $\frac{1}{16}, \frac{1}{32}, \frac{1}{64}$ **g)** Every three terms, the pattern is ab, abc, abcd, ab, abc, abcd **h)** Each step, one more happy face is added to the beginning of the pattern.

3. a)–c) 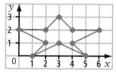 **d)** a five-pointed star

5. Ferris Wheel (0, 5), First Aid (2, 3), Washrooms (3, 2), Basketball Throw (4, 0), Battle of the Bands (7, 7), Petting Zoo (8, 1), Food Court (10, 6)

6.1 Investigate and Describe Patterns, page 182

3. Answers may vary. **a)** Starting at 1, multiply each counting number by 5. **b)** 30, 35, 40
5. a) 0, 0 + 6 = 6, 6 + 8 = 14, 14 + 10 = 24, 24 + 12 = 36, 36 + 14 = 50,… **b)** 66, 84
7. Answers will vary.
9. The top half of each triangle is cut horizontally in half. The new triangle formed alternates between a yellow and green colour.
11. A spiral pattern with rectangles 1 unit longer than the previous rectangle added in the pattern.
13. a) 5, 10, 15, 20, … **b)** 3, 6, 12, 24, … **c)** 18, 16, 14, 12, … **d)** 400, 200, 100, 50, …
15. a) 9, 11, 13, 15 **b)** Start with 9 and increase by 2. **c)** 17, 19, 21 **d)** Yes they will.
17. Answers may vary.
19. a) Start with an equilateral triangle pointing up and add a triangle pointing the opposite way of the previous triangle. **b)** A triangle is added at each stage, so the figure is growing larger. **c)**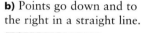
21. a) Rows of dots, arranged in order of length 5, 4, 3, 2, 1. Then the longest row is removed. **b)**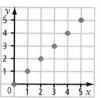
23. No, there are many answers.

6.2 Organize, Extend, and Make Predictions, pages 187–189

3. a) 4, 8, 12, 16 **b)** Each perimeter is 4 times each natural number, starting at 1.
5. a) Each number is 5 times each natural number.
b) Each number is the square of each natural number.
c) The pattern is the odd natural numbers.

d) The pattern is $\frac{1}{\text{each natural number} + 1}$.

7. a)

Number of Cubes	Number of Vertical Faces	Operations
1	4	1 × 2 + 2
2	6	2 × 2 + 2
3	8	3 × 2 + 2

b) $2n + 2$ **c)** 202
9. a) 8 **b)** 15 **c)** 103 **d)** $k + 3$
11. a) Starts at 4 and increases by 3. 4, 7, 10 **b)** 13
c) 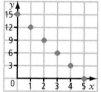 **d)** 16 **e)** Answers may vary. Build a model.

17. a) 602 **b)** 114

6.3 Explore Patterns on a Grid or in a Table of Values, pages 193–194

3. a) Points go up and to the right in a straight line. **b)** Points go down and to the right in a straight line.

5. a)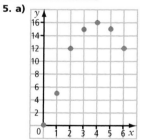

b) The points form a hill, or upside-down U-shape. The other graphs were all straight lines.

7. length = 2 × width
9. The first design has 9 toothpicks. Each one after that has 3 toothpicks more than the one before it. The pattern rule is six plus three times the stage number, $6 + 3n$.
11. Start at (5, 7) and move in each direction: North (5, 7), (5, 8), (5, 9), (5, 10); North-East (5, 7), (6, 8), (7, 9), (8, 10); East (5, 7), (6, 7), (7, 7), (8, 7); South-East (5, 7), (6, 6), (7, 5), (8, 4); South (5, 7), (5, 6), (5, 5), (5, 4); South-West (5, 7), (4, 6), (3, 5), (2, 4); West (5, 7), (4, 7), (3, 7), (2, 7); North-West (5, 7), (4, 8), (3, 9), (2, 10)
15. a) The x-coordinate is 5 less than the y-coordinate.
b) Answers may vary. **c)** $y = x + b$, $x = y - b$

6.4 Express Simple Relationships, pages 197–199

3. a) (1, 5), (2, 10), (3, 15), (4, 20); y-value is 5 times the x-value.
b) (0, 4), (1, 5), (2, 6), (3, 7) ; y-value is 4 more than the x-value.

5. a) (2, 3), (3, 4), (4, 5), (5, 6) **b)** The C-value is 1 more than the z-value. The cost (in dollars) is the number of zones + 1.

7. a)

x	y
0	0
1	2
2	4
3	6

c) Each y-value is twice the x-value.

9. Answers may vary.

11. a)

Number of Boxes, n	Cost, C ($)
1	10
2	20
3	30
4	40
5	50
6	60
7	70
8	80
9	90
10	100

b)

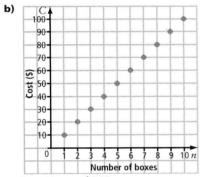

c) $C = 10n$ **d)** $150

13. a) Each y-value is 1 less than the x-values. $y = x - 1$ **b)** Each s-value is 6 times the x-value. $y = 6x$ **c)** The x-values are always 4 greater than the y-values. $y = x - 4$ **d)** Each y-value is 3 less than the x-value. $y = x - 3$

15. a)

Side Length, s (cm)	Area, A, (cm²)
1	1
2	4
3	9
4	16
5	25
6	36

b) Area is the square of the side length.

Review, pages 200–201

1. b) natural numbers
3. a) relationship
5. a) variable expression **b)** ordered pair **c)** variable **d)** algebraic equation
7. a) Start with an equilateral triangle. Join the midpoint of each side. This makes 4 congruent triangles. Colour the centre triangle. Repeat the process to divide each small white triangle into 4 smaller triangles. This process can continue indefinitely.
9. a) 9 cm **b)** 17 cm **c)** $n + (n - 1)$ cm **d)** 35 cm

11. a)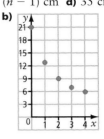

13. a) (1, 3), (2, 6), (3, 9), (4, 12) **b)** Each y-value is 3 times the x-value.
15. a) (1, 13), (2, 26), (3, 39), (4, 52), (5, 65), (6, 78) **b)** The number of computers sold equals 13 times the month number. **c)** 91

Practice Test, pages 202–203

1. A **3.** B **5.** B
7. a)

b) Each y-value is 2 more than the x-value. **c)** $y = x + 2$ **d)** 17
9. a) 4 **b)** 1 **c)** 5
11. a) diagram 1: 4 unit squares + 1 two by two square; diagram 2: 9 unit squares + 4 two by two squares + 1 three by three square. Extending the pattern, in diagram 3: 1 + 4 + 9 + 16 = 30. **b)** 30
13. a)

b) There is an increase of 175 kg of paper collected for each week. **c)** Assume 44 weeks in a school year. 7700 kg

CHAPTER 7

Get Ready, pages 208–209

1. a) 1, 2, 4, 8 **b)** 1, 17 **c)** 1, 2, 3, 4, 6, 8, 12, 24

3. a) 4, 8, 12, 16 **b)** 8, 16, 24, 32 **c)** 6, 12, 18, 24
5. a) 13 < 14 **b)** 13.6 > 13.5 **c)** 8 × 3 = 2 × 12
7. a) 2000 m **b)** 2 m **c)** 300 mm
9. 64 m²
11. 27 cm³

7.1 Understand Exponents, pages 212–213

5. a) 36 square units **b)** 144 square units
c) 121 square units
7. a) 125 cubic units **b)** 1728 cubic units
c) 8000 cubic units
9. a) 9 × 9 **b)** 7 × 7 × 7 **c)** 12 × 12 × 12
11. a) 1.69 **b)** 5.76 **c)** 16.81 **d)** 1.728 **e)** 32.768 **f)** 15.625
13. 10 × 10 × 10, 25 × 25, 8^3, 20^2
15. Because the units are squared along with the number.
17. 96 cm²
19. a) 1^2, 3^2, 6^2, 10^2, … or 1^2, $(1 + 2)^2$, $(1 + 2 + 3)^2$, $(1 + 2 + 3 + 4)^2$, …
b) 2^3, 3^3, 4^3, 5^3, … or $(3 - 1)^3$, $(6 - 3)^3$, $(10 - 6)^3$, $(15 - 10)^3$, … Each number in the sequence is the next natural number cubed.

7.2 Represent and Evaluate Square Roots, pages 216–217

5. a) 3 **b)** 5
7. a) 1.3 cm **b)** 3.5 m **c)** 0.2 mm **d)** 1.4 cm
9. a) 8 **b)** 12 **c)** 20
11. a) 1.2 **b)** 1.5 **c)** 2.4 **d)** 0.5
13. Answers may vary. **a)** Enter 81. Press the square root key.
15. a) 12 m **b)** 48 cm
17. $\sqrt{41}, \sqrt{38}, \sqrt{45}$
21. You might try systematic trial. **a)** 3 **b)** 5 **c)** 100

7.3 Understand the Use of Exponents, pages 221–223

5. a) base: 2, exp: 4 **b)** base: 1, exp: 6 **c)** base: 4, exp: 3
7. a) 32 **b)** 1 **c)** 216
9. a) 4^6 **b)** 9^3 **c)** 2^8
11. a) 3^6 **b)** 5^4 **c)** 9^7
13. a) 0.000 32 **b)** 3.8416 **c)** 0.000 729
15. a) 4^3 **b)** 4^4 **c)** 4^1
17. 19.5, 18^1, 17, 2^4, $\sqrt{225}$, 1^{18}
19. a) Notice that $10^3 = 1000$, and $10^6 = 1\,000\,000$, the number of zeros is the same as the exponent. **b)** 100
21. 81, 729

7.4 Fermi Problems, page 227

3. A loonie has a diameter of 2.6 cm. A square of side length 2.6 cm has area of approximately 7 cm². For a classroom 11m by 10 m about 160 000 loonies are needed.
5. Estimate that there are 200 words on a page, so on about 500 pages there are 10 000 words.
7. Assume the football field is 60 m by 136 m, and the textbooks are 20 cm by 25 cm. This means you need approximately 163 000 textbooks.

9. If the flat bag is 100 cm by 60 cm, it has two sides so the total area is 12 000 cm². This gives a cube of side about 45 cm and volume about 91 000 cm³. Approximate the banana by a rectangular prism 4 cm by 4 cm by 20 cm. Its volume is 320 cm³. So, about 280 bananas fit in the bag.

Review, pages 228–229

1. POWER
3. BASE
5. SQUARE ROOT
7. a) 9 square units **b)** 36 square units
9. a) 256 **b)** 1.69 **c)** 512 **d)** 1331
11. a) $3^3 > 5^2$ **b)** $14^2 < 6^3$ **c)** $3.2^2 < 2.2^3$
13. a) 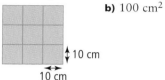 **b)** 100 cm²

15. 3 × 3 = 9
17. 120 cm
19. 128 cm
21. a) 5^4 **b)** 10^6 **c)** 3^5
23. a) 1024 **b)** 1 **c)** 0.0001
25. $2^8 = 256$
27. Answers may vary. Assume the basketball court is 15 m × 29 m, and a phonebook is 21.5 cm by 28 cm. You need approximately 7000 phonebooks.
29. Answers may vary. **a)** Assume you receive 5 coins in change a day. This means you will receive approximately 1825 coins in a year. **b)** Count the number of mixed coins needed to fill a small box. Divide 1825 by this number of coins. Multiply this answer by the volume of the box.

Practice Test, pages 230–232

1. D **3.** B **5.** A
7. a) 64 cubic units **b)** 512 cubic units
9. a) 8 **b)** 20 **c)** 1.2 **d)** 1.5
11. a) 2 × 2 × 2 × 2 × 2 × 2 × 2 × 2 × 2 = 512
b) 3 × 3 × 3 × 3 × 3 × 3 = 729
c) 4 × 4 × 4 × 4 × 4 = 1024
13. 72 cm³
15. a) Yes, 2^6 is greater than 2^5 because when a number is multiplied by 2 it gets bigger.
b) No, $1^6 = 1^5$ because any number times 1 equals itself.
17. 20 cm
19. Answers may vary. **a)** Estimate the distance from Earth to the moon and your average walking speed, then use these to find the time to get to the moon. **b)** Estimate the volume of a cell phone, then estimate the volume of a backpack, then use these to estimate the number of cell phones that would fit in the backpack. **c)** Estimate how many soft drinks one student drinks in a day, then multiply by the number of students and the number of days in a year.

CHAPTER 8

Get Ready, pages 234–235

1. a) **b)**

3. a) 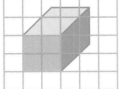 **b)**

5. vertex, edge, face

7. $A = l \times w$ **a)** 15 m² **b)** 32 cm²
9. 1 cm by 64 cm, 2 cm by 32 cm, 4 cm by 16 cm, and 8 cm by 8 cm

8.1 Explore Three-Dimensional Figures, pages 239–241

3. a) rectangular prism **b)** square-based pyramid
c) triangular prism **d)** cylinder
5. Answers may vary. **a)** sugar cube **b)** hockey puck
c) brick **d)** skateboard ramp **e)** tent **f)** globe **g)** skyscraper
7. a) one hexagon and six triangles **b)** hexagonal pyramid
9. a) Answers may vary. AB = AC = DE = DF **b)** Answers may vary. △ABC and △DEF are congruent, ACFD is congruent to ABED. **c)** triangular prism
11. A sphere, cylinder, and rectangular prism. Answers may vary. Drinking mugs can be a cylinder or hexagonal prism.
13. a) cube, any prism, or any pyramid except for a triangular pyramid **b)** cylinder **c)** triangular prism
d) cube, triangular pyramid (tetrahedron), octahedron, icosahedron, dodecahedron **e)** sphere
15. Rectangles have four right angles, opposite sides with equal length, and two pairs of opposite sides that are parallel. Squares have four right angles, all four sides with equal length, and two pairs of opposite sides that are parallel. Since a square fits all the descriptions of a rectangle, it can be thought of as a rectangle with the same length and width. So, a rectangular prism can also be thought of as a square-based prism.
17. Answers may vary.

8.2 Sketch Front, Top, and Side Views, page 245–246

3. a)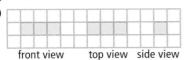

b)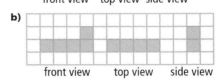

5. a)

front view   top view   side view

b)

front view   top view   side view

7. a) Diagrams may vary. **b)** Answers will vary.
9.
11. a)–d) Answers may vary. **e)** yes

8.3 Draw and Construct Three-Dimensional Figures Using Nets, pages 249–251

3. Answers may vary.

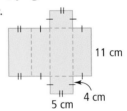

11 cm, 4 cm, 5 cm

5. rectangular prism
7. a) **b)** triangular prism
c) Answers may vary. A Toblerone chocolate bar.

9. triangular pyramid

11.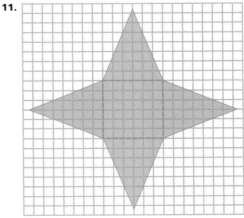

Answers • MHR **471**

13. a) trapezoid **b)**

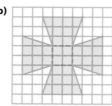

8.4 Surface Area of a Rectangular Prism, pages 255–257

3. 16 cm²
5. a) 230 cm² **b)** 412 cm²
7. 1000 cm²
9. 532 cm²
11. 1.198 m²
13. cube; S.A. = $6s^2$
15. No. You can only form a 2 by 4 by 1 prism, a 1 by 8 by 1 prism, or a 2 by 2 by 2 prism. Their surface areas, in square units, are 28, 34, and 24 respectively.
19. 456 cm²

8.5 Volume of a Rectangular Prism, pages 260–261

3. a) 175 cm³ **b)** 240 cm³
5. a) 288 cm³ **b)** 3600 m³
7. B, because washing machines are close to 1 m by 1 m by 1 m cubes.
9. 4320 cm³
11. a) 30 000 cm³ **b)** 15 L
13. 4000 cakes
15. 11.5 cm

Review, pages 262–263

1. three, polygons
3. pyramid
5. space
7. a) square-based pyramid **b)** cube

c) pentagonal pyramid or triangular prism

9. a) handle and wheels are cylinders, wagon box and handle pole are rectangular prisms **b)** Answers may vary.
11.

13. Answers may vary.
a) **b)**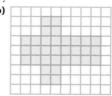

15. 1192 cm²
17. 4500 cm³
19. No. The volume of drawer is 107 250 cm³ but the volume of the suitcase is only 90 000 cm³.

Practice Test, page xxx

1. C **3.** D
5.

7. 972 cm²
9. a) **b)** no

11. 0.4 m; 2 × 2 × 0.4 = 1.6

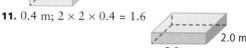

Chapters 5–8 Review, pages 268–269

1. a) 70%, 75% **b)** science test
3. 25%
5. a) 6, 10, 14, 18 **b)** 45, 34, 23, 12 **c)** 7, 14, 28, 56
d) 128, 64, 32, 16
7. a) 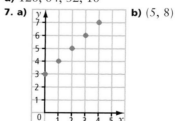 **b)** (5, 8)

9. a) 3^4 **b)** 8^3
11. a) 5 **b)** 7 **c)** 30
13. a) 16 **b)** 144
15. a) rectangular prism **b)** cylinder **c)** triangular pyramid **d)** cube
17. a) 952 cm² **b)** 1760 cm³

CHAPTER 9

Get Ready, pages 272–273

1. a) 4 **b)** 6 **c)** golf; 2 people
3. Answers will vary.

5.

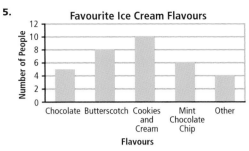

Favourite Ice Cream Flavours	
Chocolate	☺ ☺ ☺
Butterscotch	☺ ☺ ☺ ☺
Cookies and Cream	☺ ☺ ☺ ☺ ☺
Mint Chocolate Chip	☺ ☺ ☺
Other	☺ ☺

☺ represents 2 people

7. a) b) c)

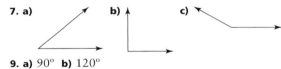

9. a) 90° **b)** 120°

9.1 Collect and Organize Data, pages 277–279

5.

Favourite Insect	Tally	Frequency
Butterfly	|||| ||||	10
Spider	||||	5
Fly			1
Ant	||||	4

7. a) five **b)**

Method	Tally	Frequency
Bus	|||| |||| ||||	14
Car	|||| |||	8
Bike	|||| |||	8
Walk	||||	4
Other	||	2

9. a) Frequency: 4, 6, 3, 8, 10
b)

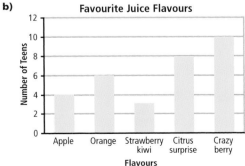

c) Crazy berry **d)** Strawberry-kiwi **e)** 31

11.

Favourite Ice Cream Flavours	
Chocolate	🍦 🍦 🍦 🍦
Strawberry	🍦 🍦 🍦
Cookies and Cream	🍦 🍦 🍦 🍦 🍦
Bubble Gum	🍦 🍦
Vanilla	🍦

🍦 represents 2 people

13. a) primary **b)** secondary **c)** primary **d)** secondary
15. a) Since each number is equally likely to be rolled, each number would probably occur about the same number of times. **b) c) d)** Answers will vary.
17. Answers will vary.
19. Answers will vary. Primary data consists of information you collect by surveying or counting. Secondary data consists of information obtained from other sources.

21.

Region	Sales ($1000s)
Ontario	20
Québec	15
Atlantic Provinces	12
Western Provinces	18
Territories	6
TOTAL	71

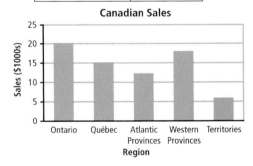

9.2 Stem-and-Leaf Plots, pages 283–285

3. a) four **b)** 1, 3 **c)** 2 **d)** 14, 17, 20, 23, 23, 28, 31, 35, 42
5. a) two children, ages 7 and 9 **b)** three teenagers, ages 13, 13, 17 **c)** 58, 59, 62, 63
7. a) tens **b)** ones **c)**

Stem (tens)	Leaf (ones)
1	0 2 4 8
2	1 2 5
3	2 4 6
4	7 7

d) Answers will vary.

9. a)

Stem (tens)	Leaf (ones)
5	9
6	4 6 7
7	3 3
8	1 2

b) $73 **c)** $565

11. a)

Stem (tens)	Leaf (ones)
4	5 9
5	3 7
6	1 8
7	3 4 7
8	0 0 2 5 5 5 8
9	0 2 5
10	3

b) three **c)** 85

13. a)

Stem (ones)	Leaf (tenths)
0	7 8 8
1	0 0 2 3 4

b) The stem represents the ones digit, and the leaf represents the tenths digit. **c)** 0.7 g

15. a) 11 **b)** $290 **c)** $348 **d)** $3303 **e)** $337; no, one employee is paid $348.

9.3 Circle Graphs, pages 290–291

5. a) **b)**

7. a)

Person	Hours	Fraction	Decimal	Section Angle
Melissa	5	$\frac{5}{12}$	$0.41\overline{6}$	150°
Zach	4	$\frac{4}{12}$	$0.\overline{3}$	120°
Cecilia	3	$\frac{3}{12}$	0.25	90°
TOTAL	12			360°

b)

Science Project Work

9. a) meat and alternatives **b)** 400 g **c)** 200 g
11. a) squirrels; $\frac{1}{4}$ **b)** Squirrels, chipmunks, and raccoons.
13.

Probability of Rolling Each Number on a Number Cube

15. Answers will vary.

9.4 Use Databases to Find Data, pages 295–297

5. a) alphabetically by type of business **b)** business name, location, phone number, sometimes advertising of services offered
7. – 19. Answers will vary.

9.5 Use a Spreadsheet to Display Data, pages 302–303

5. The bar graph clearly identifies the top students. In the pie graph, it is difficult to tell which student has the highest standing because the sections are all very close in size.
7. The bar graph clearly shows how Marisa spent her $50 because each bar identifies exactly how much was spent on each item. The pie graph only gives a rough idea of what portion of the whole $50 was spent on each item.

9. b)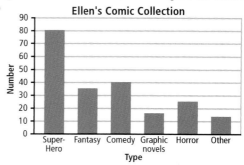

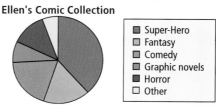

c) Answers will vary. The pie graph is a good choice because it compares the number of comic books of each type with the total number of comics Ellen owns. The bar graph is a good choice because the bars display exactly how many comic books of each type Ellen owns.
11. Answers will vary. A database is any organized collection of information. A spreadsheet is a software tool used for organizing and displaying numeric data
13. Answers will vary.

Review, pages 304–305

1. database, secondary data
3. frequency table, primary data
5. pie chart
7. a) 60 **b)** O'Connor; about 75 **d)** about 47; Ziffareto has about 28 hours and O'Connor has about 75.
9. a) 34 **b)** 43; The highest number in the data set is 43. **c)** 15
11.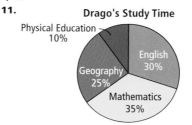

Drago's Study Time

13. a) 1, B1 **b)** 25; Add up the numbers in the B column.

15. a) Answers will vary. A spreadsheet is just one way of organizing data in a database. A database is any collection of information. **b)** and **c)** Answers will vary.

Practice Test, pages 306–307

1. D **3.** D **5.** B

7. a)

Stem (tens)	Leaf (ones)
1	9
2	2 4 5 7 7 7 8 9
3	0 1 1 3 4

b) 27; three

9. a) Frequency: 8, 4, 6, 3, 3

b)
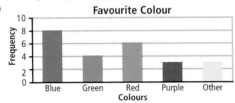

c) Favourite Colour
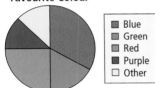

d) Answers will vary. Both charts display the data well. The best one to use depends on what you want the graph to show. If you want the graph to show how each colour compares to the whole then use the pie chart. If you want to compare the colours to each other and know their exact values, then use the bar graph.

CHAPTER 10

Get Ready, pages 310–311

1. a) 7, 4, 4, 3 **b)** play a sport **c)** other
3. a) Population is increasing. **b)** 7 years **c)** Answers may vary. Since it has increased for 7 years, it will probably keep increasing.
5. a) 12 **b)** 34.29 **c)** 70 **d)** 2.2

10.1 Analyse Data and Make Inferences, pages 315–317

3. Answers may vary. You can see the most common, least common, and range of temperature.
5. a) It increases, stays the same, then decreases.
b) It stays the same, decreases, then stays the same.
7. a) Moe and Sable 2 weeks, Lucky 3 weeks **b)** Lucky **c)** Moe
9. The heights are between 153 cm and 157 cm. 155 cm is the most common height.

11. a)

Earnings ($)	Tally	Frequency
18	I	1
19		0
20	IIII	4
21		0
22	I	1
23		0
24	I	1

b) Dale earns between $18 and $24 each week. Dale earns $20 most often.
13. a) Its position increases, stays the same, then decreases. **b)** Week 3 **c)** 2 weeks **d)** Number 1 is the highest (best) position.
15. a) Riverside: The population is increasing. Short Branch: The population stayed the same and is now decreasing. **b)** Short Branch **c)** Riverside **d)** 2001 **e)** Answers may vary. Riverside: about 1600, Short Branch: about 700.
17. Answers will vary.

10.2 Measures of Central Tendency, pages 322–325

3. Diagrams may vary. **a)** median = 4, mode = 4, mean = 5 **b)** median = 10, mode = 11, mean = 9 **c)** median = 5, mode = 5, mean = 6
5. a) Diagrams may vary. **b)** median = 7, mode = 7, mean = 6.5; median: half the girls shoot worse and half shoot better than 7 baskets out of 10, mode: most girls sink 7 baskets out of 10, mean: on average, a girl will sink 6.5 (about 6 or 7) baskets out of 10.
7. a) median = 47, mode = 41 **b)** median = 76, mode = 83

9. a)

Stem (tens)	Leaf (ones)
3	0 5 7
4	2 5 7

b)

Stem (tens)	Leaf (ones)
4	1 1 9
5	2 3
6	3 7

11. a)

Stem (tens)	Leaf (ones)
5	5 8
6	2 3 4 5 6 8
7	0 1 3 3 3 5 6
8	2 3 4 8
9	2

b) median = 72, mode = 73, mean = 72.05 **c)** Any one can be used because they are very close in value.

13. a) median = 36, mode = 36, mean = 36.7
b) The mode is most important since those are the jackets sold most often.
15. a) 39.25 **b)** 43 **c)** Answers may vary.

10.3 Bias, pages 328–330

3. Since the students might not want to hurt Wes' feelings, the results might contain bias.
5. The question says "Do you *really* like...", and Faye is asking, so the results might contain bias.
7. a) cat and dog **b)** Since cat and dog are listed, it is easier to answer them.

9. Yes; students would probably not want to do homework.
11. a) No; it only lists a few groups. **b)** No; people will have a different definition of what *rocks*. **c)** Answers may vary. "What is your favourite rock band?", with no choices given.
13. a) contains bias; Answers may vary. "Do you think the manager should be fired?" **b)** no bias **c)** contains bias; Answers may vary. "Who is the most famous Prime Minister of all time?", with no choices given.
15. a) It only lists a few shows. **b)** Answers may vary. If a broadcasting station's shows are listed, it makes them seem more popular than shows that are not listed. **c)** Answers may vary. "What is your favourite T.V. show?", with no choices given.
17. Answers will vary.
19. Answers will vary.

10.4 Evaluate Arguments Based on Data, pages 333–335

3. a) The first graph's scale starts at 0 and its scale goes up by 5s. The second graph's scale starts at 20 and goes up by 4s. **b)** The second graph make the price increase look greater than on the first.
5. a) The first graph's scale starts at 15 and increases by 1s. The second graph's scale starts at 0 and increases by 4s. **b)** The first graph make the difference in raked leaves look greater than on the second.
7.

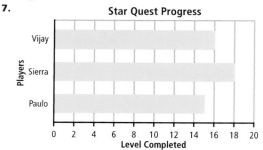

9. a) The first graph compares the ratings of the two shows for year 4. The second graph compares the ratings of the two shows over a 4-year period. **b)** The first graph because it makes the ratings look much better for Happy Times. **c)** The second graph because it shows how Buddies' ratings are increasing and Happy Times' ratings are decreasing.
11. No. The scale starts at 0 and is evenly spaced.
13. No. The scale starts at 0 and is evenly spaced.

Review, pages 336–337

1. median
3. bias
5. mode
7. Forrest Hills Collegiate; it is decreasing, but should stay above the other two schools

9. a)

Tomatoes	Tally	Frequency
6	I	1
7	I	1
8	IIII	4
9	II	2
10	I	1
11	I	1

b) Oswald picks between 6 and 11 tomatoes from each plant. The plants produce 8 tomatoes the most often.
11. a) mean 10, median 13, mode 3; mean: the average number of appointments per day is 10, median: the dentist is busier half of the time, and less busy half of the time than 13 appointments, mode: the most common number of appointments on each day is 3. **b)** Answers may vary. Since there are usually many more than 3 appointments, either the mean or median can be used.
13. a) It assumes that *Big Barney Burger* is good and popular. **b)** The owner of *Big Barney Burger*. **c)** Answers may vary. "What is your favourite fast-food burger restaurant?"
15. a) It only lists a few breakfast foods. **b)** Answers may vary. "What is your favourite breakfast food?", with no choices given.
17. a) Frequency: 11, 12, 8, 7 **b)** Graphs may vary.
c) Graphs may vary.
d) Answers may vary. The owner might use the bias graph to makes Hungry Cat look more popular.

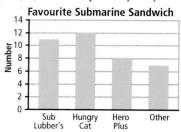

Practice Test, pages 338–339

1. D **3.** A
5. a) Hawks: decreasing; Dancers: increasing, decreasing, then increasing **b)** Answers may vary. Hawks about 600, Dancers about 800
7. a) Yes; the difference in the graph size exaggerates the difference in price by using a vertical scale that does not start at 0. **b)** People will think that Bubbles pop is much cheaper.

9.

Test Score	Tally	Frequency
64	I	1
66	I	1
68	I	1
72	I	1
73	II	2
74	I	1
75	III	3
80	I	1
81	I	1

b) 73 **c)** median = 73.5, mode = 75

CHAPTER 11

Get Ready, pages 344–345
1. a) –6 **b)** –5 **c)** 0 **d)** +2
3. A –6, B –3, C –1, D 0, E +4, F +6
5. a) 12 < 15 **b)** 32 > 23 **c)** 20 > 0 **d)** 33 < 42
e) 29 < 30 **f)** 4 < 40
7. a) 3 **b)** 30 **c)** 25 **d)** 8.4
9. a) 3 **b)** 12 **c)** 50 **d)** 60

11.1 Compare and Order Integers, pages 349–351
5. a) +1, +5 **b)** –5, –2 **c)** –5, –2, +1, +5 **d)** –5, +5
7. a) –5°C, 0°C, 8°C, 15°C, 20°C
b) –30°C, –21°C, 0°C, 8°C, 12°C, 17°C
c) –8°C, –2°C, –1°C, 11°C, 19°C, 32°C
9. a) –2, +1 **b)** –9, 0, +5, +11
11. a) –30 **b)** –12 **c)** –5 **d)** –7 **e)** –43 **f)** –14
13. a) –2, +7

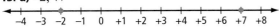

b) –1, +6

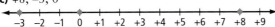

c) +8, –3, 0

![number line -3 to +9]

d) 0, +6, –4
15. Answers may vary. **a)** –10°C today, –8°C yesterday
b) –10 < –8

![number line -10 to 0]

17. a) The 3 means the third game. The 0 is his plus/minus rating. **b)** In the first game, Jake had a plus/minus rating of –1. **c)** The ordered pairs (6, 2), (7, 2), and (8, 1) represent games 6, 7, and 8. Jake can say, "In the last three games, I have improved to a positive plus/minus rating."
19. b) February 3, 5, 6, and 7 **c)** February 1, 2, 3, and 4
d) It was 6°C colder. **e)** –3°C on day 3 **f)** The trend was toward colder temperatures.

11.2 Explore Integer Addition, pages 354–355
3. a) (+2) + (–2) = 0 **b)** (+5) + (–5) = 0
5. a) (+3) + (–5) = –2 **b)** (–6) + (+3) = –3
7. a) (+3) + (–4) = –1 **b)** (+2) + (–1) = +1
c) (–7) + (+4) = –3
9. a) 0 **b)** 0 **c)** 0 **d)** 0; This happens because the integers in each pair are opposite. Opposite integers add to 0.

11.3 Adding Integers, pages 359–361
5. a) (+2) + (+2) = +4 **b)** (–1) + (–3) = –4
c) (–2) + (+4) = +2 **d)** (+1) + (–4) = –3
7. a) +9 **b)** –8 **c)** –6 **d)** –3 **e)** 0 **f)** +6
9. a) positive **b)** 0 **c)** positive **d)** negative **e)** negative
11. a) +16 **b)** 0 **c)** 0 **d)** +2 **e)** –40 **f)** –5
13. a) (–20) + (+15) = –5 **b)** (+89) + (–95) = –6
c) (–3) + (+10) = +7 **d)** (+83) + (–23) = +60
15. Answers may vary. (–1) + (–2) = –3; (+2) + (–5) = –3; (+3) + (–6) = –3
17. a) (+12) + (–15) = –3 **b)** (+25) + (–32) = –7
c) (+18) + (–11) = +7
19. a) +21 **b)** –6 **c)** –2 **d)** –11 **e)** –20
23. a)

3	–4	1
–2	0	2
–1	4	–3

b)

0	–7	–2
–5	–3	–1
–4	1	–6

11.4 Explore Integer Subtraction, pages 366–367
3. a) (–5) – (–3) = –2 **b)** (+3) – (+2) = +1 **c)** (–4) – (–4) = 0
5. a) +4 **b)** –2 **c)** –12 **d)** +5
7. a) –2 **b)** –4 **c)** –7 **d)** –8 **e)** +9 **f)** –3
9. a) +7 **b)** +4 **c)** +13 **d)** –4 **e)** –5 **f)** 0
11.
Start with –3
● ● ●
Add two zero pairs
● ● ● ● ●
○ ○
Take away –5
○ ○
The answer is +2

Start with –5
● ● ● ● ●
Take away –3
● ●
The answer is –2

13. a) Vancouver 9°C, Edmonton 5°C, Ottawa 8°C, Trois-Rivières 7°C, Fredericton 2°C, Saint John 7°C.
b) Vancouver **c)** Fredericton
15. Answers may vary. **a)** (–3) – (+4) – (–5) + (+7) = +5
b) (–3) + (–5) – (+4) – (–7) = –19

11.5 Extension: Subtracting Integers, pages 371–373
5. a) +7 **b)** (+5) + (+2) = +7
7. a) +4 **b)** –7 **c)** –14 **d)** –17 **e)** +12 **f)** +20 **g)** +4 **h)** –5
9. a) –1 **b)** –14 **c)** –4 **d)** –1 **e)** –8 **f)** –6 **g)** –1 **h)** +9
11. a) +3 **b)** –5 **c)** +12 **d)** +16 **e)** –4 **f)** –4 **g)** +1 **h)** –5
13. a) (+8) – (+15) = –7 **b)** (–8) – (+15) = –23
c) (+4) – (+7) = –3 **d)** (–5) – (+4) = –9
e) (–100) – (+600) = –700 **f)** (–20) – (–5) = –15
15. a) +8 **b)** –12 **c)** –30 **d)** –20
21. a) $100 **b)** Answers may vary. You forgot the minus sign.

11.6 Integers Using a Calculator, pages 376–377
3. (–11) + 23 + 79 + (–18) = 23 + 79 – 11 – 18; Answer: 73
5. a) +1 **b)** +16 **c)** –5 **d)** 0
7. a) +233 **b)** –40 **c)** –57 **d)** –252 **e)** –196 **f)** 232
9. +1, –1, +2, –2, +3, –3, … ; You obtain an alternating sequence of positive and negative integers.
11. 10 under par
13. a) (–5) + (–4) + (–3) = –12 **b)** (–11) + (–10) + (–9) = –30
c) (–1) + (0) + (+1) = 0
15.

1	–6	–1
–4	–2	0
–3	2	–5

17. a) –26°C **b)** 108°C
19. a) 2 – 3 + 4 = 3 **b)** 2 + 4 – 3 = 3 **c)** 4 + 2 – 3 = 3
d) –3 + 2 + 4 = 3 **e)** 2 + (–3) + 4 = 3

f) 4 + 2 − (+3) = 3; The order in which the additions and subtractions are carried out does not affect the answer.

Review, pages 378–379

1.

Term	Example
a) positive integers	D) +1 and +2
b) negative integers	A) −1 and −2
c) opposite integers	C) +2 and −2
d) zero principle	B) (+1) + (−1) = 0

3. a)

b) −10, −8, −5, −3, −2, 0, +2, +8
5. a) (+9) + (−3) = +6; The stock finished trading $6 higher. **b)** (−8) + (+6) = −2; The final temperature was −2°C. **c)** (+3) + (−4) = −1; Elizabeth's total score was 1 under par. **d)** (+7) + (−10) = −3; Keith still owed $3. **e)** (+6) + (−8) = −2. The candidate had 2 more votes against them than for them.
7. $390
9. a) −2 **b)** −10 **c)** −5 **d)** +8
11. Both have result +12. Diagrams may vary.
13. a) +10 **b)** −10 **c)** +9 **d)** −5
15. +1
17. a) (−6) + (+2) + (+3) + (−4) + (−7) + (−4) **b)** −16 kg

Practice Test, pages 380–381

1. C **3.** A **5.** C **7.** B
9. a) −9 **b)** +5 **c)** −7 **d)** +16 **e)** −10
11. a) −20 **b)** Add −10 (or subtract +10).

CHAPTER 12

Get Ready, pages 384–385

1. a) 13 **b)** 31 **c)** 26 **d)** 27
3. a) 3 + 13 = 16 **b)** 7 − 4 = 3 **c)** 2 × 6 = 12
 d) 13 − 8 = 5 **e)** 8 × 9 = 72
5. $P = 46$ cm, $A = 120$ cm^2
7. a) begin with 4, increase by 4; 16, 20 **b)** begin with 6, increase by 4; 18, 22 **c)** begin with 5, increase by 4; 17, 21

12.1 Variables and Expressions, pages 389–391

5. Let C represent the variable. **a)** C + 6 **b)** 2C + 2 **c)** 3C + 1 **d)** 2C + 4

7. a) **b)**

9. a) **b)**
c)
d)

11. a) 11 **b)** 2 **c)** 15 **d)** 13 **e)** 7 **f)** 19

13. a)

b)

c)

d)

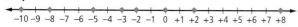

15. Variables may vary. **a)** 10 + p **b)** 8a **c)** A + 10 **d)** 2l
17. a)

b) Sonja is paid $3 to come plus $5 for every hour she babysits. **c)** $28
19. a) 7, 10, 13, 16, 19 **b)** Begin at 7, increase by 3. Answers may vary. For example, points in a basketball game where the team has 7 points and keeps sinking 3 pointers.
23. a) 9, 10 **b)** 2x gives the greater result. This will not be true for all values of x. For example: when $x = 3$, $x + 4 = 7$, $2x = 6$

12.2 Solve Equations by Inspection, pages 395–397

5. a) 9 **b)** 6 **c)** 12
7. a) 7 **b)** 2 **c)** 3
9. a) 19 **b)** 50 **c)** 9 **d)** 4 **e)** 9 **f)** 5
11. 21
13. a) yes **b)** no **c)** no **d)** no **e)** no **f)** no
15. a) 3 **b)** 6 **c)** 38 **d)** 77 **e)** 5 **f)** 4
17. a) Answers will vary. For example, $x + 4 = 8$ or $3x − 7 = 8$ **b)** Answers will vary.
19. 15 min
23. a) $3 + x = −10$ **b)** −13

12.3 Model Patterns With Equations, pages 401–403

5. a)

Number of Rows	Perimeter	Number Equation
1	1 + 1 + 1	3 × 1 = 3
2	2 + 2 + 2	3 × 2 = 6
3	3 + 3 + 3	3 × 3 = 9

b) $3r = 27$
7. Equations may vary. **a)** $5 + n = 6$, $5 + n = 7$, $5 + n = 8$ **b)** $5 + n = 17$, there are 17 marbles at the nth diagram in the pattern. Solving for n shows that $n = 12$, or that the twelfth diagram in the pattern will have 17 marbles.
9. a) $37 = 5 + 4w$, $4w + 5 = 37$, or $5 + 4w = 37$
 b) $2a + 6 = 26$, $26 = 2a + 6$, or $26 = 6 + 2a$
11. $2n − 2 = 20$, n is the number of blocks
13. Answers may vary. **a)** 6 cubes minus 2 cubes leaves 4 cubes. **b)** 11 cubes, each cube has 2 stickers on it but one of the cubes has an extra sticker on it. **c)** 4 cubes in the shape of a square, one sticker on each exposed face, but the two top faces have no stickers.
15. a) $102 = 3m + 2n$ **b)** Answers will vary.

12.4 Solve Equations by Systematic Trial, pages 408–409

5. a)

Diagram Number	Number of Dots	Pattern
1	2 on the left, 1 in the middle, 2 on the right = 5	$2 + 1 \times 1 + 2$
2	2 on the left, 2 in the middle, 2 on the right = 6	$2 + 2 \times 1 + 2$
3	2 on the left, 3 in the middle, 2 on the right = 6	$2 + 3 \times 1 + 2$
d	2 on the left, d in the middle, 2 on the right = $2 + d + 2$	$2 + d \times 1 + 2$

b) $2 + d \times 1 + 2 = 11$ **c)** $d = 7$; the seventh diagram has 11 dots.
7. a) 3 **b)** 3
9. a) 22 **b)** 4 **c)** 5 **d)** 7
11. a) $3 + 7n$ **b)** 6
13. a) 28 **b)** 26 **c)** 9 **d)** 6
15. a) $5n + 13 = 48$ **b)** $n = 7$ **c)** Answers may vary. Use systematic trial.
17. a) $1 + 3s = 52$, $s = 17$ **b)** $2 + 2s = 48$, $s = 23$
19. Answers will vary. **a)** $3n + 2 = 26$ **b)** $(n - 2) \div 3 = 2$
23. 49 no, 106 yes. The number of smiley faces is $4n + 2$ (the 4 side faces + 2 ends). This expression always gives an even number, so 49 is not possible. When there are 26 cubes in the rod, the number of smiley faces is $4 \times 26 + 2$, or 106.

12.5 Model With Equations, pages 413–415

5. a) $C \div 4 = 10$, $C = \$40$ **b)** $2A + 14 = 42$, $A = 14$
c) $S + 15 = 32$, $S = 17$ **d)** $h \div 2 - 10 = 70$, $h = 160$ cm
7. 17 cm
9. $h = 5$ cm
11. Answers will vary. **a)** You have 5 books but you want to have 7, how many more do you need? **b)** Each book costs $7, the total bill came to $28, how many books were bought? **c)** There are 3 shelves in a book case. On the first two shelves there are 14 books, there are 19 books in total on the bookcase, how many books are on the third shelf?
13. They are the same equation, just written in a different order.
15. a) $300 + 5s = C$ **b)** 140 **c)** Answers may vary. Use systematic trial.
17. a) Answers will vary. **b)** 11 cm
19. a) 80 m **b)** Answers may vary. Use systematic trial.
23. 56°

Review, pages 416–417

1. eq(**u**)a(**t**)i(**o**)n, var(**i**)able
3. solution
5. a) Three times a number is equal to 6. **b)** $3x = 6$
7. a) 5 **b)** 9
9. 70
11. $6t + 6 = 46$
13. a) 6 **b)** 12 **c)** 8 **d)** 2
15. a) $1 + 2n$ **b)** $1 + 2n = 51$ **c)** 25
17. a) $J = 2 \times 31 - 1$ **b)** 61

Practice Test, pages 418–419

1. A **3.** B **5.** C
7. a) $5h = 35$ **b)** $3h = 42$ **c)** $4a + 2 = 38$ **d)** $3a + 10 = 55$
9. a) 12 **b)** 3 **c)** 23 **d)** 12 **e)** 11
11. a) $10 + V = 75$ **b)** $65
13. $4(c - 1) + 6 = 86$, $c = 21$

Chapters 9–12 Review, pages 422–423

1. a)

Colour	Tally	Frequency
Red	||||		6
Green	|||| |||	8
Blue	||||	4
Yellow	|||| ||||	9
Orange	|||	3

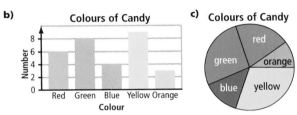

d) The bar graph makes it clear exactly how many candies of each type were in the package. The pictograph looks like a candy and gives an impression of what fraction of the total each colour is.

3.

5. a) Vinyl records: decreasing, then staying the same; CDs: increasing; Video tapes: staying the same, then decreasing. **b)** CDs **c)** 13 000; each year sales increase by 5000.
7. a) Sunny Time; the graph makes their product look more popular. **b)** The scale starts at 44.

c)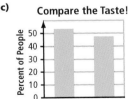

d) The two juices are almost the same in popularity, but the first graph makes it look like Sunny Time is 3 times more popular.

9. a) −8 **b)** −5 **c)** −5 **d)** −6
11. a) −30 **b)** +40 **c)** −10 **d)** 0 **e)** −150
13. 42 cm^2
15. a) 6 **b)** 4
17. 12 cm

CHAPTER 13

Get Ready, pages 426–427

1. a) No; corresponding sides and angles are not equal.
b) Yes; corresponding sides and angles are equal. **c)** No; corresponding sides are not equal.
3. a) irregular; all angles not equal **b)** regular; all sides equal and all angles equal **c)** irregular; all angles not equal **d)** irregular; all sides not equal
5. Yes; corresponding sides and angles are equal.
7. W

13.1 Explore Transformations, pages 430–433

5. Answers will vary.
7. a)

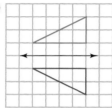

b)

9. They do not change the size or the shape of the figure. After the transformation, the image is congruent to the original.
11. You use rotations when you turn the dial. You use translations when you open and close the lock.
13. Answers may vary. **a)** A window with two panes of glass. One pane is pushed over the other to open the window. **b)** A window with one pane of glass. As you rotate a handle, the glass rotates outward, opening the window.
15. a) In the middle of the common side. **b)** 180°
17. a) Design A: squares related by translation, reflection, or rotation; triangles related by rotation or reflection; trapezoids related by reflection or rotation. Design B: parallelograms related by rotation or translation.
b) Answers will vary. Design A: reflection, rotation 180°. Design B: rotation 90°.
21. The location of the original and the final image is the same.

13.3 Extension: Translations on a Coordinate Grid, pages 440–441

5. a) 4 units right **b)** 5 units up
7. a) (3, 6) 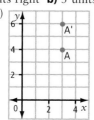   **b)** (1, 1)

9.

11. a) Michel said, "Translate the image 2 units right and 1 unit up. This brings the image back onto the original."
b) He knew this by reversing Fareeha's instructions.

13.4 Identify Tiling Patterns and Tessellations, pages 444–445

5.–9. Answers will vary.

Review, pages 456–457

1. image
3. angle of rotation
5. tessellation
7. reflection
9. Answers will vary.
11. On any of the 4 axes of symmetry, that is, along the diagonals of the square or perpendicular to and halfway along any side of the square.
13. a) rectangles: translation, reflection, or rotation; trapezoids: reflection or rotation; triangles: reflection or rotation **b)** rotation: rotate about its centre 180° or 360°; reflection: place a mirror horizontally through the centre; reflection: place a mirror vertically through the centre.
c) It does not matter. The window is symmetrical so it looks the same upside down as it does right side up.
15. Translate the figure 1 unit left and 3 units up.
17. The image was translated horizontally either left or right.
19. translates the figure 2 units left and 3 units up
21. yes
23. a) Each brick is made up of three regular hexagons with the common sides removed. Regular hexagons tile the plane. **b) c) d) e)** Answers will vary.

Practice Test, pages 458–459

1. C   **3.** A
5. Answers will vary.
7. Parallelogram ABCD is translated 2 units left and 3 units down.
9. no
11. Answers will vary.
13. a) no **b)** large triangles (F and G): turn centre is in the middle of the tangram puzzle, rotate F counterclockwise 90° or rotate G clockwise 90°; small triangles (C and E): turn centre is 1 unit right of the middle, rotate C clockwise 90° or rotate E counterclockwise 90° **c)** large triangles (F and G): reflect along the line between them; small triangles (C and E): reflect through a mirror line in centre of square parallel with the two triangles

A

acute angle An angle whose measure is less than 90°.

acute triangle A triangle in which each of the three interior angles is less than 90°.

algebraic equation An equation or formula that describes a relationship. Uses numbers and variables. $3x = 6$ and $C = 3d$ are algebraic equations.

angle The figure formed by two rays or two line segments with a common endpoint called the vertex.

angle of rotation The angle through which a figure turns.

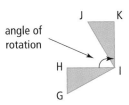

area The number of square units contained in a two-dimensional region.

B

bar graph A graph that uses horizontal or vertical bars to represent data visually.

base (exponential form) The factor you multiply.

In 5^2, the base is 5.

base (2-D geometry) A side of a polygon. Short form is b.

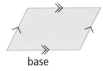

bias An emphasis on characteristics that are not typical of an entire population. Certain responses can be encouraged by the wording of a question.

C

circle The set of all points in the plane that are the same distance from a fixed point called the centre.

circle graph A graph in which a circle is used to represent a whole and is divided into sectors that show how data are divided into parts by percent. Also called a pie chart.

common denominator A number that is a common multiple of the denominators of a set of two or more fractions.

10 is a common denominator for $\frac{1}{2}$ and $\frac{1}{5}$.

composite shape A two-dimensional figure that can be split into two or more simpler figures.

concept map A diagram that places concepts or ideas in balloons. The balloons are linked together to show how concepts are related.

concrete materials Objects that can be used to help in understanding mathematical concepts and skills. Also called manipulatives.

Examples are base 10 blocks, centimetre cubes, pattern blocks, geoboards, number lines, hundred charts, spinners, number tiles, and so on.

cone A three-dimensional object with a circular base and a curved surface.

congruent figures Figures that have the same size and shape.

△ABC and △DEF are congruent.

Glossary • MHR **481**

coordinate grid The two-dimensional or (x, y) plane. Also known as the coordinate or Cartesian plane.

corresponding angles Angles that have the same relative position in geometric figures.

Corresponding pairs of angles are
$\angle A$ and $\angle D$
$\angle B$ and $\angle E$
$\angle C$ and $\angle F$

corresponding sides Sides that have the same relative position in geometric figures.

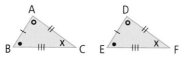

Corresponding pairs of sides are
AB and DE
BC and EF
AC and DF

cube A polyhedron with six congruent square faces.

cube (cubic number) The product of three equal factors. Represents the volume of a cube.
$2 \times 2 \times 2 = 2^3$

cylinder A three-dimensional object with two parallel circular bases.

D

data Facts or information.

database An organized collection of information. Often stored electronically.

denominator The number of equal parts in the whole or the group.

$\frac{3}{4}$ has denominator 4.

difference A number resulting from subtraction.

dodecahedron A polyhedron with 12 pentagonal faces.

E

edge Where two faces meet.

equation A mathematical statement that has equal expressions on either side of the equal sign.

equilateral triangle A triangle with all three sides equal.

equivalent fractions Fractions such as $\frac{1}{3}$ and $\frac{2}{6}$ that represent the same part of a whole or group.

estimate An approximate answer obtained using mental mathematics strategies. An estimate is used when an exact answer is not required or to check the reasonableness of a calculation.

exponent The number of factors you multiply.

exponential form A shorter method for writing numbers expressed as repeated multiplication.

expression Numbers and variables, combined by operations.
$3x + 2y$ is an expression.

F

face A flat or curved surface of an object.

factors The numbers that are multiplied to produce a specific product.
2 and 3 are factors of 6, since $2 \times 3 = 6$.

favourable outcome An outcome that counts for the probability being calculated.

Fibonacci sequence The sequence 1, 1, 2, 3, 5, 8, 13, ... Describes many patterns in nature.

formula A set of ideas, words, symbols, figures, characters, or principles used to state a general rule.

The formula for the area, A, of a rectangle with length l and width w is $A = l \times w$.

fractal A pattern that gets smaller as it repeats forever.

fraction A number that represents a part of a whole or a part of a group.

frequency table A table used to show the total numbers of occurrences in an experiment or survey.

frieze pattern A design pattern that repeats in one direction.

H

height The perpendicular distance from the base of a polygon to the opposite side. Short form is *h*.

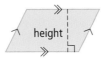

heptagon A polygon with seven sides.

hexagon A polygon with six sides.

hexagonal prism A prism whose bases are congruent hexagons.

I

icosahedron A polyhedron with 20 triangular faces.

image A figure resulting from a transformation.

improper fraction A fraction in which the numerator is greater than the denominator, such as $\frac{8}{5}$.

integer A number in the sequence ..., −3, −2, −1, 0, +1, +2, +3, ...

irregular polygon A polygon that is not regular.

isosceles triangle A triangle with exactly two equal sides.

K

kite A quadrilateral with two pairs of adjacent sides equal.

L

line of symmetry A line that divides a shape into two parts that can be matched by folding the shape in half.

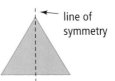

line segment The part of a line that joins two points.

M

manipulatives Objects that are can be used to help in understanding mathematical concepts and skills. Also called concrete materials.

Examples are base 10 blocks, centimetre cubes, pattern blocks, geoboards, number lines, hundred charts, spinners, number tiles, and so on.

mean The sum of a set of values divided by the number of values in the set.

measure of central tendency A value that represents the centre of a set of data. It can be the mean, median, or mode.

median The middle number in a set of data when the data are arranged in order from least to greatest. If there is an even number of pieces of data, the median is the average of the two middle values.

The median of 1, 1, 3, 5, 6, is 3.
The median of 1, 1, 3, 5, is 2.

mixed number A number made up of a whole number and a fraction, such as $3\frac{1}{2}$.

mode The value that occurs most frequently in a set of data. There can be more than one mode, or no mode.

For 1, 2, 3, 3, 8, the mode is 3.

model (noun) A physical model that can be used to represent a situation.

model (verb) To represent the facts and factors of, and the results of, a situation.

multiple The product of a given number and a natural number.

Multiples of 2 are 2, 4, 6, 8, and so on.

N

natural numbers The numbers 1, 2, 3, ... and so on. Also called positive integers.

negative integer One of the numbers −1, −2, −3, ...

net A two-dimensional drawing that can be folded to form a three-dimensional object. A single pattern piece that shows all the faces of the figure.

number line A line that matches a set of points and a set of numbers one to one.

numerator The number of equal parts being considered in the whole or the group.

$\frac{3}{4}$ has numerator 3.

O

obtuse angle An angle that measures more than 90° but less than 180°.

obtuse triangle A triangle containing one obtuse angle.

octagon A polygon with eight sides.

octahedron A polyhedron with eight triangular faces.

opposite integers Two integers with the same numeral but opposite signs.

+2 and −2 are opposite integers.

order of operations Correct sequence of steps for a calculation. Use BODMAS to remember.

B	Brackets, then
O	Order:
D	⎱ Division and Multiplication,
M	⎰ from left to right
A	⎱ Addition and Subtraction,
S	⎰ from left to right

ordered pair A pair of numbers, such as (2, 5), used to locate a point on a coordinate grid.

outcome One possible result of a probability experiment.

P

parallel lines Lines in the same plane that never meet.

parallelogram A four-sided figure with both pairs of opposite sides parallel.

pattern An arrangement of shapes, lines, colours, numbers, symbols, and so on, for which you can predict what comes next.

pattern rule A simple statement that tells how to form or continue a pattern.

pentagon A polygon with five sides.

pentagonal prism A prism whose bases are congruent pentagons.

pentagonal pyramid A pyramid with a pentagonal base.

percent Out of 100.

50% means $\frac{50}{100}$ or 0.5.

percent circle A circle divided into 100 equal sections. Each section represents 1%.

perfect square A number whose square root is a natural number.

4 is a perfect square. Its square root is 2.

perimeter The distance around the outside of a two-dimensional shape or figure.

perpendicular lines Two lines that cross at 90°.

pictograph A graph that illustrates data using pictures and symbols.

place value The value given to the place in which a digit appears in a number.

In the number 2345, 2 is in the thousands place, 3 is in the hundreds place, 4 is in the tens place, and 5 is in the ones place.

polygon A two-dimensional closed figure whose sides are line segments.

polyhedron A three-dimensional figure with faces that are polygons.

positive integer One of the numbers +1, +2, +3, ….

power A number in exponential form. Includes a base and an exponent.

primary data Data you collect yourself.

Data from a survey are primary data.

prism A three-dimensional object with two parallel, congruent polygonal bases. A prism is named by the shape of its bases, for example, rectangular prism, triangular prism.

probability The chance that something will happen.

product A number resulting from multiplication.

proper fraction A fraction in which the denominator is greater than the numerator, such as $\frac{5}{8}$.

pyramid A polyhedron with one base and the same number of triangular faces as there are sides on the base.

Q

quadrilateral A four-sided polygon.

quotient A number resulting from division.

R

random A type of choice or pick in which each outcome is equally likely.

ray A part of a line with one endpoint.

rectangle A quadrilateral with two pairs of equal opposite sides and four right angles.

rectangular prism A prism whose bases are congruent rectangles.

rectangular pyramid A pyramid with a rectangular base.

reflection A flip over a mirror line.

regular polygon A polygon with all sides equal and all angles equal.

relationship A pattern formed between two sets of numbers. Can often be seen by plotting ordered pairs on a coordinate grid.

repeating decimal A decimal with a digit or group of digits that repeats forever. Write the repeating digits with a bar: $0.333… = 0.\overline{3}$.

rhombus A quadrilateral in which the lengths of all four sides are equal.

right angle An angle that measures 90°.

right triangle A triangle containing a 90° angle.

rotation A turn about a fixed point.

S

scalene triangle A triangle with no sides equal.

secondary data Data obtained from someone else. An encyclopedia is an example of secondary data.

similar figures Figures that have the same shape but different size.

△RST and △UVW are similar.

simulation A probability experiment used to model a real situation.

solution A number that makes an equation true.

solving by inspection A method of solving equations using mental math.

solving by systematic trial A method of solving equations by substituting values for the variable until the correct answer is obtained.

sphere A round ball-shaped object. All points on its surface are the same distance from a fixed point called the centre.

spreadsheet A software tool for organizing and displaying numeric data.

square A rectangle in which the lengths of all four sides are equal.

square-based pyramid A pyramid with a square base.

square (number) The product of two equal factors. Represents the area of a square.

$3 \times 3 = 3^2$

square root (of a number) A factor that multiplies by itself to give that number.

Since $8 \times 8 = 64$, the square root of 64 is 8.

statistic A value calculated from a set of data.

stem-and-leaf plot A way of organizing numerical data by representing part of each number as a stem and the other part of the number as a leaf.

sum A number resulting from addition.

surface area The number of square units needed to cover the outside of an object.

survey A sampling of information. Can be conducted by asking people questions or interviewing them.

symmetry A balanced arrangement on either side of a centre line. This line is called a line of symmetry.

T

table of values A table listing two sets of numbers that may be related.

tally chart A table used to record experimental results or data. Tally marks are used to count the data.

tessellation A pattern that covers a plane without overlapping or leaving gaps. Also called a tiling pattern.

tetrahedron A polyhedron with four triangular faces.

tiling pattern A pattern that covers a plane without overlapping or leaving gaps. Also called a tessellation.

tiling the plane Using repeated congruent shapes to cover a region completely.

transformation A change in a figure that results in a different position, orientation, or size. Three types of transformations are translations, rotations, and reflections.

translation A slide along a straight line.

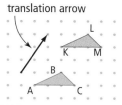

translation arrow An arrow that shows the distance and direction of a translation.

trapezoid A four-sided figure with exactly one pair of opposite sides parallel.

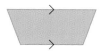

tree diagram A diagram that shows outcomes as sets of branches. Useful for organizing combined outcomes.

triangle A closed, three-sided figure.

triangular prism A prism whose bases are congruent triangles.

triangular pyramid A pyramid with a triangular base.

turn centre A fixed point about which a figure rotates.

V

variable A letter that represents an unknown number.

In $2x + 4$, the letter x is a variable.

variable expression An expression that contains variables and operations with numbers.

$2x + 4$ is a variable expression.

Venn diagram A diagram that uses nested and/or overlapping shapes to show relationships.

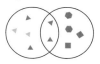

vertex A point at which two sides of a figure meet.

volume The number of cubic units contained in a space.

X

x-axis The horizontal number line on a coordinate grid.

x-coordinate The first number in the ordered pair describing a point on a coordinate grid.

The point P(2, 5) has x-coordinate 2.

Y

y-axis The vertical number line on a coordinate grid.

y-coordinate The second number in the ordered pair describing a point on a coordinate grid.

The point P(2, 5) has y-coordinate 5.

Z

zero principle The principle that opposite integers cancel each other out. The sum of a pair of opposite integers is zero.

For example, $(+1) + (-1) = 0$.

Index

A

Acute angle, 53
Acute triangle, 56
Adding integers, 352–361
Algebraic equations, 195–199
Analysing data, 312–314
Angles, 52–53
 in circle diagrams, 273, 287
 classified by size, 53
 in triangles, 55, 56
Animating a pattern, 435
Architecture and mathematics, 266
Area, 5
 of a parallelogram, 12–14, 18–21
 of a rectangle, 235
 of a square, 209
 of a trapezoid, 24–25, 30–33
 of a triangle, 16–17, 22–25
 calculating percent of, 168
 of composite shapes, 40–41, 42, 43
 explained, 11

B

Bar graph, 272–273, 276, 285
Base, 211
Bias, 326–330
Board games and math, 113, 121–125, 144
Brackets in the order of operations, 20–21, 26–27

C

Calculators and integers, 374–375
Calendars, 180, 184
CANSIM database, 292–294
Card games and math, 126–130
Chance *see* Probability
Circle diagram, 273
Circle graph, 286–291, 299–300, 301
Collecting data, 274–279
Common denominators
 adding and subtracting fractions, 98–103
 finding, 94–96
Composite shapes, 34–37, 40–45
 area, 40–41, 42, 43
 perimeter, 42, 43
Computer memory, 218
Computer software *see The Geometer's Sketchpad®*
Cone, 236
Congruent and similar figures, 70–74
Congruent figures, 66–69, 237, 426
Coordinate grid, 179, 190–194, 427
 and translations, 429, 436–439
Cube, 236
 evaluating, 211–212
 representing, 210
 volume of, 209
Cubic number, 210
Cubic units, 259
Cylinder, 236

D

Data
 analysing, 312–317
 collecting and organizing, 274–279
 displaying, 293–294, 298–303
 evaluating arguments from, 331–335
 fraction, decimal, percent, 162–168, 171
 reading sports data, 136
Databases, 292–297
 and sports reporting, 303
Decimals
 applying, 166–171
 comparing and ordering, 150, 155
 converting fractions to, 115
 data as, 163–165, 167–168
 and fractions, 150, 152–157, 163–165
 and percents, 151, 159, 163–165
 place value, 150, 155
 repeating decimals, 154
Dominoes, 177
Drawing
 different views, 242–246
 trapezoids, 28–29, 34–36
 using software *see The Geometer's Sketchpad®*

E

Edge, 235
Electronic databases *see* Databases
Equations
 modelling patterns, 398–403, 405, 406
 modelling problems, 410, 412–415
 solving by inspection, 392–397
 systematic trial, 404–409
Equilateral triangle, 56
Equivalent fractions, 85, 94, 114, 136
Evaluating statements based on data, 331–335
Exponential form
 defined, 211
 understanding the use of, 218–223
Exponents
 square roots, 214–216
 understanding, 210–213
 understanding the use of, 218–223
 see also Cube; Square; Square roots

F

Face, 235
Factors, 208
Favourable outcomes, 117
Fermi problems, 224–227
Fibonacci sequence, 178

488 MHR • Index

Fishbone organizer, 285
Floor plans, 246
Formulas, 385, 411
Fraction strips, 136
Fractions
  applying, 166–171
  changing to decimals, 115
  comparing and ordering, 84–85, 115, 153
  data as, 162–165, 167–168, 171
  and decimals, 150, 152–157, 163–165
  equivalent fractions, 114, 136
  finding common denominators, 85, 94–97
  manipulatives for adding, 86–89, 98
  manipulatives for subtracting, 90–93, 99
  more fraction problems, 104–107
  and percents, 122, 135, 159, 162–165
  using a common denominator, 98–103
  writing, 84, 100
Frequency table, 116, 274–276, 310
Frieze pattern
  exploring transformations, 428–429
  The Geometer's Sketchpad®, 434–435

G

Geometer's sketchpad *see The Geometer's Sketchpad*®
Geometric pattern, 181
Graphing ordered pairs, 179, 190–194
Graphing skills, 427
Graphs
  identifying trends, 312–317
  line graphs, 310–311, 312–317
  misleading graphs, 332–335
Grid, 179, 190–194
  plotting ordered pairs on, 427
  translations on, 436–441

H

Hexagon, 12

I

Image, 429
Improper fractions, 84
Integer chips, 352–355, 362–365
Integers, 344
  addition, 352–361
  comparing and ordering, 346–351
  opposite, 348, 369
  positive and negative, 347
  subtraction, 362–373
  using a calculator, 374–377
  zero principle, 354
Irregular polygon, 426
Isosceles triangle, 56

K

Kite, 61

L

Line graphs, 310–311, 312–317
Line of symmetry, 59
Line segments, 52
Lists and probability, 127

M

Magic squares, 421
Manipulating trapezoids, 39
Mean, 114, 311, 319, 320, 321
  calculating, 345
Measures of central tendency, 318–323
  finding with a spreadsheet, 324–325
  using a stem-and-leaf plot, 319
Measuring trapezoids, 39
Media bias, 330
Median, 318, 319–321
  calculating, 345
Metric units, 4–5, 10–11
Mixed numbers, 84
Mode, 318, 319–321
Modelling, 153
  for calculating probability, 122
  equations, 398–403, 405, 406, 410–415
  integer chips, 352–355
  math expressions, 386–389
  a translation, 438
Multiples, 85, 95, 96, 208

N

Natural numbers, 186
Negative integer, 347
Nets, 247–251
Number line, 356, 357–358, 368–371
Number pattern, 181
Nutrition and math, 144

O

Obtuse angle, 53
Obtuse triangle, 56
Octagon, 12
Opposite integers, 348, 369
Order of operations, 20–21, 26–29, 384
Ordered pair, 179, 190–199, 427
Organizers, 322
Organizing and extending patterns, 185–189
Organizing data, 274–279
Outcomes, 117
  organizing outcomes, 121–125
  to predict probabilities, 126–130

P

Page, 158
Parallelogram, 12–14, 60
  area, 18–21
  definition, 12
Pattern blocks, 79
  and fractions, 86–89, 90–93
Pattern rule, 191, 192
Patterns
  and equations, 398–403, 405, 406
  exploring, 178
  on grid or in table of values, 190–194
  identifying and extending, 385
  investigating and describing, 180–184

number or geometric, 181
organize, extend, predict, 185–189
in simple relationships, 195–199
Percent circle, 291
Percents
applying, 166–171
calculating, 158–161
data as, 162–165, 167–168, 171
and decimals, 151, 159, 163–165
and fractions, 122, 135, 159, 162–165
representing, 151, 159
Perfect squares, 215, 216
Perimeter, 4, 6–9
of composite shapes, 42, 43
explained, 10
of two-dimensional shapes, 12–17
Pictograph, 272–273, 275, 276, 285
Pie chart, 286–291, 299–300, 301
Place value, 150, 155
Plotting ordered pairs, 179, 190–194
Polygon, 234
regular, 12
regular and irregular, 426
Polyhedron, 237, 238
regular, 241
Positive integer, 347
Powers
evaluating, 219
reading, 219
writing, 220
Predictions from patterns, 185–189
Primary data, 274–276
Prime symbols to name a point, 436
Prism, 236, 237, 238, 248
surface area, rectangular prism, 252–257
volume of a rectangular prism, 258–261
Probability

calculating, 117
defined, 116
introducing, 116–120
organizing outcomes, 121–125
simulating, 131–133
in sports and games, 134–139
using outcomes to predict, 126–130
Proper fractions, 84
Pyramid, 236, 237, 238, 248, 250

Q

Quadrilateral
classifying, 60–65
definition, 12, 60

R

Random choice, 117
Rays, 52
Rectangle, 60
Reflections, 53, 427, 429
Regular polygon, 234, 426
Regular polyhedron, 241
Regular shapes
hexagon, 12
octagon, 12
polygon, 12, 234, 426
Relationships
in a table of values, 192
expressing, 195–199
Repeating decimals, 154
Rhombus, 60
Right angle, 53
Right triangle, 55, 56
Roots, 215
Rotation, 53
Rotational tessellations, 452–455
Rotations, 427, 429

S

Scalene triangle, 55, 56
Secondary data, 275, 276
Section angles, 287
SI *see* Metric units
Similar figures, 70–74, 75
Simulation of probability, 131–133
Solving problems

inspection and mental math, 392–397
modelling with equations, 410, 412–415
systematic trial, 404–409
Sphere, 236
Sports and games
board games, 113, 121–125, 144
card games, 126–130
geometry, 80, 217
probability, 134–139
sports data, 136, 280–283, 303, 356
statistics, 149, 152–153, 166–167
Spreadsheet, 298–303
measures of central tendency, 324–325
Square
area of, 209
evaluating, 210–211
a quadrilateral, 60
representing, 210
Square-based pyramid, 236, 237
Square number, 210
Square roots, 214–217
defined, 214
reading, 215
Statistics, 149, 152–153, 166–167
in sports, 356
Statistics Canada (CANSIM), 292–294
Stem-and-leaf plot, 280–285
mode and median, 319
Subtracting integers, 362–373
Surface area of a rectangular prism, 252–257
Surveys and bias, 326–330
Symbols, 208
Symmetry, 59
Systematic trial, 404–409, 411

T

Table of values, 191–199
Tables, 32
Tally chart, 116, 132, 272, 276
Tangram, 60
Tessellation

explained, 425
identifying, 442–445
rotational tessellation, 452–455
translational tessellation, 448–451
The Geometer's Sketchpad®, 31–33
  animating a pattern, 435
  creating tiling patterns, 446–447
  drawing trapezoids, 37–39
  frieze patterns, 434–435
  identifying similar triangles, 75
  rotational tessellation, 454–455
  translational tesselation, 450–451
Three-dimensional figures
  classifying and exploring, 236–241
  different views, 242–246
  and nets, 247–251
Three-dimensional objects, 237
Tiling pattern, 442–445
  The Geometer's Sketchpad®, 446–447

Tiling the plane, 442–445
Transformations, 53
  and congruence, 80
  exploring, 428–433
  types, 427, 429
Translation, 53, 427, 429
  on a coordinate grid, 436–441
Translation arrow, 437, 439
Translational tessellations, 448–451
Trapezoid, 24–25, 28–29, 61
  area, 30–33
  definition, 12
  drawing, 34–36
  measuring and manipulating, 39
  The Geometer's Sketchpad®, 37–39
Tree diagram, 122, 123, 127, 135
Trends in data, 312–317
Triangle area formula, 16–17
Triangles
  area, 22–25
  classifying, 54–59
Triangular prism, 236, 237
True statements, 208

U

Unit conversions, 208–209

V

Variable, 186–187, 386–387, 389
Variable expression, 186–187, 387–388, 389
Vertex, 52, 235
Views of objects, 242–246
Visuals in reports, 170
Volume
  of a cube, 209
  of a rectangular prism, 258–261
Volume formula, 415

X

x-coordinate, 179

Y

y-coordinate, 179

Z

Zero principle, 354

Credits

Photo Credits

t=top; b=bottom; c=centre; l=left; r=right

vi (r) Roland W. Meisel, (l) www.comstock.com/ca; **vii** NASA/GPN-2000-001444; **viii** (r) PhotoDisc/SP000240, (l) NASA/MSFC-9022272; **ix** (l) Photodisc/AA038262, (r) www.comstock.com/ca; **xx** Roland W. Meisel; **5** © Richard Hutchings/CORBIS/MAGMA; **7** Roland W. Meisel; **8-9** © Roy Ooms/Masterfile; **12** Roland. W. Meisel; **18** Roland W. Meisel; **22** National Oceanic and Atmospheric Administration/National Environmental Satellite, Data, and Information Service/Department of Commerce; **23** National Oceanic and Atmospheric Administration/National Environmental Satellite, Data, and Information Service/Department of Commerce; **25** Roland W. Meisel; **30** Roland W. Meisel; **33** Roland W. Meisel; **34** www.comstock.com/ca; **36** Roland W. Meisel; **37-39** Screen shots from The Geometer's Sketchpad®, Key Curriculum Press, 1150 65th St., Emeryville, CA 94608, 1-800-995-MATH; **43** Roland W. Meisel; **50-51** © Royalty-free/CORBIS/MAGMA; **52** Roland W. Meisel; **54** Roland W. Meisel; **58** Roland W. Meisel; **60** Copyright © 2003 Jerry Slocum. From The Tangram Book by Jerry Slocum, Sterling Publishing Co., Inc., 2001; **65** Roland W. Meisel; **69** Roland W. Meisel; **70** Don Ford; **74** Roland W. Meisel; **75** Screen shots from The Geometer's Sketchpad®, Key Curriculum Press, 1150 65th St., Emeryville, CA 94608, 1-800-995-MATH; **76** Roland W. Meisel; **77** Roland W. Meisel; **80** A.G.E. Foto Stock/firstlight.ca; **90** Roland W. Meisel; **94** Roland W. Meisel; **95** Roland W. Meisel; **98** Roland W. Meisel; **104** Roland W. Meisel; **112-113** David Madison/Getty Images; **116** © CORBIS/MAGMA; **117** Roland W. Meisel; **121** Roland W. Meisel; **124** Roland W. Meisel; **126** Roland W. Meisel; **129** Roland W. Meisel; **130** Roland W. Meisel; **134** Roland W. Meisel; **139** David Young-Wolff/Getty Images; **140** Roland W. Meisel; **144** CP(Aaron Harris); **148-149** © Kim Kulish/Corbis/MAGMA; **149** (br) CP(Tom Hanson); **158** CP(Fred Chartrand); **161** NASA/GPN-2000-001444; **166** © David Young-Wolff/PhotoEdit; **171** Roland W. Meisel; **176-177** Rubberball Productions/rbv2_87; **177** (br) SANDER ERIC/PONOPRESSE; **185** Canadian Museum of Civilization/D2004-11201; **186** (b) Roland W. Meisel; **188** Roland W. Meisel; **190** Chris Cole/Getty Images; **195** ©Jiang Jin/SuperStock; **196** (r)Photo Credit: Helmut Hasse. Source: *Emmy Noether*, Auguste Dick, Translated by HI Blocher, Birkhäuser, Basel, 1981; **203** SANDER ERIC/PONOPRESSE; **205** Roland W. Meisel; **206-207** © Ric Ergenbright/CORBIS/MAGMA; **213** Roland W. Meisel; **214** (t) Photodisc/SP002866; **218** Roland W. Meisel; **222** © Ric Ergenbright/CORBIS/MAGMA; **223** © Lowell Georgia/CORBIS/MAGMA; **225** Roland W. Meisel; **226** (l) Roland W. Meisel; **232-233** © CORBIS/MAGMA; **233** (br) Roland W. Meisel; **236** Roland W. Meisel; **237** Roland W. Meisel; **241** (c) © Lloyd Sutton/Masterfile, (b) Roland W. Meisel; **243** Roland W. Meisel; **247** (t) © PAVLOVSKY JACQUES/CORBIS SYGMA/MAGMA, (b) Roland W. Meisel; **248** Roland W. Meisel; **251** Roland W. Meisel; **252** (t) © Andrew Wenzel/Masterfile, (b) Roland W. Meisel; **253** Roland W. Meisel; **258** © Walter Hodges/CORBIS/MAGMA; **264** © MAGMA/CORBIS; **266** www.comstock.com/ca; **270-271** © CORBIS/MAGMA; **273** Roland W. Meisel; **280** (t) CP(Adrian Wyld), (b) CP(PC/AOC); **286** Michael Krasowitz/Getty Images; **287** Roland W. Meisel; **292-295** Screen shots. Source: Statistics Canada, E-STAT, <http://estat.statcan.ca>, March 2004; **298** Roland W. Meisel; **299-300** Screen shots from AppleWorks 6. Copyright © 2004 Apple Computer, Inc. All rights reserved; **303** © ORBAN THIERRY/CORBIS SYGMA/MAGMA; **308-309** REUTERS/Charles Platiau; **312** Roland W. Meisel; **318** Roland W. Meisel; **324-325** Screen shots from AppleWorks 6. Copyright © 2004 Apple Computer, Inc. All rights reserved; **326** Roland W. Meisel; **332** © CORBIS/MAGMA; **340** (t) Photodisc/WL003062, (b) © CORBIS/MAGMA; **342-343** Photodisc/200024142-001; **346** NASA/ MSFC-9022272; **351** Photodisc/200024142-001; **352** Roland W. Meisel; **356** PhotoDisc/SP000240; **361** Roland W. Meisel; **362** Roland W. Meisel; **368** Courtesy of HST; **374** Scott Halleran/Getty Images; **379** Roland W. Meisel; **382-383** © AbleStock / Index Stock Imagery; **386** Roland W. Meisel; **388** Roland W. Meisel; **392** Roland W. Meisel; **397** © CORBIS/MAGMA; **398** Roland W. Meisel; **400** (cl) Roland W. Meisel; **404** Roland W. Meisel; **409** Roland W. Meisel; **410** Roland W. Meisel; **414** © David Muir/Masterfile; **424-425** © Archivo Iconografico, S.A./CORBIS/MAGMA; **428** © Dave G. Houser/CORBIS/MAGMA; **430** (t) Photodisc/SP001283, (b) Photodisc/AA009519; **432** Don Ford; **433** Photodisc/AA038262; **434** Screen shots from The Geometer's Sketchpad®, Key Curriculum Press, 1150 65th St., Emeryville, CA 94608, 1-800-995-MATH; **438** Roland W. Meisel; **442** © Collier Campbell Lifeworks/CORBIS/MAGMA; **446-447** Screen shots from The Geometer's Sketchpad®, Key Curriculum Press, 1150 65th St., Emeryville, CA 94608, 1-800-995-MATH; **448** M.C. Escher's "Symmetry Drawing E105" © 2004 The M.C. Escher Company - Baarn - Holland. All rights reserved; **450-451** Screen shots from The Geometer's Sketchpad®, Key Curriculum Press, 1150 65th St., Emeryville, CA 94608, 1-800-995-MATH; **452** M.C. Escher's "Symmetry Drawing E99" © 2004 The M.C. Escher Company - Baarn - Holland. All rights reserved; **454-455** Screen shots from The Geometer's Sketchpad®, Key Curriculum Press, 1150 65th St., Emeryville, CA 94608, 1-800-995-MATH.

Statistics Canada information is used with permission of Statistics Canada. Users are forbidden to copy this material and/or redisseminate the data, in an original or modified form, for commercial purposes, without the expressed permission of Statistics Canada. Information on the availability of the wide range of date from Statistics Canada can be obtained from Statistics Canada's Regional Offices, its World Wide Web site at http://statcan.ca, and its toll-free access number 1-800-263-1136.

Illustration Credits

xii Tina Holdcroft Enterprises Inc. **xv** Ben Hodson **xvi** Ben Hodson **xix** Ben Hodson **3** Tina Holdcroft Enterprises Inc. **18** Tina Holdcroft Enterprises Inc. **26** Steven Schulman **49** Adam Wood **82–83** Ben Hodson **86** Tina Holdcroft Enterprises Inc. **113** Ben Hodson **131** Tina Holdcroft Enterprises Inc. **135** Tina Holdcroft Enterprises Inc. **152** Ben Hodson **162** Tina Holdcroft Enteprises Inc. **187** Tina Holdcroft Enterprises Inc. **196** Juliana Kolesova **224** Ben Hodson **242** Ben Hodson **267** Ben Hodson **274** Tina Holdcroft Enterprises Inc. **276** Tina Holdcroft Enterpises Inc. **326** Tina Holdcroft Enterprises Inc. **331** Amy Occhipinti **337** Claire Louise Milne **412** Tina Holdcroft Enterprises Inc. **436** Adam Wood

Technical Art

Tom Dart, Greg Duhaney, Kim Hutchinson, Claire Milne, and Adam Wood of First Folio Resource Group, Inc.